The Basics of Chemistry

The Basics of Chemistry

Richard Myers

Basics of the Hard Sciences
Robert E. Krebs, Series Editor

GREENWOOD PRESS
Westport, Connecticut • London

Library of Congress Cataloging-in-Publication Data

Myers, Richard, 1951–
   The Basics of chemistry / Richard Myers.
      p. cm.—(Basics of the hard sciences)
   Includes bibliographical references and index.
   ISBN 0–313–31664–3 (alk. paper)
  1. Chemistry. I. Title. II. Series.
QD33.2.2.M94   2003
540—dc21   2002028436

British Library Cataloging in Publication Data is available.

Library of Congress Catalog Card Number: 2002028436
ISBN: 0–313–31664–3

First published in 2003

Greenwood Press, 88 Post Road West, Westport, CT 06881
An imprint of Greenwood Publishing Group, Inc.
www.greenwood.com

Printed in the United States of America

The paper used in this book complies with the
Permanent Paper Standard issued by the National
Information Standards Organization (Z39.48–1984).

10 9 8 7 6 5 4 3 2

Every reasonable effort has been made to trace the owners of copyrighted materials in this book, but in some instances this has proven impossible. The author and publisher will be glad to receive information leading to more complete acknowledgments in subsequent printings of the book and in the meantime extend their apologies for any omissions.

To Chris.
For all your support and companionship during our thirty years.

Contents

Preface   xiii

Acknowledgments   xv

Chapter 1   Introduction: Chemistry and Its Divisions   1
Introduction   1
The Scientific Enterprise   2
What Is Chemistry?   3
Divisions of Chemistry   4

Chapter 2   A Brief History of Chemistry   7
Introduction   7
Early History   7
Greek Science   9
Alchemy   11
Chemistry in the Middle Ages   13

Chapter 3   The Birth of Modern Chemistry   17
Introduction   17
The Beginnings of Modern Chemistry   17
The Phlogiston Theory   19
Pneumatic Chemistry   20
Lavoisier   25

Chapter 4   The Atom   31
Introduction   31
The Law of Definite Proportions   31
Dalton and the Birth of the Atomic Theory   33
Gay-Lussac and Avogadro's Hypothesis   34
The Divisible Atom   35
Atomic Structure   38

	Quantum Mechanics	40
	A Modern View of the Atom	44
	Atomic Mass, Atomic Numbers, and Isotopes	45
	Ions	45
	Summary	46
Chapter 5	Atoms Combined	49
	Introduction	49
	Mixtures, Compounds, and Molecules	49
	Naming Elements	50
	Naming Compounds	51
	Writing Chemical Formulas	54
	Chemical Reactions	54
	Types of Reactions	55
	From Molecules to Moles	56
Chapter 6	Elements and the Periodic Table	61
	Introduction	61
	Discovery of the Elements	61
	The Modern Periodic Table	64
	The Main Group Elements	64
	d and f Block Elements	67
	Metals, Metalloids, and Nonmetals	67
	Periodic Problems	67
	Summary	69
Chapter 7	Chemical Bonding	71
	Introduction and History	71
	Dot Formulas, the Octet Rule, and Ionic Bonds	74
	Covalent Bonds	76
	Electronegativity and the Polar Covalent Bond	76
	Polar Molecules and Hydrogen Bonds	78
	Bond Energies	79
	The Metallic Bond	80
	Bonding and Molecular Geometry	80
	Modern Bonding Theory	82
Chapter 8	Intermolecular Forces and the Solid and Liquid States	85
	Introduction	85
	Characteristics of Solids, Liquids, and Gases	86
	Intermolecular Forces	86
	Crystalline Solids	90
	Amorphous Solids	92
	Liquids	93
	Material Science	95

Chapter 9	Gases	99
	Introduction	99
	Pressure	99
	The Gas Laws	102
	The Ideal Gas Law	106
	Partial Pressure and Vapor Pressure	107
	Applications of the Gas Laws	109
Chapter 10	Phase Changes and Thermochemistry	113
	Introduction	113
	Changes of State	113
	Energy and Phase Changes	113
	Heating Curves	115
	Calorimetry	118
	Heat of Reaction	120
	Applications of Heat of Reactions	122
Chapter 11	Solutions	125
	Introduction	125
	The Solution Process	126
	Electrolytes	128
	Concentration of Solutions	129
	Solubility of Gases	130
	Colligative Properties	131
	Reactions of Solutions	134
	Colloids and Suspensions	136
Chapter 12	Kinetics and Equilibrium	139
	Introduction	139
	Reaction Rates	139
	The Collision Theory	140
	Factors Affecting Reaction Rates	142
	Reaction Mechanisms and Catalysis	144
	Chemical Equilibrium	147
	Le Châtelier's Principle	149
	The Equilibrium Constant	151
	Ammonia and the Haber Process	152
Chapter 13	Acids and Bases	155
	Introduction and History	155
	Chemical Definitions of Acids and Bases	156
	Strengths of Acids and Bases	159
	Acid Concentration	160
	pH and pOH	161
	Neutralization Reactions	163
	Buffers	166
	Summary	169

Chapter 14	Electrochemistry	171
	Introduction	171
	History	171
	Oxidation and Reduction	176
	Oxidation Numbers	177
	Electrochemical Cells	179
	Electrolysis	184
	Batteries	185
	Corrosion	189
	Electroplating	190
	Commercial Applications and Electrorefining	191
	Summary	193
Chapter 15	Organic Chemistry	195
	Introduction and History	195
	Organic Compounds	198
	Aliphatic Hydrocarbons	199
	Aromatic Hydrocarbons—Benzene	205
	Functional Groups and Organic Families	207
	Petroleum	216
	Organic Products	219
	Summary	220
Chapter 16	Biochemistry	221
	Introduction	221
	Carbohydrates	221
	Lipids	226
	Proteins	229
	Nucleic Acids	233
	Biotechnology	236
Chapter 17	Nuclear Chemistry	241
	Introduction	241
	Nuclear Stability and Radioactivity	241
	Radioactive Decay and Nuclear Reactions	243
	Half-Life and Radiometric Dating	244
	Nuclear Binding Energy—Fission and Fusion	246
	Fusion and Stellar Evolution	251
	Transmutation	252
	Nuclear Medicine	254
	Radiation Units and Detection	255
	Biological Effects of Radiation	257
	Summary	259

Chapter 18   Environmental Chemistry   261
              Introduction   261
              Stratospheric Ozone Depletion   262
              Acid Rain   266
              The Greenhouse Effect   270
              Water Quality   273
              Air Pollution   278
              Pesticides   281
              Summary   287

Chapter 19   The Chemical Industry   289
              Introduction   289
              Acids and Alkalis   290
              Explosives   292
              Synthetic Dyes   293
              Synthetic Fibers and Plastics   297
              Synthetic Rubber   299
              Petrochemicals   301
              Industry Histories   302
              Summary   307

Chapter 20   Chemistry Experiments   309
              Introduction   309
              The Scientific Method   309
              Chemistry Experimentation   312
              Chemical Activities   313

Chapter 21   A Future in Chemistry   327
              Introduction   327
              Chemist's Job Description and Training   327
              Careers in Chemistry   330
              Professional Development and the American Chemical Society   332

              Glossary   335

              Brief Timeline of Chemistry   351

              Nobel Laureates in Chemistry   355

              Table of the Elements   361

              Selected Bibliography   363

              Author Index   365

              Subject Index   369

Preface

Perhaps not everyone is a student of chemistry in the formal sense of being enrolled in a chemistry course, but everyone uses chemistry on a daily basis. Every time a package label is read to check ingredients, food is tasted to determine whether it needs spices added, or mixtures are prepared, chemistry is practiced. Without even knowing it, all of us have accumulated a basic understanding of chemistry by observing the world around us. Observations such as salt melting ice, milk souring, leaves turning color, iron rusting, and wood burning demonstrate that changes are constantly occurring. Because chemistry is essentially the study of change and we are constantly observing these changes, we all are students of chemistry.

The Basics of Chemistry is written for students beginning a formal study of chemistry. These readers are primarily high school and college students enrolled in their first chemistry course. In addition to these students, individuals who are not enrolled in a chemistry course but would like a general overview of the subject should find this book helpful. Teachers of all grades may use *The Basics of Chemistry* as a general reference on the subject.

The Basics of Chemistry is a general reference book that presents the basic scientific concepts of chemistry in addition to providing information on several related subjects. Chapters progress through several areas. The first several chapters focus on chemistry's roots as a modern science. Although the first chapters focus heavily on the historical development of chemistry, most of the other chapters give a historical overview of the chapter's content. The heart of this book is devoted to explaining basic chemistry concepts. In these chapters, I review the content found in a beginning chemistry course. Subjects such as nomenclature, chemical bonding, acids and bases, equilibrium, kinetics, solutions, and gases are covered. Building on these general concepts, the chapters that follow explore several subdivisions of chemistry and present additional concepts especially important in these subdivisions. Chapters on organic chemistry, biochemistry, electrochemistry, nuclear chemistry, and environmental chemistry are included. Chapter 19 presents an overview of the chemical industry and includes information on industrial chemistry. Chapter 20 presents the experimental method and twenty chemistry activities. The final chapter provides an overview of the chemistry profession, possible careers in chemistry, and chemical education.

This book is enhanced by other features. It includes a glossary of chemistry terms. Glossary terms are typed in boldface when they first appear in the text. The year of birth and year of death of important individuals are given in parentheses following the first appearance of a person's name. Ample illustrations, tables, and analogies are used to clarify basic chemistry concepts and highlight important chemical information. The chapter on the chemistry industry provides background information on the founding of a number of well-known chemical companies.

The most difficult task in writing a book called *The Basics of Chemistry* is deciding what material to include in the book. What may be "the basics" for one person may not be for another. Rather than present a list of basic facts, I present chemistry from a broad perspective. An introductory book on such a broad subject should give the student a glimpse of the discipline, provide basic concepts and information, and offer a foundation for further study. In this respect, an introductory text is like wiping the moisture from a fogged window. Many objects become clear through the transparent opening, but the rest of the glass remains foggy. Objects viewed through the fog are unclear. Even after wiping away all of the moisture, only a small portion of the outside world can be seen. My hope is that *The Basics of Chemistry* provides the student with a small, cleared opening into the world of chemistry. Once a glimpse is gained through this opening, the student can continue to enlarge the opening. After clearing the window, then it's time to move to the next window.

Acknowledgments

Many individuals helped in the preparations of this book. Jennifer Rochelle provided timely typing whenever needed. Sean Tape and the team at Impressions Book and Journal Services did an exceptional job of editing the manuscript. Greenwood Press initiated this book as part of a series of general reference books in the sciences. I would like to thank the staff at Greenwood for the opportunity to write this book. Greenwood's Debby Adams provided guidance and a firm hand at the wheel as she shepherded the project through production. Rae Déjur's illustrations demonstrate that one picture is truly worth a thousand words, and in some cases much more. A special thanks goes to Robert Krebs who mentored me throughout this project by providing encouragement and valuable feedback.

A number of chemical structures were drawn by ACD/ChemSketch version 4.55 developed by Advanced Chemical Development, Inc. For more information please check the ACD Web site at *www.acdlabs.com.*

Introduction: Chemistry and Its Divisions

Introduction

What do you think of when the word "chemistry" is mentioned? For many, chemistry projects images of a crazed scientist dressed in a white lab coat mixing a bubbly, spewing, stinking concoction. The absentminded professor is the classic stereotype of a chemist. The original Disney movie (*The Absentminded Professor*) released in 1961 tells the story of a chemistry professor who accidentally creates "flubber," a substance with gravity-defying levitative properties. A few years later in 1965, chemists produced what seemed to be real-world flubber. It was and continues to be marketed as the super-ball®. A more modern version of a chemist is FBI agent Stanley Goodspeed portrayed by Nicholas Cage in the movie *The Rock*. Goodspeed is the FBI's point man on chemical weapons. In the movie, he secretly penetrates Alcatraz Island, which is held by a group of ultrapatriotic military extremists. Once on Alcatraz, he must beat a deadline to disarm rockets loaded with a deadly chemical weapon aimed at San Francisco.

Hollywood's portrayal of chemistry stretches the imagination, but probably no more than that of a person living at the beginning of the twentieth century who could foresee life at the start of the twenty-first century. A quick look around is all it takes to see how chemistry has transformed our modern world. The plastic disposable ballpoint pen I use to write with has been in existence only since 1950. Your clothes may be made of natural materials, but chances are you are wearing plenty of synthetics such as nylon, dacron, and polyurethane. You drink tap water that was chlorinated in the last few hours and will be chlorinated again at the treatment plant before it continues its journey through the water cycle. Our vehicles burn refined oil, and catalytic converters reduce the amount of pollution entering the atmosphere. Just opening a medicine cabinet gives ample evidence of the advances in chemistry during the last one hundred years. Cosmetics, soaps, medicines, and cleansers are all products of the modern chemical industry.

Much of the twentieth century has been characterized by the use and abuse of chemistry. There is no doubt that advances in chemistry and chemical technology have improved and extended life, but there is also a price for living in a modern chemical society. Environmental costs such as ozone depletion, acid rain, water pollution, air pollution, and hazardous wastes are part of the price we pay for the benefits of living in a chemical society. New drugs and medicines have eradicated disease, and advances in chemistry have allowed possible fertilizers and other agricultural advances to increase the world's food supply. These developments have led to a booming world population currently at six billion and growing. To sustain a growing world population of over six billion means converting raw materials into products. In recent years, we have recognized the need to "reduce, reuse, and recycle." We know that one of the fundamental principles of chemistry is that matter is not created or destroyed, and chemical processes themselves play a major role in recovering raw materials from used products. Many items once considered waste are now recycled supplementing raw materials obtained from the Earth. As more countries such as China develop their industrial infrastructure, the demand for energy and material resources is bound to increase.

The laws of thermodynamics tell us that the conversion of raw materials into finished goods requires energy and ultimately results in an ever-increasing amount of heat released into the environment. This waste heat is accompanied by carbon dioxide produced from the burning of fossil fuels. There seems to be a consensus among scientists that our planet's temperature is increasing due to the **greenhouse effect**, although a minority of scientists do not believe that our planet's temperature is increasing. Many scientists also feel that increased temperatures are part of natural fluctuations and that humans have little impact on the recent trend of elevated global temperatures. International efforts are now underway to curb carbon dioxide emissions worldwide, such as revegetation, curbing slash and burn practices, and sequestering carbon dioxide in the sea, are stalled by disagreements between developed and underdeveloped countries, as well as Europe and the United States.

The Scientific Enterprise

So how do we strike a balance between the benefits of better living through chemistry and the costs associated with this lifestyle? One answer to this question is knowledge. Greater knowledge equips us with the tools to make informed decisions when we live in a free democratic society. Chemical knowledge is based on a long history of chemical research. The methodological procedure to attain scientific knowledge is often defined as the scientific method. The classical view of the scientific method presents science as a neat step-by-step procedure starting with formulating a question, posing a **hypothesis**, performing a **controlled experiment** to address the hypothesis, and drawing conclusions from the experiment. Although this view is accurate in an idealized sense, in reality, true science is filled with pitfalls, blind alleys, dead ends, false starts, fortuitous accidents, and human idiosyncrasies. Books on science often give the impression that science proceeds in a linear, uniform fashion. More often, scientific progress follows a circuitous route with periods of accelerated discoveries and long, relatively dormant periods.

The theories and facts we cling to today are only an experiment away from revision. As you read these words, chemical research is occurring in thousands of labs across the globe. The modern scientific enterprise is vastly different from that of the quintessential scientist working in isolation in a small laboratory. Modern research is characterized by research teams led by scientists who specialize in very specific fields. Teamwork is essential as research teams across the world attempt to answer fundamental questions. These teams work both cooperatively and competitively with other teams working on similar problems. Work is continually reviewed, critiqued, revised, and published in scientific journals. The advent of the Internet has led to both a wider audience and a faster distribution for research results. Much of today's research takes place in universities and is supported by governmental agencies that distribute millions of dollars annually to those teams whose work is deemed worthy of support. In addition to this research, numerous private companies operate their own private research labs to develop products and processes to give them an advantage over their competitors. Chemical research is both basic and applied. Basic research addresses the fundamental laws of nature and is undertaken to advance the theoretical foundation of chemistry. Applied research is performed with a specific application in mind, for example, better products or to improve the quality of life. Research should not be viewed in terms of a basic or applied research dichotomy. Most research contains elements of both basic and applied research.

As new knowledge is gained through research, additional questions are raised and new areas of research are continually exposed in a neverending process. Because knowledge is power, as we embark on the twenty-first century, a knowledge of chemistry is essential to shape a sustainable future. Chemical knowledge may lead to a simple decision to recycle or choose one product over another. Perhaps this knowledge helps you make a decision to undertake a certain medical treatment or vote for or against an issue. Chemical knowledge can help guide you in any number of choices you make today. More important, this

knowledge will help you with decisions concerning novel areas we can only imagine but know will develop as the twenty-first century unfolds.

What Is Chemistry?

Chemistry is the branch of science that deals with the composition and structure of **matter** and the changes that matter undergoes. Matter is anything that has mass and occupies spaces, which means just about anything you consider. This book, your body and the air you breathe are all examples of matter. Matter is simply the stuff that makes up our universe.

Chemistry, like all branches of science, is a method that attempts to simplify and organize. Every object we might consider is a separate piece of matter, and matter can be classified in any number of ways. One simple classification scheme for matter is based on the three states of matter: solid, liquid, and gas (four if we include **plasma**). Another classification scheme, and one that is fundamental to chemistry, is classifying matter by chemical composition. Rather than speaking of water as a form of matter, we can speak of water composed of approximately 11% hydrogen and 89% oxygen by mass. Similarly, air is a mixture of matter containing approximately 78% nitrogen, 21% oxygen, a little less than 1% argon by volume, and an assortment of other gases in small percentages.

Much of the history of chemistry has concerned itself with determining the ultimate composition of matter. The Greeks considered all matter to be composed of different combinations of earth, air, fire, and water. It is only during the last two hundred years that the modern idea of **chemical elements** developed and only in the last one hundred years that we have determined that elements themselves are composed of protons, neutrons, and electrons. Hence, we can think of all matter to be composed of 91 naturally occurring elements, and protons, neutrons, and electrons are the building blocks that make up the elements. Chemistry not only deals with the composition of matter, but how the pieces of matter fit together, that is, the structure of matter. The structure of matter can have a major impact on its properties. For example, carbon can exist as a two-dimensional flat structure, or a soft, black graphite, or in the pyramidal three-dimensional tetrahedron shape we know as diamond (Figure 1.1).

So far, we have noted that chemistry involves the composition and structure of matter. But beyond knowing what the composition of matter is and how the pieces fit together, chemistry is about how matter changes from one substance to another. When we refer to **chemical change**, we mean that the composition of matter has changed. A chemical change occurs when a substance or substances change into other substances.

For example, when hydrogen and oxygen are brought together under the right conditions, they can combine and change into water. When the carbon in paper reacts with oxygen in the air during combustion, it changes into carbon dioxide. The key when considering whether a chemical change has taken place is to ask the question: do I end up with something different from what I started? Chemical changes result in the production of something entirely different from the original substances. The element sodium is a highly reactive metal that reacts violently with water. Chlorine is a highly toxic gas. Yet, put sodium and chlorine together under the right conditions and you have table salt.

Not all changes are necessarily *chemical changes*. Changes may also be **physical changes**. When water changes from ice to liquid to steam, change has certainly taken place, but not a chemical change. Water in its solid ice state is H_2O. After melting and vaporizing, it is still H_2O. The substance water has not changed. The physical state has changed, but not the substance. Similarly, when sugar dissolves in water, no chemical change takes place. We start with sugar and water and end with sugar and water. Again, only a physical change occurs, because we still have sugar and water after the sugar dissolves. To summarize, chemistry can be thought of as the science of the study of matter: what matter is made of, how it is made, and the changes in matter from one substance into another.

Figure 1.1
Diamond and Graphite Structure
(Rae Déjur)

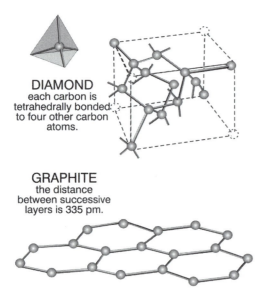

DIAMOND
each carbon is tetrahedrally bonded to four other carbon atoms.

GRAPHITE
the distance between successive layers is 335 pm.

Divisions of Chemistry

The definition given in the last section is very general. Although all chemists are involved in the study of matter, the field is so broad that it helps to divide chemistry into divisions. These divisions characterize different aspects of the study of chemistry using some common feature. Chemistry may be divided into the two very large divisions of inorganic and organic chemistry.

Inorganic chemistry involves the study of the chemical elements, their **compounds**, and reactions excluding those that contain carbon as the principal element. **Organic chemistry** is the study of the composition and the reaction of carbon-containing compounds. The distinction between inorganic and organic results from the days when chemical compounds were classified according to their origin. Inorganic means not derived from life. Inorganic substances were thought to originate only from mineral, nonliving sources. Conversely, organic compounds were thought to have come from living plant or animal sources. All organic substances contain carbon, and it was once believed that these carbon substances could originate only from a living source. We know today that many organic compounds are synthesized from inorganic, nonliving sources. Some examples include polypropylene, acrylics, and nylon used in clothing; polystyrene (styrofoam) used for insulation; and ethylene glycol (antifreeze). Even though the original distinction separating inorganic and organic does not hold, the terms are still used to distinguish two broad areas of chemistry. For our purposes, we can think of organic chemistry as simply the chemistry of carbon compounds and inorganic chemistry as the study of all other compounds. It should be noted, though, that certain carbon compounds, most notably carbon dioxide and carbonates, are classified as inorganic.

In addition to the large divisions of organic and inorganic chemistry, several other large divisions of chemistry exist. Biochemistry is the study of chemical substances associated with living organisms. Biochemists study the nature of biological substances. The study of DNA, viruses, and immune systems are examples of biochemical research. Physical chemistry deals with the structure of matter and how energy affects matter. Physical chemists are concerned with the physical characteristics of matter. Areas of research in physical chemistry might include how chemicals absorb light or how much energy is released or absorbed when a chemical reaction occurs. Closely related to physical chemistry is the study of nuclear chemistry. Nuclear chemistry focuses on the study of atomic nuclei, **nuclear fission** reactions, and **nuclear fusion** reactions. The nucleus is the positive central portion of an atom composed of positively charged protons and uncharged neutrons. Fission reactions involve the physical splitting of heavier elements into smaller elements, and fusion involves lighter elements physically combining to form heavier elements, for example, the fusion of hydrogen to form helium. Analytical chemistry deals with techniques used to identify and quantify the composition of matter. Analytical chemists, as the name implies, analyze substances for their content. An analytical chemist might perform tasks such as determining the sugar content of a fruit juice, the amount of pollutant in a water body, or the purity of a drug. Environmental chemistry focuses on the occurence of natural and synthetic substances in the environment and their impact on the environment. The study of the chemistry of pollution is a major area of concern for environmental chemists. Problems involving water pollution, air pollution, solid waste, hazardous materials, and toxic substances involve a study of environmental chemistry. Each of the major areas of chemistry mentioned above, however, are not rigid divisions.

The nature of chemistry is such that the fields of study in many areas naturally cross over into other areas. A biochemist naturally works in the area of organic chemistry, and an environmental chemist who studies radiation is concerned with nuclear chemistry.

Many branches of chemistry exist in conjunction with other areas of science. For example, the study of the chemistry of rocks and minerals falls under the category of geochemistry. The combination of medicine and chemistry leads to medicinal chemistry. Other divisions include agricultural chemistry, food chemistry, petroleum chemistry, soil chemistry, and polymer chemistry.

As chemical knowledge continues to expand, we are certain to develop new areas of study in chemistry. Just recently, chemists have performed simple computer functions using chemical reactions rather than performing these functions electronically. Chemical reactions that perform computer operations bring computers closer to modeling human thought processes. Our thoughts result from chemical reactions that occur in our brains. As chemists work at the forefront of this area of technology and others, new interdisciplinary areas will certainly develop. Those new areas add to existing branches of chemistry and at the same time create new areas of study.

In the chapters to follow, the basic concepts that guide chemists in their study of matter are examined. In addition to introducing these basic concepts, the historical foundation of chemical thought, as well as new areas of chemical research, are explored. This knowledge is intended to give the reader a clearer view of the wonderful world of chemistry that is all around us; and with this, a foundation to build upon. All progress by scientists is dependent on the work of those who precede them. In a letter to Robert Hooke, Isaac Newton stated, "If I have seen farther than others, it is because I was standing on the shoulders of giants." The goal of this book is to help you see a little farther.

2

A Brief History of Chemistry

Introduction

Chemistry has always existed. The formation of the Earth and the development of life involved innumerable chemical processes. Humans, as part of the natural world, are part of these natural processes. As humans evolved and observed chemical processes constantly occurring all around them, they attempted to control these processes in their daily existence. For example, combustion in the form of fire could be used to clear land, light the darkness, cook food, and provide protection. Eventually, humans learned that fire could be use to harden clays and melt ores to refine metals. Other chemical discoveries were used to improve living conditions. In this respect, our ancient ancestors were no different than modern humans. Chemical technology has been used throughout history to improve our lives. As a modern science, chemistry has only existed for two hundred years. Modern chemistry developed out of a tradition of chemical technology applied throughout history. This chapter examines some of that history, and the following chapter focuses on the events that established chemistry as a modern science.

Early History

Chemistry is often called the central science. It derives this name because of its importance to all the other sciences. Although chemistry did not exist as a modern science until two hundred years ago, humans have used chemistry from prehistoric times. Evidence of the early uses of chemistry includes cave paintings dating from 25,000 B.C. It is not hard to imagine our ancestors obtaining natural pigments by squeezing berries or mixing crushed rocks with water to produce different colors to use as paints. Similar processes could have been used to obtain dyes for clothes or for body decorations. Fragrances obtained from flower extracts or fats rendered from animals were the prehistoric version of the cosmetic and repellant industry. Evidence for the use of **fermentation** to produce wine and beer dates from the earliest civilizations. Fermentation involves the conversion of glucose to ethyl alcohol (ethanol) in the presence of yeast:

$$C_6H_{12}O_{(aq)} \xrightarrow{\text{yeast}} 2C_2H_5OH_{(aq)} + 2CO_{2(g)}$$

glucose   ethyl alcohol carbon dioxide

One of the earliest uses of fire was in pottery making. Early humans undoubtedly observed that when clay was heated its water was driven out and a hard rock substance remained. Our ancestors made clay implements and art fixtures by heating their work in open pit fires, and clay implements from 20,000 years ago have been found.

Early human civilizations used stone, bone, and wood for objects. Approximately ten thousand years ago, metals first appeared. The first metals used were those found in their native form, or in a pure, uncombined state. Most metals today are acquired from an ore containing the metal in combination with other elements such as oxygen. The existence of native metals is rare, and only a few metals exist in native form. Iron and nickel were available in limited supply from meteorites. The first metals utilized widely by humans were copper, silver, and gold. Pure nuggets of these metals were pounded, in a process known as "cold hammering," with stones into various shapes used for weapons, jewelry, art, and various domestic implements. Eventually, smiths discovered if a metal was heated it could be shaped more easily. The heating process is known as **annealing**. Because the supply of native metals was limited, metal items symbolized wealth and status for those who possessed them.

The discovery that metals could be obtained from ore-bearing minerals signaled the true start of the metal age. The first metal to be extracted from its ore was copper. The technology used for firing clay could be directly applied to the **smelting** of metals. Kilns and furnaces were adapted with troughs to capture metals obtained during the smelting process. Smelting involved heating the metal-bearing ore with wood or charcoal. The carbon from the charcoal combined with the oxygen in the ore separating the metal from the oxygen. During the smelting process, the sulfur, which was often associated with metal ores, was driven off as sulfur dioxide. Sulfur, with its characteristic odor and yellow fumes that are released when it burns, came to be closely associated with the transformation of metals and subsequently played a key role in alchemical reactions. The earliest evidence of the smelting of copper dates from around 6000 B.C.

By 4000 B.C., smelters were combining copper and tin to produce the **alloy** bronze. An alloy is simply a mixture of metals. The bronze produced in ancient times consisted mostly of copper mixed with between 1% and 10% tin. Because the production of bronze required a source for both copper and tin, major trade centers and trade routes developed to mine and produce bronze. One of the major sources of tin from the Bronze Age until recent times was the Cornwall region in southwest England. The Romans and other groups settled the area and exported tin to other parts of Europe. Bronze was the first metal used over wide geographic regions. Advances in kilns, annealing, alloying, **oxidation**, and smelting continued to advance the craft of metalsmithing as the production of bronze flourished.

Iron, which requires a higher temperature to smelt than does bronze, is found in relics dating from approximately 1500 B.C. Early evidence of the use of iron was provided upon the opening of King Tut's tomb when an iron dagger that dated from 1350 B.C. was found buried with the king. The first irons produced were of inferior quality. Repeated heating and hammering (**tempering**) were needed to produce an iron of acceptable quality. A major advancement made in iron production was the use of bellows to obtain temperatures high enough (approximately 1,500°C) to smelt iron, although even irons created with the higher temperatures and the tempering process were inferior to bronze. It was only when iron was heated with carbon that an iron with superior strength qualities was obtained. This process foreshadowed the modern steel industry. Steel is iron to which carbon has been added.

The metal age brought about a number of advances in chemical technology. Yet, the smelting of metals and their incorporation into products involved more art and craft than science. Smiths were thought to possess magical powers to be able to obtain refined metals from rock. Civilizations and their rulers closely guarded their trade secrets. After all, metals were the strategic weapons of ancient civilizations. Still, progress made in metallurgy had a direct impact on the development of chemistry. Advances in heating devices such as furnaces and kilns, fuels to fire the kilns, and processes such as mixing, extracting, and mining were directly applicable to latter advances in chemistry.

Greek Science

The technologies of early civilizations used chemical technology **empirically**. That is, different substances and methods were discovered through experimentation. Applied science took precedence over any theoretical understanding of the general laws and principles governing chemical processes. As early as 600 B.C., Greek philosophers sought a more basic understanding of matter. Anaximander of Miletus (circa 610–545 B.C.) believed air was the primary substance of matter and that all other substances came from air. For example, fire was a form of thin air, and earth was thick air. Heraclitus of Ephesus (544–484 B.C.) believed fire was the primary substance and viewed reality in terms of change. Matter was always in a state of flux, and nature was always in a process of dissolving and reforming. His ideas on change are summarized in his saying: "you cannot step twice into the same river." Heraclitus' philosophy also considered reality in terms of opposites. The way up and the way down were one and the same depending on one's perspective.

The Eleatic school of Greek philosophy was centered in the southern Italian city-state of Elea, which flourished in the sixth and fifth centuries B.C. The Eleatics directly opposed the ideas of Heraclitus and believed the universe did not change. They felt knowledge acquired through sensory perception was faulty because reality was distorted by our senses. One of the leading Eleatics was Parmenides (circa 515–450 B.C.), who viewed the world as a unity and believed change was an illusion. To make their ideas about the absence of change conform to observations, the Eleatics expanded upon Anaximander's idea of one primary substance. Rather than considering air as the primary substance, the Eleatic philosopher Empedocles of Agrigentum (circa 495–435 B.C.) proposed four primary elements: air, earth, fire, and water.

In response to the Eleatics, the atomist school led by Leucippus (circa 450–370 B.C.) and his student Democritus (circa 460–370 B.C.) developed. The atomists refuted the idea that change was an illusion and developed a philosophy that supported the sensory observations of the physical world. The atomist school proposed that the world consisted of an infinite void (vacuum) filled with atoms. According to Democritus, "nothing exists except atoms and empty space, everything else is opinion." Atoms were eternal, indivisible, invisible, minute solids that comprised the physical world. Atoms were homogeneous but existed in different sizes and shapes. The idea of atoms developed on logical grounds. If a piece of gold was continually divided into smaller and smaller units, the gold would eventually reach a point where it was indivisible. The ultimate unit of gold, or any other substance, was an atom. According to Democritus, an infinite number of atoms moved through the void and interacted with each other. The observable physical world was made of aggregates of atoms that had collided and coalesced. The properties of matter depended on the size and shape of the atoms. Change was explained in terms of the motion of the atoms through the void or how atoms arranged themselves. Aristotle (384–322 B.C.) rejected the idea of atoms, but Democritus' atomist philosophy was later adopted by Epicurus (341–270 B.C.) and the Roman poet Lucretius (circa 95–55 B.C.). Almost two thousand years would pass until atoms were resurrected by Dalton in the development of modern chemistry.

Of all ancient philosophers, Aristotle had the greatest influence on scientific thought. Aristotle's writings and teaching established a unified set of principles that could be used to explain natural phenomenon. Although today we would consider his science rather foolish, it was based on sound, logical arguments that incorporated much of the philosophical thought of his time. The teleological nature of Aristotle's explanations led to the subsequent acceptance by the Church in western Europe. This contributed to Aristotle's two thousand year longevity in Western thought.

Aristotle's *Meterologica* synthesized his ideas on matter and chemistry. Aristotle believed that four qualities could be used to explain natural processes. The four qualities were hot, cold, dry, and moist. Hot and cold were active qualities. Hence, the addition or subtraction of heat leads to the transformation of things. Moist and dry were the result of the action of the active qualities. Only four possible combinations of these qualities could exist in substances. These were hot and moist, hot and dry, cold and moist,

and cold and dry. Opposite qualities could not coexist. The combinations of hot and cold and moist and dry were impossible in Aristotle's system. The four allowable combinations determined the four basic elements. Air possessed the qualities hot and moist, fire possessed hot and dry, earth possessed dry and cold, and water possessed wet and cold. Aristotle's system is summarized in Figure 2.1, which is found in a number of ancient chemical writings.

Using the four qualities of matter and four elements as a starting point, Aristotle developed logical explanations to explain numerous natural observations. Both the properties of matter and the changes in matter could be explained using Aristotle's theory.

Aristotle explained the process of boiling as a combination of moisture and heat. If heat is added to a substance that contains moisture, the heat draws the moisture out. The process results in the substance dividing into two parts. When the moisture leaves, the substance becomes thicker, whereas the separated moisture is lighter and rises. Aristotle used this type of reasoning to explain numerous physical and chemical processes including evaporation,

Figure 2.1
Aristotle's Diagram on Matter and Change

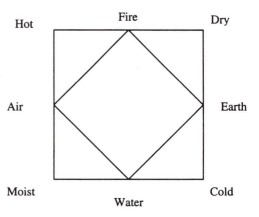

decaying, **condensation**, melting, drying, putrefaction, and **solvation**. Although Aristotle's philosophy explained much, it did have its shortcomings. A simple example helps to illustrate how problems arise with any scientific theory. According to Aristotle, hard substances were hard because they had an abundance of earth and possessed the qualities of dry and cold. Aristotle claimed earth was heavy and so it moved down. Soft substances would not contain as much earth, but contained more water and air. These substances would not be attracted as strongly downward. Now consider ice and water. Ice is dry and cold compared to water, which is moist and cold. We would expect ice to have more of the element earth and a stronger downward attraction than water. Yet, hard solid ice floats in water. Even though many other problems existed with Aristotle's theory, it provided reasonable explanations for many things.

In closing this section, remember that Aristotle rejected the concept of atoms. Aristotle could not accept the idea of a void space and believed that "nature abhors a vacuum." Furthermore, Aristotle did not consider internal structure. Substances contained their qualities and elements as a **homogenous** mixture. An Aristotelian would explain the reaction of hydrogen gas and oxygen gas to produce liquid water as

$$\text{Air} + \text{Air} \rightarrow \text{Water}$$

This reaction shows how one basic element could convert directly to another. An entirely new element is produced from an original element using Aristotle's system. A modern chemist would write the reaction:

$$2H_{2(g)} + O_{2(g)} \rightarrow 2H_2O_{(l)}$$

In this reaction, water is formed from the gases hydrogen and oxygen. We end up with water, but we still have hydrogen and oxygen. Rather than producing new elements, the elements hydrogen and oxygen combine and rearrange themselves to form water.

Alchemy

The teachings of Aristotle and events in Egypt, India, China, and Mesopotamia stimulated the practice of alchemy. The early roots of alchemy, which was practiced for nearly two thousand years, are difficult to trace because much of the practice was shrouded in mystery and transmitted by oral tradition. The two main goals of the alchemists were to produce gold from base metals and to develop potions that would confer health and even immortality. The alchemists were crafts people who combined serious experimentation with astrology, incantations, and magic in hopes of finding the **philosopher's stone**.

The philosopher's stone or *material prima* was a substance (a powder, tincture, or stone) that was more pure than gold itself. It was thought that by using a minute quantity of the philosopher's stone a base metal could be elevated to gold through an alchemical process. Besides perfecting common substances into gold, the philosopher's stone was thought to convey immortality, cure all common ills, and cleanse the spiritual soul. The philosopher's stone was sought by all and possessed by few. Its power conveyed perfection and the ability to reach a pure state to whoever held the stone. The origination of the idea of a philosopher's stone is unclear, but reference is made to the idea in several ancient cultures. An elixir of immortality is present in Indian writings as early as 1000 B.C. Alchemical practice in China several centuries B.C. through A.D. 500 referred to a "pill of immortality" and the use of gold to confer immortality. The Chinese desire for gold, as opposed to that in the West, was strictly concerned with physical and spiritual health rather than acquiring wealth. Hermetic philosophy and the writings of Hermes (a pseudonym) incorporated both Egyptian and Greek mythology in the first few centuries A.D. and mentioned a solidified spirit derived from the Sun capable of perfecting all things. Ancient ideas on the philosopher's stone were undoubtedly passed among cultures. Ideas from India and China could have easily found their way into medieval Europe through Arab cultures.

Alchemy's foundation was based on the teachings of Aristotle. Aristotle held that rocks and minerals were alive and grew in the interior of the Earth. Like humans, minerals attempt to reach a state of perfection through the growth process. Perfection for minerals was reached when they ripened into gold. Based on these premises, the alchemist sought to accelerate the ripening process for metals by subjecting them to a series of physical and chemical processes. One typical series might include heating the metal with sulfide to remove impurities. Then a starter seed of gold was added to the metal. After the seed was added, the metal was treated with arsenic sulfide. This treatment resulted in whitening of the metal, which could be interpreted as a production of silver, or a stage halfway to the perfect gold stage. Finally, the whitened metal could be treated with polysulfides to produce the characteristic yellow gold color. Of course, the process could take weeks or months, and it all had to be performed under the right sign of the stars.

The early center of alchemy was the intellectual capital of ancient Greece, Alexandria. Very little remains of the original alchemical manuscripts from ancient Greece. The rise of Christianity and concerns about disrupting the economy eventually led the Roman Emperor Diocletian to outlaw alchemy in A.D. 296. Diocletian's edict resulted in the destruction of much of the knowledge accumulated by the alchemists. Two original works that did survive are known as the *Leyden Papyrus X* and the *Stockholm Papyrus*. These two related works were discovered in a tomb of a buried Egyptian in ancient Thebes. The papyri date from the end of the third century, although it is believed the papyri are copies of an original written sometime between A.D. 100 and 300. They are written in Greek and consist of a series of recipes for alchemical preparations. The recipes include instructions for making alloys, determining the purity of metals, making fake gold and silver, producing imitation gemstones, and dyeing textiles.

The rise of the Arab Empire shifted advances in learning to the East. The Arab Empire at its peak spanned from its eastern borders with China and India all the way across northern Africa into southern Spain. The Arabs translated and dispersed the works of the Greeks, and for the next one thousand years learning and advances in science, medicine, astronomy, and mathematics were made by the Arabs. The Arab influence can be seen in the word "alchemy" itself. The word "al" is the Arab prefix for "the." Adding this prefix to "chyma," Greek for melting of metals, gives "al chyma." The term "alchemy" first appeared in the fourth century A.D. Another term to appear was "al iksirs." Iksirs were coloring agents used as an ingredient in the procedure to produce gold. Today, the term "elixir" is a reminder of its alchemical root.

The period between A.D. 700 and 1100 was the peak of the Arab Empire. The vast geographic expanse of the Arab influence meant that a great variety of materials were available and that there was ample opportunity for the cross-fertilization of knowledge. Several individuals from this period had a lasting influence on the development of the chemical arts. Jabir ibn Hayyan (Latin name Geber, 721–815) accepted many of the ideas of Aristotle but also modified Aristotle's ideas. Jabir proposed two primary exhalations. One exhalation was smoky and believed to contain small particles of earth. The other was vaporous and contained small particles of water. The principal smoky exhalation was sulfur, and the primary vaporous exhalation was mercury. Jabir believed pure sulfur and mercury combined in different proportions to produce all metals. Most of the writings attributed to Jabir are actually a compilation from numerous Arab alchemists and early European alchemists written over several hundred years. These writings described processes for producing acids, metals, dyes, inks, and glass.

Al-Razi (Rhazes, 854–925) was a Persian who studied in Baghdad. Al-Razi wrote extensively on medicine, philosophy, astronomy, and alchemy, but he was primarily a physician. Al-Razi was less mystical than his contemporary alchemists and classified chemicals by their origin. According to Al-Razi, chemicals came from either animals, plants, and minerals or were derived from other chemicals. Al-Razi wrote *The Comprehensive Book*, which was an enormous medical encyclopedia that synthesized medical practices of ancient Greeks, Syrians, Arabs, and Persians. Al-Razi was the first person known to describe the disease smallpox. Most of his alchemical writings have been lost, but Al-Razi believed in the atomic nature of matter. Al-Razi took a systematic approach to science and rejected the idea of divine intervention. His rational methods and descriptions were more consistent with modern science than most individuals of his time. Ali al Husayn ibn Sina (Avicenna, 980–1037) was another Persian physician whose voluminous works, including *The Canon of Medicine*, guided the practice of medicine for 500 years after his death. Avicenna rejected the idea that a base metal could be transformed into gold. Avicenna claimed correctly that diseases were spread through air and water. Much of Avicenna's teachings questioned the status quo and teachings of Aristotle.

By Avicenna's time, around 1000, the Arab Empire was in decline from both internal and external forces. Factions of the Islamic faith battled one another. A general intolerance of science pervaded Arab culture, and scientists were not free to publish their ideas. Christian Crusaders from the West and Mongol invaders from the East exerted pressure on the Arabic world. As traditional Arab regions were recaptured by Europeans, the classical knowledge that had been preserved and advanced by the Arabs influenced European thinking. Major Arab learning centers, such as Toledo in Spain, provided works to rekindle European science. From the twelfth century, major advances in the chemical arts shifted from Arab lands to western Europe.

Chemistry in the Middle Ages

Scholars, especially clergy from the Catholic Church, were primarily responsible for translating Arabic, Syrian, and Persian writings into Latin. The translators not only reintroduced traditional Aristotelian philosophy, but they also explained numerous technological advances made during the period in which the Arabs flourished. Translated works brought to light advances in the practice of distillation, glass making, filtration, **calcination, sublimation**, and crystallization. From these translations it was apparent that the Arabs and Persians had advanced chemical knowledge significantly, especially during the period from A.D. 700

to 1100. Contributions from the Arab and Persian regions followed the work of Jabir. Jabir developed chemistry by focusing on practical applications using the method of experimentation.

Arab and Persian work in the area of chemistry was principally conducted by individuals practicing medicine. In addition to preparing medicines, experimenters sought improvements in various crafts such as refining metals, preparing inks, dyeing fabrics, waterproofing materials, and tanning leather. Modern distillation originated with the Arabs and was improved by Avicenna, who invented the refrigerated coil. Advanced distillation techniques spread throughout the Middle East, and the Arabs used the refrigerated coil to distill ethanol from fermented sugar. The ethanol was used as a solvent for extracting plant oils, which were substituted for animal fats in preparations. Because essential oils were considered the basis of odors and flavors, any improvement in extraction processes were valued. The Arabs also incorporated other industries into their empire as they expanded. The manufacture of glass was dominated by Syria and Alexandria until A.D. 850. The Arabs conquered these areas and incorporated and improved the art of glass making. Glass was used as a building material and aesthetically in ornamentation and religious designs. After the fall of the Arab Empire, the center of glass making moved west to Venice. Besides technical advances in processes and apparatus, the Arabs had developed and improved the purity of substances such as alcohols, acids, and gunpowder, which were not available to the Europeans.

Several Europeans, while not alchemists themselves, contributed to setting the stage for modern chemistry. Albertus Magnus (1200–1280), also known as Albert the Great, played an important role in intro-

ducing Greek and Arabic science and philosophy in the Middle Ages. Albert produced numerous commentaries on the works of Aristotle. These commentaries helped establish the value of studying natural philosophy along with the philosophical logic of Aristotle. Albert had a major influence on his most famous student, Thomas Aquinas (1225–1274). Aquinas reconciled Aristotelian logic and Christian teaching. Albert was declared a saint in 1931 and is considered the patron saint of all who study science.

Roger Bacon (1214–1294) was an English contemporary of Albert Magnus. Bacon was a Franciscan clergy who taught at Oxford and conducted studies in alchemy, optics, and astronomy. He conducted major studies on gunpowder. In his *Opus Majus* (Major Work), *Opus Minus* (Minor Work), and *Opus Tertium*, Bacon argued that the study of the natural world should be based on observation, measurement, and experimentation. Bacon proposed that the university curriculum should be revised to include mathematics, language, alchemy, and experimental science. Because of his teachings, Bacon often had difficulties with his superiors, although today, we accept many of Bacon's ideas as the foundation of modern science.

Another influential figure was the Swiss self-proclaimed physician Paracelsus (1493–1541). Paracelsus' real name was Phillippus Aureolus Theophrastus Bombastus von Hohenheim. The name "Paracelsus" was a nickname he chose for himself, meaning "superior to Celsus." Celsus was a second-century Greek poet and physician. Paracelsus was an itinerant scholar. He traveled throughout Europe undertaking both formal studies and absorbing folk remedies. At various times in his life, he worked as a surgeon (which at that time was a low-class medical craft), a physician to miners, and a medical admistrator. Paracelsus applied the principles of alchemy to medicine. As such, he rejected the traditional medical practices of his time, which were based on the ancient principles of Galen (circa 129–199). Paracelsus adopted the ideas of Aristotle and applied these to medicine. He believed that disease was the result of an outside agent invading the body. According to Paracelsus, disease was the result of seeds being spread between individuals, and it was the physician's job to find a specific cure to negate each disease seed. Paracelsus experimented and searched for remedies by distilling materials to produce purified medicines. In addition to preparing an assortment of remedies, Paracelsus knew it was important to determine the dosage necessary to produce maximum benefit. Paracelsus used mercury salts to cure syphilis and as disinfectants in working against infections. Although much of his approach involved a systematic experimental search for cures, Paracelsus was a mystic who also used astrology in his practice.

Paracelsus attacked the established medical community of his time. His criticism was harsh, and he alienated many of his established fellow physicians with his bombastic writings. Yet, many of his practices were sound, and more importantly, those physicians who did ascribe to his practices achieved better results than when using traditional methods. Paracelsus' works were not published and distributed widely until some twenty years after his death. By the end of the sixteenth century, Paracelsus' methods had attracted a number of followers, and his methods formed the new basis of standard medical practice. Paracelsus' followers became known as the **iatrochemical** physicians. They used a variety of drugs and treatments that depended on specific dosages of medicine prepared with specific purity. His ideas on chemical purity and for-

mulation also found use in other crafts such as metallurgy and soap making. Chemists became known as dealers in medicinal drugs who gained knowledge about preparations from skilled masters (Figure 2.2). Although Paracelsus was an alchemist, his work led to the term "alchemy" being restricted to the practice of trying to convert base metal into gold. Through the work of Paracelsus and his followers, the term "chymia" was now reserved for the study of matter.

Several other individuals made important contributions at about the same time as Paracelsus, further establishing chymia as a science apart from alchemy. Vannoccio

Figure 2.2
Woodcut print of chemical lecture from 1653. Small furnace sits on table and various glassware used for distillation rests on shelves above stove in background. Image from Edgar Fahs Smith Collection, University of Pennsylvania Library.

Biringuccio (1480–1539) wrote *Pirotechnia* in 1540. This work was a basic text written in Italian on the science of metallurgy. Biringuccio's work provided techniques for processing metal, and he rejected the alchemists' idea of transmutation of metals. A similar work focusing on the processing of metals was *Concerning Metals* written by Georgius Agricola (1494–1555). Andreas Libavius (1540–1616) critiqued Paracelsus. His work *Alchemia* published in 1597 was a comprehensive review of chemical knowledge available at the time. Libavius' work organized chemistry by summarizing the methods, operations, chemicals, and properties of chemicals. Libavius entitled the section of *Alchemia* describing the properties of substances *Chymia*. The term "chymia" eventually became the term "chemistry." Libavius contributed to the establishment of chemistry as a unique discipline deserving study in its own right.

By the start of the seventeenth century, the stage had been set for transformation in the study of chemistry. Up to this time, chemistry played a central role in the arts and crafts, but as a science, it had made only modest advances from the days of Aristotle. This started to change as individuals adopted a scientific approach and subjected the ideas of Aristotle to experimental testing. By 1700, the walls of unquestioned authority concerning scientific thought had been shattered by individuals such as Galileo, Isaac Newton, and Francis Bacon. The time was now ripe for rapid changes in the study of chemistry.

3

The Birth of Modern Chemistry

Introduction

Through the sixteenth century, the chemical arts had advanced due to developments in metallurgy, medicine, and alchemy. Physicians, metal smiths, tanners, textile workers, soap makers, and a host of other crafts people and artisans employed chemical knowledge to carry out their trades. Although alchemy and the teaching of the ancient Greeks persisted well into the nineteenth century, increasingly individuals started to conduct science by performing experiments. Natural philosophers critically examined ideas such as Aristotle's four-element theory and proposed alternative explanations to describe the behavior of matter. Chemistry was now ready to establish its own identity. Throughout Europe during the two hundred-year period from 1600 to 1800, natural philosophers posed questions, experimented, published their findings, and modified their ideas until the basic principles of chemistry started to take shape. The story of the developments leading up to the late eighteenth-century chemical revolution must be told through the individuals responsible for this revolution.

The Beginnings of Modern Chemistry

Jan Baptista van Helmont (1577–1644) was one of the early iatrochemists who followed the practices of Paracelsus. Van Helmont developed the idea of a gas, which he distinguished from ordinary air. He used the term "spiritus sylvestrius" to describe the gas produced during the combustion process, and he realized that this same gas also was produced during fermentation and when **acids** reacted with sea shells. To van Helmont, a gas was a substance that could not be retained in a vessel and was invisible. Van Helmont considered air and water as the basic elements of matter, and so he never associated different gases as new substances. Because he could condense water vapor but not ordinary air, he used the Greek word for chaos, "khaos." The Greek "kh" is phonetically pronounced as "g" in Flemish, and the result was the English word gas. To van Helmont, a gas was modified water, and water was the basis for chemical change. In one of his experiments van Helmont attempted to prove that plants were transformed water. Van Helmont planted a willow in a measured weight of soil. Reweighing the soil and willow after five years of watering, van Helmont discovered no change in the weight of the soil, but the willow had gained approximately 80 kilograms of weight. From this, van Helmont concluded the water had been converted into plant mass.

The most influential figure of seventeenth-century chemistry was Robert Boyle (1627–1691). Boyle was born in Ireland to a wealthy English family; his father was the Earl of Cork. The Boyle's family wealth allowed Robert to pursue studies in natural philosophy, free from worry about income to support himself. While still in his teens, Boyle traveled to Italy and was in Florence when Galileo died. Boyle was greatly influenced by Galileo's approach to science, and throughout his life based his approach to science on an experimental, mathematical-mechanical study of nature. Because of civil unrest in his native Ireland, Boyle returned to Oxford where he was a member of the group known as the Invisible College. This group was an informal, independent association of thinkers devoted to developing new ideas in science and philosophy.

Boyle, like many scholars of his day, studied and published in a number of areas including theology, philosophy, science, and political thought. In the area of chemistry, Boyle, in the tradition of van Helmont, studied gases. Aided by his assistant Robert Hooke, Boyle used a vacuum pump to conduct experiments in which he discovered air was necessary for life, sound does not travel in a vacuum, and that the volume of a gas is inversely proportional to pressure. This last discovery is one of the basic gas laws, and today is known as **Boyle's Law** (see Chapter 9). Boyle applied his work on gases to a study of the atmosphere and determined the density of air, and how atmospheric pressure changes with elevation.

During the seventeenth century, an atomic view of nature was employed to explain the physical properties of matter. Daniel Sennert (1572–1627) proposed four different types of atoms corresponding to the ancient elements of earth, air, fire, and water. Boyle took an atomic view of nature believing that matter consisted of corpuscles (an idea borrowed from Descartes), which were particles of different size and shape. Like van Helmont, Boyle considered air to be one substance, but considered gases to be various varieties of air rather than a different form of water. Boyle's studies included appreciable work on chemical analysis. In this area, Boyle rejected the idea that water was an element. In one of his many experiments, he boiled water in a flask for many days, adding more water to replace water that boiled away. Boyle discovered sediment in the flask after the boiling process, and he suggested this finding might indicate water was a compound. He speculated that the sediment may have come from the glass and suggested weighing the glass before and after the boiling process to determine the source of the sediment. Boyle never repeated the experiment to determine if the sediment came from the glass. One hundred years later Antoine Lavoisier (1743–1794) repeated Boyle's experiment and demonstrated that indeed the sediment came from the glass flask and had not been derived from the water. Boyle's lasting contribution to chemistry was his book *The Sceptical Chymist* published in 1661. In this work, Boyle sought to expose the inconsistencies and paradoxes of chemical theory of his day. Using the same approach that Galileo used in his *Dialogue*, Boyle presented his ideas in the form of a conversation between several individuals. One person speaks for Aristotle, another for Paracelsus, and one who questions both systems by providing new explanations for the behavior of matter. Boyle refuted Paracelsus' idea that salt, sulfur, and mercury were basic substances. He also advanced the corpuscular theory of matter and used this theory to differentiate mixtures from compounds. Boyle also performed experiments on combustion with Robert Hooke. Boyle and Hooke showed that combustion of charcoal and sulfur required the presence of air, and that these substances could ignite in a vacuum when mixed with saltpeter (potassium nitrate, KNO_3). Hence, Boyle concluded there must be something common to both air and saltpeter (which we now know was oxygen). Boyle also did work on acids and **bases** and he developed indicators to distinguish between them.

John Mayow (1643–1679), a contemporary of Boyle, did important work on **combustion** and respiration. Mayow isolated gases emitted from combustion and respiration by using a siphon to capture the gases in a vessel underwater. Mayow discovered that only part of a volume of air is utilized during combustion and respiration. Mayow called this part nitro-aerial particles. The work, conducted in the 1600s, pointed the way for advances that would take place in chemical theory in the 1700s. The study of gases, respiration, combustion, and calcination were areas that sparked interest. While we are familiar with the process of combustion, to understand chemical developments in the 1700s, it is important to understand the process of calcination. Calcination is a process in which a metal is heated without melting. During the procedure, the metal's volatile components escape. Calcination is often employed as an initial step in the extraction of a metal from its ore. The material left after a substance has undergone the calcination process is known as a calx. Combustion and calcination have been known since ancient times. Combustion was considered to be a process in which something was removed from a substance. The ancients thought that fire was liberated when a substance burns. Paracelsus associated sulfur with combustion, and Boyle proposed special fire particles to explain combustion and calcination. In this line of thinking, fire was the primary element in the combustion/calcination process and air played only a secondary role. Air was the media that absorbed and transported the fire away from a substance during combustion. Using this interpretation, a candle goes out when a jar is placed over it because the air in the jar becomes saturated with fire. Because the air cannot absorb any more fire, the candle goes out.

Using the ancient idea of fire and its modification over the ages, reasonable explanations were developed for chemical processes. Yet, a fundamental fact continued to plague individuals working in chemistry. Combusted substances weighed less after combustion because fire had left the substance. This fact was consistent with the four-element theory. Conversely, when metals were heated during calcination, they gained weight. If fire was supposed to leave a material, why did calcinified metals gain weight? As more emphasis was placed on quantitative methods in the 1700s, this problem grew increasingly troublesome.

The Phlogiston Theory

The phlogiston theory was an attempt to formulate a comprehensive explanation for chemical observations made up to the start of the eighteenth century. John Becher (1635–1682) is generally given credit for laying the foundation of the phlogiston theory. Born in Germany, Becher was a self-educated scientist, economist, and businessman. Becher spent his life serving in the administrations of European royalty, promoting different industrial projects, and pursuing various other business schemes. He never remained in place long, quickly wearing out his welcome with his unfulfilled promises. Included in the latter were a perpetual motion machine and a contraption to change sand into gold. Becher traveled throughout Europe, always keeping one step ahead of his enemies and detractors. Although considered a charlatan by many, Becher conducted and published serious scientific studies. Becher proposed five fundamental elements: air, water, fusible earth, fatty earth, and fluid earth. Fatty earth was combustible earth, fusible earth did not burn, and fluid earth was distillable. Becher's three earths were analogous to Paracelsus' sulfur (fatty earth), mercury (fluid earth), and salt (fusible earth) principles of matter. Becher's work summarized many of the ideas of his predecessors and included many ideas that the Chinese held on chemical processes.

Becher did not develop his theory completely; this was left to his principal disciple, the German physician George Stahl (1660–1734). Stahl changed Becher's term for fatty earth (terra secunda) to phlogiston and popularized the phlogiston theory. Phlogiston is derived from the Greek word *phlogistos* meaning flammable. Stahl accepted phlogiston as a real substance that was responsible for combustion. Phlogiston had no color or taste. Materials that burned almost to completion, such as coal, were believed to be almost completely phlogiston. Substances that did not burn possessed little or no phlogiston. Although we might be quick to dismiss the phlogiston theory today, we must remember that it was the accepted theory for explaining a wide range of phenomenon by the mid-1700s. The phlogiston theory made sense of a number of chemical observations and among these were:

• Combustibles lose weight when they burn because they lose phlogiston.
• Charcoal almost burns completely because it is almost entirely phlogiston.
• A flame is extinguished because the surrounding air becomes saturated with phlogiston.

• Animals die in airtight spaces because the air becomes saturated with phlogiston.
• Metal calxes turn to metal when heated with charcoal because phlogiston transfers from the charcoal to calx.

While the phlogiston theory explained a number of observations, the problem of calcination still remained. Why did metals increase in weight when heated? After all, if phlogiston left the metal during calcination, the metal should weigh less. To accommodate the phlogiston theory, different solutions were proposed to make the theory work. This is precisely the way science works. A theory develops and claims widespread acceptance. As scientists use the theory, observations and experimental results occasionally arise that are not consistent with the theory. Scientists are then forced to modify the theory in some fashion to explain their results. In the case of the calcination problem, some individuals proposed different types of phlogiston. One type had negative weight; therefore, when it was liberated from the substance, the substance weighed more. The idea of negative weight may sound a bit far-fetched, but we must remember people were still considering only a few basic elements at that time. It's easy for us to reason a balloon stays suspended in air because it contains helium, and the balloon sinks as it loses helium. In essence, negative-weight phlogiston was the eighteenth-century equivalent to helium. It was an attempt to resolve a major inconsistency in accepted phlogiston theory, but as further experimentation on gases ensued, a new theory developed to take its place.

Pneumatic Chemistry

The stage was now set for a critical examination of phlogiston theory and the birth of modern chemistry. By the mid-

1700s, the study of chemistry as a distinct discipline had been established. Although the research of Boyle, van Helmont, Stahl, and others provided a sound foundation to build on, it was primarily a series of experiments on gases that took place in the last half of the eighteenth century that created the science of modern chemistry. Several men stand out as prominent figures who contributed to the demise of the phlogiston theory. Black, Priestley, and Cavendish conducted important studies in Great Britain, but it was a Frenchman who revolutionized chemistry. Antoine Lavoisier synthesized his work and that of other natural philosophers into a modern theory of chemistry.

The **pneumatic chemists** did experiments on gases. To do this work, a method was required to generate and isolate the gases produced in a chemical reaction. It was important that the gases generated during reactions not become contaminated with air. An invention that facilitated the study of gases was the **pneumatic trough** (Figure 3.1). Stephen Hales (1677–1761) perfected this device in his studies of animal and plant physiology. Hales' trough was nothing more than a gun barrel bent and sealed at one end. Hale placed a substance in the sealed end and heated it. The open end of the barrel was directed upward into a flask filled with water and partially submerged in a bucket of water. Any gas generated when a substance was heated would be directed through the tube and would displace the water in the flask. Using Hales' pneumatic trough, a group of English chemists conducted important experiments that assisted Lavoisier in formulating his ideas.

Joseph Black (1728–1799) conducted an important series of experiments during work on his doctoral dissertation in medicine. Black was searching for a material to dissolve kidney stones. He chose magnesia alba (magnesium carbonate), but

Figure 3.1
Diagram of Pneumatic Trough (Rae Déjur)

he found it could not be used for his primary topic on kidney stones so he turned his attention to stomach acidity. His studies of magnesia alba were based on a number of experiments on carbonates. Black observed that magnesia alba reacted with acid to produce a gas and a salt. Black also worked with limestone (calcium carbonate, $CaCO_3$) and quicklime (calcium oxide, CaO) at the same time, and so he thought that magnesia alba, which behaved similarly to limestone, might be another form of limestone. Black subjected magnesia alba to calcination and found that it lost a substantial amount of weight during calcination. The residue left after calcination did not behave at all like quicklime. Because quicklime is obtained when limestone is calcined, Black concluded magnesium alba was not a form of limestone. Black continued his investigation by treating both the magnesia alba and the calcined magnesia alba with acid. He found that magnesia alba and acid produced a salt and gas, while the calcined magnesia alba produced the same salt but not the gas. Blacks two reactions were

$$\text{acid} + \text{magnesia alba} \rightarrow \text{salt} + \text{gas}$$
$$\text{acid} + \text{calcined magnesia alba} \rightarrow \text{salt}$$

Black explained his results by determining that calcination of magnesia alba gives off a gas, and this gas is the same gas produced when magnesia alba is treated with acid. It should be remembered that Black interpreted his results based on phlogiston theory. According to phlogiston theory, the magnesia alba should have gained weight when it was calcinated due to phlogiston leaving the fire and flowing into the magnesia alba. Therefore, during the calcination of magnesia alba, gas was given off and phlogiston was gained. Black then devised a method to determine the weight of the gas evolved from magnesia alba. Hales' pneumatic trough would not work, because the gas dissolved in water. Therefore, Black weighed the acid and magnesia alba before and after they reacted. Black measured the difference in weights as the weight of gas evolved. Once he did this, Black found that the change in weight was close to the weight loss when magnesia alba was calcined. This result indicated that there was no weight gain due to phlogiston entering the magnesia alba. Black could explain the changes using conservation of mass without resorting to a gain of phlogiston. The changes in weight in magnesia alba could be explained by the loss of a gas during both calcination and reaction with an acid.

Black decided that the gas given off in the reaction between magnesia alba and acid was similar to one described by van Helmont. He coined the term "fixed air" to indicate that this gas was fixed or trapped in magnesia alba. Black also recognized that **fixed air** was the same gas produced in respiration, combustion, and fermentation. Today, we know Black's fixed air was car-

bon dioxide. Black published his work in 1756 in England, but it took a number of years for word of his study to spread. Black's work inspired other British scientists to study gases. He had demonstrated that gases had to be accounted for in chemical reactions, especially when looking at changes in weight. Black also showed that chemical reactions could be explained without resorting to phlogiston. According to the phlogiston theory, limestone, when heated, absorbed phlogiston to become quicklime. Black showed that the production of quicklime could be explained by the loss of fixed air from limestone. Phlogiston was not needed to explain the process, although Black never abandoned the phlogiston theory.

Joseph Black, upon completion of his study on magnesia alba, received his medical degree. He published very little after this study. Black taught at universities in Glasgow and Edinburgh, and continued to do solid research, which he presented in his lectures. One area in which he did important work was heat and the latent heat of steam. His work in this area inspired one of his students James Watt to apply Black's ideas in making improvements to the steam engine.

Similar to Black, Daniel Rutherford (1749–1819) studied gases for his medical degree dissertation. Rutherford found that common air contained a part that supported respiration and a part that did not. Initially, Rutherford assumed the part that did not support respiration was contaminated by fixed air. Rutherford experimented and removed the fixed air, and he discovered the uncontaminated air still did not support life or combustion. Rutherford assumed the gas he had isolated was ordinary air saturated with phlogiston; hence, it was phlogisticated air, which he referred to as noxious air. What Rutherford had isolated was nitrogen, and he is given credit for its discovery.

Another important figure in the chemical revolution was Joseph Priestley (1733–1804). Priestley was a minister by training who earned his living at various times as a pastor, tutor, and school administrator. He had no formal science training, and throughout his life he was branded a dissident and radical. Early in his scientific studies, Priestley met Benjamin Franklin who sparked his interest in electricity. One year after meeting Benjamin Franklin, Priestley wrote *The History of Electricity*. Priestley's first investigations in chemistry dealt with the gas produced in a brewery near his parish in Leeds. Priestley discovered he could dissolve the gas in water to produce a pleasant tasting beverage. Priestley had actually produced soda water, the equivalent of sparkling water. Priestley's process for producing carbonated water was made commercially feasible in the 1790s by Jacob Schweppe (1740–1821), the founder of the company that still bears his name. Priestley's work on electricity and gases resulted in his election to the French Academy of Sciences in 1772, and he was also honored by the Royal Society in 1773.

In 1772, the second Earl of Shelburne employed Priestley as an advisor and personal secretary. This position gave Priestley ample time and freedom to continue his studies on gases. Priestley used a strong magnifying glass to heat substances trapped under glass, and in this manner he was able to liberate gases from substances. He also employed a pneumatic trough using mercury instead of water. In this manner, Priestley was able to isolate fixed air (CO_2) and a number of other gases. He produced nitrous oxide, N_2O, which he called **dephlogisticated nitrous air**. Priestley noted dephlogisticated nitrous air caused people to laugh, and thus, this gas became known as laughing gas. Nitrous oxide was used years later for anesthesia, and it is currently used as a

propellant in cans of whipped cream. Priestley also isolated alkaline air (ammonia, NH_3) vitriolic acid air (sulfur dioxide, SO_2), heavy inflammable air (carbon monoxide, CO), and nitrous acid air (nitrogen dioxide, NO_2). His research on gases was summarized in a series of volumes published between 1774 and 1786 entitled *Experiments and Observations on Different Kinds of Air*. Priestley's most famous studies concerned the production of oxygen (Figure 3.2). Priestley heated the calx of mercury (mercuric oxide, HgO) and collected the gas. He observed that a candle burned brightly in the gas. At first, Priestley assumed the gas was dephlogisticated nitrous air (N_2O) because of results he had obtained previously. When Priestley further examined the gas and its solubility properties, he decided it was not dephlogisticated nitrous air but an entirely new gas. He noticed that mice lived longer in the gas, and the gas did not lose its ability to support combustion as quickly as dephlogisticated nitrous oxide. Priestley came to realize he was dealing with a gas similar to common air but with a greater ability to support combustion and respiration. We know that Priestley was dealing with oxygen, but Priestley was a strong believer in phlogiston theory. He explained his findings in terms of phlogiston. Priestley assumed oxygen was **dephlogisticated air**, common air in which phlogiston had been removed. To Priestley, dephlogisticated air was air that had the ability to absorb a substantial amount of phlogiston. Priestley was conservative in the interpretation of his results, which was a major difference between Priestley and Lavoisier. Additionally, Lavoisier emphasized quantitative measurements in his study of gases while Priestley placed secondary importance on quantitative relationships.

Priestley left the employment of Lord Shelburne in 1780 having made his major scientific discoveries. Priestley's political

Figure 3.2
Apparatus used by Priestley for investigations on gases. The tub is Priestley's pneumatic trough. From *Experiments on Different Kinds of Air*. Image courtesy of School of Chemical Sciences, University of Illinois at Urbana-Champaign.

and religious ideas continued to cause him trouble within England. He had supported the colonists in the Revolutionary War and supported the French Revolution. His unorthodox Unitarian views ran against those of the Church of England. In 1782, he published a book entitled *History of Corruptions of Christianity*, which was banned by the Church of England. Eventually, the situation got so bad for Priestley that his home and church were burned by an angry mob in 1791. Priestley was forced to move to the United States in 1794, where he lived the last decade of his life in Pennsylvania. In 1874, on the 100th anniversary of Priest-

ley's discovery of oxygen, a celebration was held, and the event marked the founding of the American Chemical Society. The highest honor this society gives is the Priestley Award.

While Joseph Priestley generally receives credit for the discovery of oxygen, it was Carl Sheele (1742–1786) who probably deserves credit as the first to isolate oxygen. Scheele was a Swede who became interested in chemistry during his apprenticeship as an apothecary. Scheele acquired his own pharmacy and during the process researched the production of different medicines. Scheele realized that common air contained a component that was responsible for combustion and termed this component fire air. Scheele succeeded in isolating fire air by reacting nitric acid (HNO_3) and potassium hydroxide (KOH) to obtain potassium nitrate (KNO_3) and then heated the potassium nitrate and used a salt to absorb the nitrogen oxides from the gas produced. The process resulted in fire air. Unfortunately, by the time Scheele published his results in 1777, Priestley's results were well known. In addition to his work on gases, Scheele made numerous other discoveries in the field of chemistry. He is credited with the discovery of a number of organic acids including tartaric, oxalic, uric, lactic, and citric. Scheele also discovered numerous acids such as hydrofluoric, hydrocyanic, and arsenic. The fact that Scheele lived far from the centers of scientific activity and died at the age of 44 contributed to his lack of recognition. Scheele was a firm believer in phlogiston theory and explained all his findings within the phlogiston framework.

Another of the English pneumatic chemists who made important discoveries was Henry Cavendish (1731–1810). For years, it was known that gas was produced when metals reacted with acid. Cavendish studied this gas and in 1766 used the term "inflammable air" to describe the gas. Cavendish's inflammable air was hydrogen. Because of its explosive nature, Cavendish considered inflammable air to be pure phlogiston. Cavendish was able to demonstrate water was a compound by combining inflammable air (hydrogen) and dephlogisticated air (oxygen) and using a spark to ignite the mixture to produce water.

By 1780, a number of different types of airs had been isolated and discoveries made that raised serious questions about the nature of matter. Air, earth, and water were shown to be compounds and mixtures rather than the basic elements from ancient times.

Fire was associated with the mysterious material known as phlogiston. This phlogiston could not be detected with the senses and had either positive, negative, or no weight depending on how experiments were interpreted. The discoveries of Boyle, Black, Cavendish, Scheele, Priestley, and other pioneer chemists were interpreted using the phlogiston theory. Yet, phlogiston theory left too many questions unanswered. Lavoisier, like his contemporaries, was initially a confirmed phlogistonist, but his interpretation of his own work and that of others lead him to abandon phlogiston. Lavoisier's work followed Newton's revelations in physics and preceded Darwin's theory of evolution and established the modern science of chemistry at the end of the eighteenth century.

Lavoisier

Antoine Laurent Lavoisier was a member of the French nobility; his father was a wealthy Parisian lawyer. Educated at the best French schools, Lavoisier developed a deep interest in science. Lavoisier showed early promise in geology and astronomy, but he also was interested in chemistry, which he studied under Giullame Francois Rouelle (1703–1770). Lavoisier was wealthy enough to fund his own scientific studies, but he also was involved in significant government work and business dealings. In 1768, Lavoisier purchased a position in a private French tax-collecting firm known as Ferme Générale. It was through this firm that Lavoisier met his wife, Marie-Anne, who assisted Lavoisier in his scientific endeavors throughout his life.

Lavoisier's serious work in chemistry started around 1770. Lavoisier conducted studies on the combustion of sulfur and phosphorous and focused on the weight increase during the combustion process. Lavoisier's work on combustion led him to believe that a weight gain may accompany all combustion and calcination reactions. Lavoisier related his ideas on air being fixed in substances during combustion and calcination in a note dated November 1, 1772, to the French Academy.

In 1774, Lavoisier's work *Opuscles Physiques et Chymiques* (Physical and Chemical Tracts) appeared in print. This book presented Lavoisier's findings and was based largely on repeating and expanding upon the work of Joseph Black. *Opuscles* indicated that Lavoisier believed calcination and combustion involved combination with some component of air. Initially, Lavoisier thought fixed air was the gas involved with calcination and combustion. At this point, Lavoisier was still unclear about the gas involved in combustion and calcination, and he turned to the work of other natural philosophers to point the way. This was typical of Lavoisier. Much of Lavoisier's laboratory work was based on repeating the work of other scientists. Using this method enabled Lavoisier to observe a wide range of chemical observations and tie these together without resorting to phlogiston. After publication of *Opuscles*, Lavoisier met with Priestley in Paris during the fall of 1774. Lavoisier also was familiar with the work of French chemist Pierre Bayen (1725–1798); Bayen also rejected the idea of phlogiston. Bayen, like Priestley, had reported that the calx of mercury when heated produced fixed air in the presence of carbon and another gas when carbon was not present. Lavoisier also received a letter from Scheele on his work in the fall of 1774.

Lavoisier resumed his work on combustion and calxes in the spring of 1775. At this time, Lavoisier wrote his initial paper entitled *On the Nature of Principles which Combine with Metals During Calcination and Increases Their Weight*. A revised version of this paper did not appear until 1778.

In this paper, Lavoisier said air consists of two elastic fluids and distinguished between active and inactive components of air. Lavoisier's reactive air was the same as Priestley's dephlogisticated air (oxygen). Lavoisier slighted Priestley by not giving Priestley due credit for helping him interpret his work. Priestley felt Lavoisier was claiming credit for his work, but Lavoisier was equally adamant in staking claim to an original interpretation of the role of dephlogisticated air in combustion and calcination. This claim irritated Priestley immensely and resulted in a bitter rivalry between the two for the rest of their lives.

In 1775, Lavoisier was appointed Chief of the Royal Gunpowder Administration. In this position, Lavoisier reformulated French gunpowders and improved their quality to become the best in Europe. At his arsenal lab, Lavoisier continued his chemical experiments. Lavoisier focused much of his attention on acids. He was the first to use the term "oxygine" in September of 1777, a word derived from the Greek terms *oxys* and *gen* meaning acid producing. Lavoisier believed incorrectly oxygine was a basic component in all acids.

Another problem Lavoisier wrestled with was the reaction of dephlogisticated air and inflammable air (oxygen and hydrogen) to produce water. Lavoisier repeated Cavendish's experiments. He also performed an experiment where he passed steam through a heated iron rifle barrel obtaining hydrogen and iron calx (rust). By decomposing water into hydrogen and oxygen, Lavoisier claimed credit for demonstrating water was a compound. Again, others such as Cavendish, Priestley, and James Watt had previously shown water was a compound, but it was Lavoisier who explained the true nature of water without using phlogiston. Lavoisier conducted numerous experiments with different gases, isolating them and studying their characteristics (Figure 3.3). By 1780, Lavoisier was well into developing his new philosophy to explain chemical reactions. He published *Reflections on Phlogiston* in 1783. In this work he attacked the phlogiston theory and introduced his own oxygen theory to explain chemical reactions.

Lavoisier's ideas met initial resistance from the established chemical community. After all, phlogiston had been used for a century to explain chemistry, and natural philosophers had explained their work based on phlogiston theory. The leading chemists of the day, including Richard Kirwan (1733–1812), Priestley, Cavendish, and Torbern Bergman (1735–1784), not only rejected Lavoisier's ideas, but they also were antagonistic toward him. This was not to say

that Lavoisier did not have support. Other chemists had questioned phlogiston, but they could not put forth an acceptable alternative theory. Lavoisier's strongest disciples were several of his countrymen. The French chemists Claude Louis Berthollet (1749–1822), Guyton de Morveau (1737–1816), and Antoine Francois de Fourçroy (1755–1809) joined Lavoisier in promoting his new chemistry. In addition to Lavoisier's new theoretical explanation, he and his colleagues proposed a new system of nomenclature. Led by de Morveau, the French associates proposed a new system of naming chemicals based on their elemental composition. At the time substances were named using archaic names passed down through the ages. Names such as fuming liquor of Libavius (tin tetrachloride, $SnCl_4$), Epsom

Figure 3.3
Lavoisier conducting human respiration experiment while Madame Lavoisier takes notes. Image from Edgar Fahs Smith Collection, University of Pennsylvania Library.

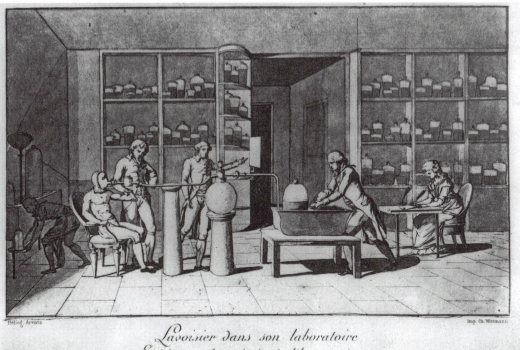

salt (magnesium sulfate, $MgSO_4$), azote (nitrogen), and acid of salt (hydrochloric acid, HCl) gave no indication of the true composition of substances. In their *Methods of Chemical Nomenclature*, published in 1787, the French chemists proposed a system consisting of Latin and Greek names using prefixes and suffixes to distinguish the composition of substances. Using the new nomenclature, chemists now had a language that told them what a substance was made of and pointed the way on how to make the substance. The supporters of phlogiston were reluctant to accept the system and initially rejected the new nomenclature.

Lavoisier summarized his ideas developed over the previous twenty years in his seminal 1789 book *Traité Élémentaire de Chimie* (Elements of Chemistry). This work presented his findings on gases and the role of heat in chemical reactions. He explained his oxygen theory and how this theory was superior to phlogiston theory. Lavoisier established the concept of a chemical element as a substance that could not be broken down by chemical means or made from other chemicals. Lavoisier also presented a table of thirty-three elements. The thirty-three elements mistakenly included light and caloric (heat). Lavoisier put forth the modern concept of a chemical reaction, the importance of quantitative measurement, and the principle of conservation of mass. The final part of Lavoisier's book presented chemical methods, a sort of cookbook for performing experiments.

Lavoisier's *Elements of Chemistry* synthesized and explained in a coherent manner the chemistry of his day. The work was readily accepted by many as a substantial improvement over the phlogiston theory. Young chemists quickly adopted Lavoisier's ideas and new nomenclature rather than trying to fit their work within a phlogiston framework. Upon reading Lavoisier's book,

many of the "old guard" became converts. Lavoisier's ideas spread quickly to other countries. English, Spanish, German, Italian, and Dutch translations of *Elements of Chemistry* were made within a few years. Furthermore, Lavoisier actively promoted his new chemical system in lectures, reviews, and by lobbying fellow chemists.

Lavoisier's greatest ally in his campaign to promote the new chemical system was Marie-Anne Lavoisier. In one episode, Marie-Anne staged a scene where she publicly burned the works of Stahl and other books promoting phlogiston. Marie-Anne served Antoine in several important capacities. As a gifted linguist and artist, she translated English works and illustrated his publications. She served as a personal secretary transcribing notes, editing papers, and assisting in his laboratory experiments.

Lavoisier, by all accounts, was not a humble person. Some suspect he took credit for the work of others, was arrogant, and did little original work. While true in some respects, Lavoisier nevertheless synthesized the vast wealth of chemical knowledge that had accumulated by the end of the eighteenth century. Lavoisier never discovered an element or isolated a new gas, but what he accomplished was far more important. Lavoisier established chemistry as a modern science and laid the foundation for the quantitative experimental methods of chemistry. Lavoisier also served France throughout his life as a public servant. His work on French gunpowder has already been mentioned, but Lavoisier also worked on reform in agriculture, the measurement system (he was an early proponent of the metric system), penal systems, and medicine. Unfortunately, Lavoisier fell victim to the Reign of Terror of the French Revolution. As a partner of Ferme Générale, he was associated with this despised tax-collecting agency. Because Lavoisier was a member of the French nobility, he received little sympathy from the revolutionaries. Although the case against him was weak, Lavoisier was convicted and executed by the guillotine on May 8, 1794.

By the end of the eighteenth century, chemistry had established itself as a modern science. Conservation of mass and the accounting of reactants and products in a chemical balance sheet, although established early in the century, was perfected by Lavoisier as a unifying principle in chemistry. Lavoisier used conservation of mass to draw conclusions, devise new experiments, and transform ideas in chemistry. In this manner, Lavoisier established a method for examining chemical reactions that applies to this day. Like Lavoisier, modern chemists work within a theoretical framework and are guided by general principles as they conduct experiments. Experimental results collectively are used to lend support or ultimately refute accepted theories. Lavoisier's experiments led him away from the phlogiston theory, and in so doing created another framework that guided chemical thought. Lavoisier established modern chemical reasoning and a model for modern chemists. Lavoisier's experiments often required the modification of equipment or fabrication of new equipment. Likewise, advances in modern science are dependent on advances in equipment and technology. Lavoisier also stressed the importance of language in presenting the results of science. Chemical nomenclature today is highly specific with the names of substances portraying their chemical composition using a precise language. Finally, Lavoisier demonstrated how chemical knowledge could be used to better society and promote industry. Lavoisier applied chemistry to real-world problems such as improving agriculture, manufacturing gunpowder, or developing lighter-than-air balloons. Most modern chemists operate in just this manner, applying chemical principles to a endless multitude of practical problems.

The Atom

Introduction

As the eighteenth century drew to a close, Lavoisier's new system of chemistry had won the day. Mysterious phlogiston was no longer needed to interpret chemical phenomenon, yet phlogiston still had a few staunch adherents such as Joseph Priestley. Priestley persisted in his belief of phlogiston until his death in 1804. By this time, however, a whole new school of chemists was busy wrestling with problems raised by Lavoisier's new chemistry. Foremost among these problems was determining the correct composition of substances and assigning the correct **relative masses** to the chemical elements. The relative mass refers to the ratio of the mass of one element to another, for example, how many times heavier is oxygen than hydrogen. The composition and mass problems were closely related. Without knowing the exact composition of a substance, it was impossible to determine accurately the mass of individual elements. Likewise, inaccurate relative masses of the elements brought into question the exact composition of a substance. The mass/composition dilemma was critical to advancing chemistry and drew the attention of chemists at the start of the nineteenth century. Atoms were adopted and gradually accepted during the first part of the century. By the end of the century, the atomic model was one of the fundamental theories in science and fundamental particles comprising the atom began to be discovered.

The Law of Definite Proportions

The problem of chemical composition was directly related to that of chemical combination. Combining chemicals to produce new substances had been a goal of individuals throughout history. Lavoisier's work emphasized the quantitative study of chemical combination. Chemists sought to determine the proportion in which chemicals combined, and how much of one substance it took to react with another. Rather than combining substances using the alchemist's trial and error method, chemists sought to determine the specific ratio of chemicals involved in chemical reactions.

Jermias Benjamin Richter (1762–1807) was one of the first chemists to focus attention on the relative weights of the elements. Richter conducted experiments in which he determined how much acid of various types it took to neutralize bases. Richter was searching for the underlying mathematical relationships inherent in chemical reactions. Richter used measurements, math, and observations to decipher chemical reactions. He coined the word "**stoichiometry**," which is still used to describe mass balances associated with chemical reactions. Richter based his calculated atomic weights on equivalences, or how much of one substance would react with another. A specific amount of sulfuric acid was used as a reference. Ernest Gottfried Fischer (1754–1831) brought attention to Richter's work when he published his German translations of Berthollet's study on chemical combinations.

According to Lavoisier's disciple Berthollet, when elements combined to form compounds, the compounds contained variable proportions of elements within defined limits. Berthollet believed the composition of a compound depended on how much of each element was present when the compound was formed. If salt happened to be produced by a process with ample sodium and little chlorine, then it would be enriched in sodium. Because of Berthollet's stature as a chemist, his views and ideas carried great weight at the turn of the nineteenth century. He had based his ideas on years of study. Berthollet had accompanied Napoleon to Egypt and noticed that in salt lakes sodium chloride reacted with limestone (calcium carbonate) deposits to give sodium carbonate and calcium chloride. Berthollet knew from his lab work in France that solutions of calcium chloride and sodium carbonate reacted to give calcium carbonate and sodium chloride. He now observed the reverse reaction occurring in Egypt's salt lakes.

Lake Reaction

$$2NaCl_{(aq)} + CaCO_{3(s)} \rightarrow Na_2CO_{3(s)} + CaCl_{2(aq)}$$

sodium chloride + calcium carbonate
$\rightarrow$ sodium carbonate + calcium chloride

Lab Reaction

$$CaCl_{2(aq)} + Na_2CO_{3(aq)} \rightarrow CaCO_{3(s)} + 2NaCl_{(aq)}$$

calcium chloride + sodium chloride
$\rightarrow$ calcium carbonate + sodium chloride

This observation provided evidence to advance the idea of reversible reactions. Berthollet analyzed many substances and found their composition did vary within a limited range strengthening his conviction that substances did not display constant composition. Joseph Louis Proust (1754–1826) opposed Berthollet's views on the composition of substances. Proust's chemical education began as an apprentice to his apothecary father in Angers in western France. Subsequently, Proust moved to Paris and left for Spain during the French Revolution. In Madrid, Proust held several academic positions and conducted studies in a well-equipped laboratory. Through his analyses of substances, Proust found that compounds had a fixed and invariable chemical composition. Proust based his findings largely on careful studies of various metal compounds of sulfates, carbonates, and oxides. Proust discovered that laboratory-prepared copper carbonate was no different than copper carbonate found in nature. Proust expressed his views clearly on the composition of compounds:

The properties of true compounds are invariable, as is the ratio of their constituents. Between pole and pole, they are found identical in these two respects; their appearance may vary owing to the manner of aggregation, but their properties never.

Proust put forth his views starting in 1799 and for the next ten years attacked Berthollet's ideas on variable composition. Proust demonstrated that Berthollet's results were based on analyzing substances with impurities and pointed out several other flaws in Berthollet's methods. Berthollet countered with his own explanations, but by 1808 he adopted Proust's view. Proust's Law, also know as the **Law of Definite Proportions**, was one of the cornerstones in the development of modern chemistry. The Law of Definite Proportions meant that compounds had a fixed chemical composition that could be used to help determine the relative masses of elements making up the compound.

Dalton and the Birth of the Atomic Theory

John Dalton (1766–1844) was a Quaker and largely self-educated schoolteacher who developed an interest in chemistry as a result of his work in meteorology. His early scientific work included essays on meteorological observations, equipment, and color-blindness (Daltonism). As a result of his interest in meteorology and the atmosphere, Dalton turned his attention to a study of gases. He discovered that the vapor pressure of water increases with temperature. Dalton considered air to be a mixture of gases rather than a chemical compound, and he formulated the concept of partial pressure. Dalton determined that the total pressure of air was the sum of the individual pressures exerted by individual gases in a mixture. Furthermore, the pressure exerted by each gas was independent of the other gases present in the mixture.

One question that Dalton considered and which guided his formulation of the atomic theory was why the gases in air formed a homogenous mixture and did not separate according to density. Dalton knew that water vapor was lighter than nitrogen, and nitrogen was lighter than oxygen. Why

then didn't these gases form stratified, or separate, layers in the atmosphere? The reason for this according to Dalton was that gases diffused in each other. According to Dalton, the forces between different gas particles were responsible for the mixing of gases in the atmosphere. Like gas particles repelled each other, and unlike gas particles were neutral with respect to each other.

In attempting to explain the behavior of gases, Dalton experimented on the solubility of gases in water. Dalton's work on solubility was done with his close colleague William Henry (1774–1836). Dalton was convinced that the different solubility of gases in water was due to the weight of the ultimate particle of each gas, heavier particles being more soluble than lighter particles. Dalton needed to know the relative weights of the different gases to support his ideas on gas solubility. This need led him to develop a table of comparative weights of ultimate gas particles. In 1803, he presented his table of relative weights (Table 4.1). Over the next five years, he continued to develop his ideas on atomism and work on atomic weights.

As a result of his work on relative weights, Dalton formulated the **Law of Multiple Proportions**, which states that when elements combine to form more than one compound, then the ratio of the masses of elements in the compounds are small whole number ratios of each other. For example, the elements carbon and oxygen form the two compounds carbon monoxide (CO) and carbon dioxide (CO_2). The ratio of

Table 4.1

Relative Weights of Ultimate Gas Particles According to Dalton

Hydrogen	1
Azot (nitrogen)	4.2
Carbon	4.3
Oxygen	5.5
Water	6.5
Gaseous oxide of carbone (CO_2)	9.8

carbon to oxygen masses using Dalton's Table would be

CO

$$\frac{C}{O} = \frac{4.3}{5.5} = 0.78$$

$$\frac{C}{O} \text{ in CO compared to } CO_2 = \frac{0.78}{0.39} = \frac{2}{1}$$

CO₂

$$\frac{C}{O} = \frac{4.3}{11.0} = 0.39$$

Dalton's Law of Multiple Proportions meant that two elements combine in simple whole number ratios. Dalton believed that compounds found in nature would be simple combinations. Hence, knowing that hydrogen combines with oxygen to give water, Dalton's formula for water would consist of 1 H and 1 O. Its formula would be HO using modern nomenclature. Both Proust's Law of Definite Proportions and Dalton's Law of Multiple Proportions are outcomes of an atomic view of nature. In 1808 Dalton published his table of relative atomic weights along with his ideas on atomism in *A New System of Chemical Philosophy*.

Because Dalton did not know the chemical formula for compounds, he assumed the greatest simplicity. This worked fine for some compounds such as CO or NO, but introduced error for other compounds, for example, assuming water was HO. Nevertheless, Dalton's ideas laid the foundation for the modern atomic theory. Dalton's ideas briefly summarized are:

1. Elements are made of tiny indivisible particles called atoms.
2. Elements are characterized by a unique mass specific to that element.
3. Atoms join together in simple whole number ratios to make compounds.
4. During chemical reactions, atoms rearrange themselves to form new substances.
5. Atoms maintain their identity during the course of a chemical reaction.

Dalton's work on relative weights, multiple proportions, and the atomic theory did not have an immediate effect on chemists of his day. Dalton's ideas did provide a framework for determining the empirical formula of compounds, but his table of relative weights was not accurate enough to give consistent results. Many scientists still debated the existence of atoms in the second half of the nineteenth century. Still, little by little, the atomic theory was adopted by chemists as a valid model for the basic structure of matter. While Dalton continued his life as a humble tutor in Manchester, other chemists used Dalton's ideas to establish the atomic theory. Foremost among these was Jöns Jacob Berzelius (1779–1848) of Sweden, the foremost chemical authority of the first half of the nineteenth century.

Gay-Lussac and Avogadro's Hypothesis

Dalton's work provided a system for representing chemical reactions, but inevitably, conflicts arose when trying to resolve Dalton's idea on chemical combination with experimental evidence. According to Dalton, one volume of nitrogen gas combined with one volume of oxygen to give one volume of nitrous gas (nitric oxide). Dalton referred to combination of atoms as compound atoms. Using Dalton's symbols, this reaction would be represented as

1 nitrogen atom + 1 oxygen atom   gives   1 compound atom of nitrous gas

But when one volume of nitrogen reacted with one volume of oxygen, the result was two volumes of nitrous oxide, not one. Joseph Louis Gay-Lussac (1778–1850) had experimentally demonstrated that two volumes of nitrous gas result from combining one volume of nitrogen with one volume of oxygen. Dalton could not use atomic theory to explain Gay-Lussac's results. Berzelius knew Gay-Lussac's experiments were correct and supported Gay-Lussac's hypothesis that volumes combined just like atoms. This meant equal volumes of gases contained equal numbers of atoms. If Gay-Lussac was correct, then volumes could be used to determine the correct formulas. All one would have to do is determine the ratio of combining volumes. This would be similar to the practice of using the ratio of combining masses. Employing gas volumes gave chemists another method for determining atomic masses. It must be remembered that at the time different formulas were obtained depending on the atomic mass used. For example, the formula for water could be HO, H_2O, HO_2, or H_2O_2 depending on which values were used for the masses of hydrogen and oxygen.

One way to resolve the problem of combining volumes was to split the combining atoms in two before the combination. But atoms were supposed to be indivisible, and so this was not an acceptable solution. Berzelius urged Dalton to reexamine his ideas in light of Gay-Lussac's findings, but Dalton would not accept Gay-Lussac's results. In 1811, Amedeo Avogadro (1776–1856) resolved the problem using both Dalton and Gay-Lussac's ideas. Avogadro hypothesized that equal volumes of gases contain equal numbers of molecules as opposed to atoms. Avogadro took the basic unit of a gas as a molecule not an atom. This meant when nitrogen and oxygen combined, the reaction could be represented as

1 molecule of nitrogen + 1 molecule of oxygen gives 2 molecules of nitrous gas

Avogadro had no experimental evidence to back his claim, yet his hypothesis solved the problem. It took fifty years for the scientific community to accept Avogadro's idea. Today, we take for granted that common gases such as hydrogen, oxygen, and nitrogen exist as molecules rather than atoms in their natural state.

The Divisible Atom

Dalton proposed that atoms are the basic building blocks of matter in the early 1800s. It is one thing to claim the existence of atoms, but it is another to develop the concept until it is readily accepted. A century after Lavoisier's chemical revolution the true nature of the atom began to unfold. Although the question about the true nature of the atom concerned chemists, physicists led the way in the study of the atom. Evidence accumulated during the 1800s indicated that Dalton's atoms might not be indivisible particles. As the twentieth century unfolded, so did the mystery of the atom.

One area of research that raised questions about the atom involved experiments conducted with gas discharge tubes. Gas discharge tubes are sealed glass tubes containing two metal **electrodes** at opposite ends of the enclosed tube (similar to small fluorescent lights). During operation, the metal electrodes are connected to a high voltage power source. The negatively charged electrode is called the **cathode**, and the positively charged electrode is the **anode**. Before a gas discharge tube is sealed, most of the air is pumped out of it. Alternatively, the tube may contain a gas such as hydrogen or nitrogen under very low pressure. During the last half of the nineteenth century, numerous improvements were made in gas discharge tubes.

William Crookes (1832–1919) observed a faint greenish glow and beam using his own discharge tubes called Crookes tube. Crookes noted that the beam in his tubes originated from the cathode, and hence, the beams became known as cathode rays. The tubes themselves took on the name cathode-ray tubes (Figure 4.1). Crookes conducted numerous experiments with cathode-ray tubes. He observed that cathode rays moved in straight lines, but they could be deflected by a magnet. From this work, Crookes concluded cathode rays were some sort of particle, but other researches believed cathode rays were a form of light.

Part of the problem involved in interpreting cathode rays resulted from the ambiguous results obtained by various researchers. For example, cathode rays were bent in a magnetic field supporting the particle view of cathode rays, yet when subjected to an electric field, cathode rays were not deflected. The fact that cathode rays were not deflected in an electric field supported the view that cathode rays were a form of electromagnetic radiation or light. The true nature of cathode rays was ultimately determined by a group of physicists led by J. J. Thomson (1856–1940) at the Cavendish Laboratory at Cambridge University.

Joseph John Thomson was appointed head of the prestigious Cavendish Laboratory (endowed by the Cavendish family in honor of Henry Cavendish) at the age of 28.

Figure 4.1
Cathode-Ray Tube (Rae Déjur)

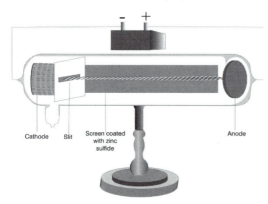

Thomson was a brilliant mathematician and experimenter who tackled the problem of cathode rays at the end of the nineteenth century. Thomson's initial experiments led him to believe cathode rays were actually negatively charged particles. Using cathode-ray tubes in which the air had been evacuated until a very low pressure was attained ($\approx 1/200$ atmosphere), Thomson demonstrated cathode-rays could indeed be deflected by an electric field. The failure of previous researchers to evacuate their discharge tubes to such low pressure had prevented the cathode-ray beams from being deflected in an electric field. Thomson was convinced that cathode rays were particles or in his terms "corpuscles," but he still had to determine the magnitude of the charge and mass of his corpuscle to solidify his arguments.

Thomson and his colleagues at Cavendish conducted numerous experiments on cathode-rays. They used different gases in their tubes, varied the voltage, and employed different metals as electrodes. As the years went by and Thomson accumulated more and more data, he was able to calculate a charge to mass ratio for the corpuscle. Thomson's results startled the scientific community. His results showed that the charge to mass ratio (e/m) was one thousand times greater than the accepted value for hydrogen ions (hydrogen ions are just protons).

$$\frac{\text{Charge}}{\text{Mass}} = \frac{1000e}{m} \qquad \frac{e}{m}$$

Thomson's Corpuscle   Hydrogen Ion

At the time, hydrogen ions were the smallest particle known to exist. Thomson's results could be interpreted as either the charge of the corpuscle being one thousand times greater than the hydrogen ion or the mass of the corpuscle being 1/1,000 that of the hydrogen ion. Thomson suspected the charge of the corpuscle was equal and opposite to that of the hydrogen ion, and therefore, assumed the corpuscle had a mass of 1/1,000 of a hydrogen ion. Thomson published his results in 1897. Thomson's corpuscles eventually came to be called electrons based on a term coined by George Stoney (1826–1911) to designate the fundamental unit of electricity.

While Thomson's work proved the existence of electrons, his studies still left open the question of the exact charge and mass of the electron. Robert Andrew Millikan (1868–1953) tackled this problem at the University of Chicago by constructing an instrument to measure the mass and charge of the electron (Figure 4.2). Millikan's instrument consisted of two parallel brass plates separated by about one centimeter. A pinhole-size opening drilled into the top plate allowed oil droplets sprayed from an atomizer to fall between the two plates. The opening between the two plates was brightly illuminated and a microscopic eyepiece allowed Millikan to observe individual oil droplets between the two plates. Millikan calculated the mass of oil droplets by measuring how long it took an oil droplet to move between the two uncharged plated. Next, Millikan charged the top brass plate positive and the bottom plate negative. He also used an x-ray source to negatively

charge the oil droplets when they entered the space between the plates. By varying the charge, Millikan could cause the oil droplets to rise or stay suspended between the plates. Using his instrument and knowing the mass of the oil droplet enabled Millikan to calculate the charge of the electron, negative. In 1913 Millikan determined that the charge of an electron was -1.591×10^{-19} coulomb. This is within 1% of the currently accepted value of $-1.602177 \times 10^{-19}$ coulomb. With Millikan's value for charge, the electron's mass could be determined. The value was even smaller than given by Thomson and found to be 1/1,835 that of the hydrogen atom. Millikan's work conducted around 1910 gave definition to Thomson's electron. The electron was the first subatomic particle to be discovered, and it opened the door to the search for other subatomic particles.

In their studies with cathode rays, researchers observed different rays traveling in the opposite direction of cathode rays. In 1907, Thomson confirmed the rays carried a positive charge and had variable mass depending on the gas present in the cathode-ray tube. Thomson and others found the positive rays were as heavy or heavier than hydrogen atoms. In 1914, Ernest Rutherford (1871–1937) proposed that the positive rays were composed of a particle of positive charge as massive as the hydrogen atom. Subsequent studies on the interaction of alpha particles with matter demonstrated that the fundamental positive particle was the proton. By 1919, Rutherford was credited with identifying the proton as the second fundamental particle.

The last of the three fundamental particles is the neutron. Experimenters in the early 1930s bombarded elements with alpha particles. One type of particle produced had the same mass of the proton, but carried no charge. James Chadwick (1891–1974), in collaboration with Rutherford, conducted

Figure 4.2
Picture of Millikan's Instrument (Rae Déjur)

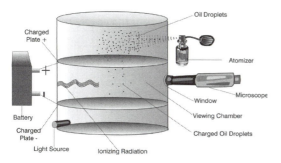

experiments to confirm the existence of the neutral particle called the neutron. Chadwick is credited with the discovery of the neutron in 1932.

Atomic Structure

With the discovery of the neutron, a basic atomic model consisting of three fundamental subatomic particles was complete. While discoveries were being made on the composition of atoms, models were advanced concerning the structure of the atom. John Dalton considered atoms to be solid spheres. In the mid-1800s, one theory considered atoms to consist of vortices in an ethereal continuous fluid. J.J. Thomson put forth the idea that the negative electrons of atoms were embedded in a positive sphere. Thomson's model was likened to raisins representing electrons in a positive blob of pudding; the model was termed the "plum-pudding" model as shown in Figure 4.3. The negative and positive parts of the atom in Thomson's model canceled each other, resulting in a neutral atom.

A clearer picture of the atom began to emerge toward the end of the first decade of the twentieth century. This picture was

Figure 4.3
Thomson proposed the atom consisted of negative electrons embedded in a positive pool, like raisins in plum pudding.

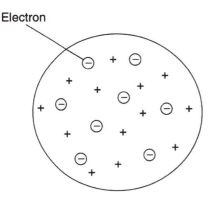

greatly aided by several significant discoveries in physics. In 1895 while observing the glow produced by cathode rays from a sealed Crookes tube, Wilhelm Conrad Roentgen (1845–1923) noticed that nearby crystals of barium platinocyanide glowed. Barium platinocyanide was a material used to detect cathode rays, but in this case, the cathode rays should have been blocked by the glass and an enclosure containing the Crookes tube. Roentgen also discovered that a photographic plate inside a desk was exposed and contained an image of a key resting on top of the desk. Evidently whatever had caused the barium platinocyanide to glow could pass through the wooden desk but not through the metal key. Roentgen coined the term "x-rays" for his newly discovered rays. He was awarded the first Nobel Prize in physics for his discovery in 1901.

Roentgen's discovery of x-rays stimulated great interest in this new form of radiation worldwide. Antoine Henri Becquerel (1852–1908) accidentally discovered the process of radioactivity while he was studying x-rays. Radioactivity involves the spontaneous disintegration of unstable atomic nuclei. Becquerel had stored uranium salts on top of photographic plates in a dark drawer. When Becquerel retrieved the plates, he noticed the plates contained images made by the uranium salts. Becquerel's initial discovery in 1896 was further developed by Marie Curie (1867–1934) and Pierre Curie (1859–1906). Marie Curie coined the word "radioactive" to describe the emission from uranium.

Three main forms of radioactive decay involve the emission of alpha particles, beta particles, and gamma rays. An alpha particle is equivalent to the nucleus of a helium atom. Beta particles are nothing more than electrons. Gamma rays are a form of electromagnetic radiation.

Table 4.2
Forms of Radiation

Type	Symbol	Mass	Charge	Equiv
Alpha	α	4	+2	^4_2He
Beta	β	~0	-1	$^0_{-1}\text{e}$
Gamma	γ	0	0	

Physicists and chemists doing pioneer work on atomic structure employed **radioactive** substances to probe matter. By examining how radiation interacted with matter, these researchers developed atomic models to explain their observations. Rutherford along with Johannes Wilhelm Geiger (1882–1945), creator of the first Geiger counter in 1908 to measure radiation, and Ernest Marsden (1889–1970) carried out their famous gold foil experiment that greatly advanced our concept of the atom. Rutherford's experimental set-up is shown in Figure 4.4.

Rutherford used foils of gold, platinum, tin, and copper and employed polonium as a source of alpha particles. According to Thomson's plum pudding model, alpha particles would pass through the gold with little or no deflection. The uniform distribution of charge in Thomson's model would tend to balance out the total deflection of a positive alpha particle as it passed through the atom. Rutherford, Geiger, and Marsden found a significant number of alpha particles that experienced large deflections of greater than 90°. This result surprised Rutherford who years later described the phenomenon of widely scattered alpha particles "as if you had fired a 15-inch shell at a piece of tissue paper and it came back and hit you." To account for these experimental results, Rutherford proposed a new atomic model in which the atom's mass was concentrated in a small positively charged nucleus. The electrons hovered around the nucleus at a relatively great distance. To place the size of Rutherford's atom in perspective, consider the nucleus to be the size of the period at the end of this sentence, the electrons would circle the nucleus at distances of several meters. If the nucleus was the size of a ping-pong ball, the electrons would be about one kilometer away. Rutherford's planetary model represented a miniature solar system with the electrons rotating around the nucleus like the planets around the sun. Most of the atom's volume in Rutherford's model consisted of empty space, and this space explained the results from the gold foil experiment. Most alpha particles passed through the foil with little or no deflection. This occurred because the net force on the alpha particle was close to zero as it passed through the foil. The widely scattered alpha particles resulted when the positively charged alpha particles approached the nucleus of a gold atom. Because like charges repel each other, the alpha particle would be deflected or even back-scattered from the foil.

Figure 4.4
Rutherford's Set-Up (Rae Déjur)

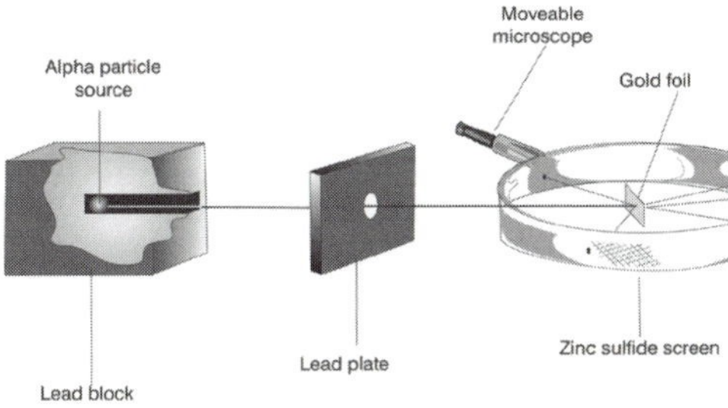

Niels Bohr (1885–1962) advanced Rutherford's model by stipulating that an atom's electrons did not occupy just any position around the nucleus, but they occupied specific orbitals to give a stable configuration. Bohr based his ideas on a study of the spectrum for hydrogen. A spectrum results when light is separated into its component colors by a prism. When visible light passes through a prism, a continuous spectrum similar to a rainbow results. Light from a hydrogen discharge tube does not produce a continuous rainbow, but it gives a discontinuous pattern of lines with broad black areas separating the lines. Bohr proposed specific electron energy levels or orbitals to account for the lines produced in hydrogen's emission spectrum. In Bohr's model, when the hydrogen in the tube became energized by applying a voltage, electrons jumped from a lower energy level to a higher level. As they moved back down, energy was released in the form of visible light to produce the characteristic lines of hydrogen in its spectrum. Figure 4.5 displays this graphically. While Bohr's model successfully explained hydrogen's line spectrum, it could not account for the spectra obtained for gases with more than one electron such as that for helium. Nonetheless, Bohr had shown that matter was discontinuous and introduced the idea of the quantum nature of the atom.

Quantum Mechanics

World War I delayed advances in basic research on the atom for several years as scientists turned their attention to national security. The end of WWI brought renewed interest and further development on atomic theory in the 1920s. Louis de Broglie (1892–1977) from France introduced the idea of wave properties of matter in his doctoral thesis in 1924. The Austrian Edwin Schrödinger (1887–1961) formulated wave mechanics to explain the behavior of electrons in terms of wave properties. Werner Heisenberg (1901–1976) from Germany developed an atomic theory using matrix mathematics. Schrödinger and Heisenberg's work between 1925 and 1928 contributed to the formulation of **quantum mechanics**. In quantum mechanics, the position and energy of electrons are described in terms of probability. Rather than saying an electron is located at a specific position, quantum numbers are used to describe each electron in an atom. The numbers give the probability of finding an electron in some region surrounding the nucleus. The quantum-mechanical model is a highly mathematical and nonintuitive picture of the atom. Rather than thinking of electrons orbiting the nucleus, the quantum mechanical picture places electrons in clouds of different shapes and sizes around the nucleus. These clouds are referred to as orbitals. Several orbital are shown in Figure 4.6.

Figure 4.5
Electrons Changing Energy Level in the Bohr Atom

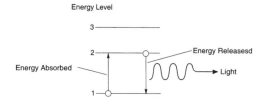

Figure 4.6
Representation of Several Different Orbitals

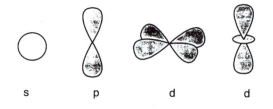

In the quantum-mechanical model of the atom, electrons surround the nucleus in orbitals, which are in regions called shells. Shells can be thought of as a group of orbitals, although the first shell contains only one orbital. The number of electrons surrounding the nucleus equals the number of protons in the nucleus forming a neutral atom. An atomic orbital is a volume of space surrounding the nucleus where there is a certain probability of finding the electron. The electrons surrounding the nucleus of a stable atom are located in the orbitals closest to the nucleus. The orbital closest to the nucleus has the lowest energy level, and the energy levels increase the farther the orbital is from the nucleus. The location of electrons in atoms should always be thought of in terms of electrons occupying the lowest energy level available. An analogy can be drawn to the gravitational potential energy of an object. If you hold an object such as this book above the floor, the book has a certain amount of gravitational potential energy with respect to the floor. When you let go of the book, it falls to the floor where its potential energy is zero. The books rests on the floor in a stable position. If we pick the book up, we used work to increase its potential energy and it no longer rests in a stable position. In a similar fashion we can think of electrons as resting in stable shells close to the nucleus where the energy is minimized.

Orbitals can be characterized by quantum numbers. One way to think of quantum numbers is as the addresses of electrons surrounding the nucleus. Just as an address allows us to know where we are likely to find a person, quantum numbers give us information about the location of electrons. We cannot be sure a person will be home at their address, and we can never be exactly sure an electron will be located within the orbital. The address of electrons in an atom is given by using quantum numbers and is referred to as the electron configuration. The electron configuration of each element provides valuable information about the element's chemical properties.

Four quantum numbers characterize the electrons in an atom. These are:

1. The principal quantum number symbolized by the letter n. The principal quantum number tells which shell the electron is in and can take on integral values starting with 1. The higher the principal quantum number, the farther the orbital is from the nucleus. A higher principal quantum number also indicates a higher energy level. Letters may also be used to designate the shell. Orbitals within the same shell have the same principal quantum number.

Principal Quantum Numbers

n	1	2	3	4	5	6
Shell	K	L	M	N	O	P

— Energy →

2. The angular momentum quantum number is symbolized by the letter l. The angular momentum quantum numbers gives the shape of an orbital. Values for l depend on the principal quantum number and can assume values from 0 to n–1. When l = 0, the shape of the orbital is spherical, and when l = 1, the shape is a three-dimensional figure eight (see representations in Figure 4.6). As the value of l increases, more complex orbital shapes result. Letter values are traditionally used to designate l values.

l value	0	1	2	3	4
Letter	s	p	d	f	g

— Energy →

The letter designation originated from spectral lines obtained from the elements. Lines were characterized as sharp, principal, or diffuse (s,p,d). Lines after d were labeled alphabetically starting from the letter f.

3. The magnetic quantum number symbolized by m_l. The magnetic quantum num-

ber gives the orientation of the orbital in space. Allowable values for m_l are integer values from -1 to $+1$, including 0. Hence, when $l = 0$, then m_l can only be 0, and when $m_l = 1$, m_l can be -1, 0, or $+1$.

4. The spin quantum number symbolized by m_s. The spin quantum number tells how the electron spins around its own axis. Its value can be $+\frac{1}{2}$ or $-\frac{1}{2}$, indicating a spin in either a clockwise or counterclockwise direction.

Using the above definitions for the four quantum numbers, we can list what combinations of quantum numbers are possible. A basic rule when working with quantum numbers is that no two electrons in the same atom can have an identical set of quantum numbers. This rule is known as the Pauli Exclusion Principle named after Wolfgang Pauli (1900–1958). For example, when $n = 1$, l and m_l can be only 0 and m_s can be $+\frac{1}{2}$ or $-\frac{1}{2}$. This means the K shell can hold a maximum of two electrons. The two electrons would have quantum numbers of $1,0,0,+\frac{1}{2}$ and $1,0,0,-\frac{1}{2}$, respectively. We see that the opposite spin of the two electrons in the K orbital means the electrons do not violate the Pauli Exclusion Principle. Possible values for quantum numbers and the maximum number of electrons each orbital can hold are given in Table 4.3 and shown in Figure 4.7.

Using quantum numbers, electron configurations can be written for the elements. To write electron configurations we need to remember that electrons assume the most stable configuration by occupying the lowest energy levels. We also must remember that no two electrons can have the same four quantum numbers. Let's start with the simplest element hydrogen and write several configurations to demonstrate the process. Hydrogen has only one electron. This electron will most likely be found in the K ($n = 1$) shell. The l value must also be 0. We don't

Figure 4.7
Maximum Numbers of Electrons per Orbital

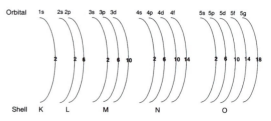

need to consider the m_l and m_s since the former is 0 and the latter can arbitrarily be assigned either plus or minus $\frac{1}{2}$. The electron configuration for hydrogen is written as

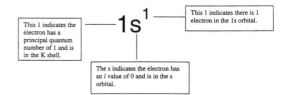

Helium has two electrons. Both electrons have n values of 1, l values of 0, and m_l values of 0, but they must have opposite spins to conform to the Pauli Exclusion Principle. Therefore, the electron configuration of He is $1s^2$. Lithium with an **atomic number** of 3 has three electrons. The first two electrons occupy the 1s orbital just like helium. Because the 1s orbital in the K shell can hold only two electrons, the third electron must go into the next lowest energy level available in the L shell. This is where the n value is 2 and the l value is 0. The electron configuration of lithium is thus $1s^2 2s^1$. As the atomic number increases moving through the periodic table, each element has one more electron than the previous element. Each additional electron goes into the orbital with the lowest energy. The electron configuration for the first twenty elements is given in Table 4.4.

An examination of the electron configuration of the first 18 elements indicates that the addition of each electron follows a reg-

Table 4.3
Quantum Numbers

Shell	n	Possible l	Possible m_l	m_s	Maximum Electrons
K	1	0	0	$+\frac{1}{2},-\frac{1}{2}$	2
L	2	0,1	-1,0,1	$+\frac{1}{2},-\frac{1}{2}$	8
M	3	0,1,2	-2,-1,0,1,2	$+\frac{1}{2},-\frac{1}{2}$	18
N	4	0,1,2,3	-3,-2,-1,0,1,2,3	$+\frac{1}{2},-\frac{1}{2}$	32
O	5	0,1,2,3,4	-4,-3,-2,-1,0,1,2,3,4	$+\frac{1}{2},-\frac{1}{2}$	50

ular pattern. One would expect that the electron configuration of element 19, potassium, would be $1s^2\,2s^22p^6\,3s^23p^63d^1$, but Table 4.4 indicates the 19th electron goes into the 4s orbital rather than 3d. The same is true for calcium with the 20th electron also falling into the 4s orbital rather than 3d. It is not until element 21, scandium, that the 3d begins to fill up. The reason for this and several other irregularities has to do with electron repulsion. It is easier for the electrons to go into the 4th shell rather than the partially filled 3rd shell.

The quantum mechanical model and the electron configurations of the elements provide the basis for explaining many aspects of chemistry. Particularly important are the electrons in the outermost orbital of

Table 4.4
Electron Configurations of the First 20 Elements

1 H	$1s^1$	11 Na	$1s^2\,2s^2 2p^6\,3s^1$
2 He	$1s^2$	12 Mg	$1s^2\,2s^2 2p^6\,3s^2$
3 Li	$1s^1\,2s^1$	13 Al	$1s^2\,2s^2 2p^6\,3s^2 3p^1$
4 Be	$1s^2\,2s^2$	14 Si	$1s^2\,2s^2 2p^6\,3s^2 3p^2$
5 B	$1s^2\,2s^2 2p^1$	15 P	$1s^2\,2s^2 2p^6\,3s^2 3p^3$
6 C	$1s^2\,2s^2 2p^2$	16 S	$1s^2\,2s^2 2p^6\,3s^2 3p^4$
7 N	$1s^2\,2s^2 2p^3$	17 Cl	$1s^2\,2s^2 2p^6\,3s^2 3p^5$
8 O	$1s^2\,2s^2 2p^4$	18 Ar	$1s^2\,2s^2 2p^6\,3s^2 3p^6$
9 F	$1s^2\,2s^2 2p^5$	19 K	$1s^2\,2s^2 2p^6\,3s^2 3p^6\,4s^1$
10 Ne	$1s^2\,2s^2 2p^6$	20 Ca	$1s^2\,2s^2 2p^6\,3s^2 3p^6\,4s^2$

an element. These electrons, known as the **valence** electrons, are responsible for the chemical properties elements display, bonding, the periodic table, and many chemical principles. The role the valence electrons play will be discussed more fully when these topics are addressed.

A Modern View of the Atom

Since the beginning of the twentieth century when Thomson identified the electron as part of the atom, hundreds of other subatomic particles have been identified. Modern particle accelerators and atom smashers have enabled scientists to catalog a wide variety of particles each characterized by their unique mass, charge, and spin. Murray Gell-Mann (1929–) and George Zweig (1937–) independently proposed the existence of **quarks** in 1963 as the basic constituents of matter. The term "quark" was coined by Gell-Mann who obtained it from a line in James Joyce's 1939 novel *Finnegans Wake:* "Three quarks for Muster Mark!" According to the quark theory, small hypothesized particles call quarks combine to produce subatomic particles.

Figure 4.8
Structure of Protons and Neutrons

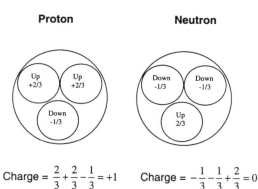

$$\text{Charge} = \frac{2}{3} + \frac{2}{3} - \frac{1}{3} = +1 \qquad \text{Charge} = -\frac{1}{3} - \frac{1}{3} + \frac{2}{3} = 0$$

Quarks carry a fractional charge of $\pm 1/_3$ or $\pm 2/_3$. Six "flavors" or types of quarks make up all subatomic particles. Each flavor of quark can be further classified as having one of three colors. These are not colors or flavors as commonly thought of, but part of a classification scheme used to explain how matter behaves. The language of quarks makes them seem like some creation of fantasy, but the quark theory can be used to explain many properties of subatomic particles. For example, a proton can be considered to be made of two up quarks and a down quark, and a neutron of two down quarks and an up quark (Figure 4.8). Quark flavors and charges are given in Table 4.5.

Table 4.5
Flavor and Charge of Quarks

Quark Flavor	Charge
Up	$+\dfrac{2}{3}$
Down	$-\dfrac{1}{3}$
Strange	$-\dfrac{1}{3}$
Charmed	$+\dfrac{2}{3}$
Bottom (beauty)	$-\dfrac{1}{3}$
Top (truth)	$+\dfrac{2}{3}$

Electrons are not thought to be made of quarks, but of another class of elementary particles called leptons. Leptons do not seem to be made of other particles and are either neutral or carry a unit charge.

Atomic Mass, Atomic Numbers, and Isotopes

The most basic unit of a chemical element that can undergo chemical change is an atom. Atoms of any element are identified by the number of protons and neutrons in the nucleus. The number of protons in the nucleus of an element is given by the **atomic number**. Hydrogen has one proton in its nucleus so its atomic number is one. The atomic number of carbon is six, because each carbon atom contains six protons in its nucleus. Besides protons, the nucleus contains neutrons. The number of protons plus the number of neutrons is the **mass number** of an element. A standard method of symbolizing an element is to write the elements with the mass number written as a superscript and the atomic number as a subscript. Carbon-12 would be written as

$$^{12}_{6}C$$

An element's atomic number is constant, but most elements have varying mass numbers. For example, all hydrogen atoms have an atomic number of one, and almost all hydrogen atoms have a mass number of one. This means most hydrogen atoms have no neutrons. Some hydrogen atoms have mass numbers of two or three, with one and two neutrons, respectively. Different forms of the same element that have different mass numbers are known as **isotopes**. Three isotopes of hydrogen are symbolized as

$$^{1}_{1}H \qquad ^{2}_{1}H \qquad ^{3}_{1}H$$

Hydrogen   Deuterium   Tritium

Isotopes of elements are identified by their mass numbers. Hence, the isotope C-14 is the form of carbon that contains eight neutrons. Isotopes of an element have similar chemical properties. Therefore, when we eat a bowl of cereal and incorporate carbon in our bodies, we are assimilating C-14 (and other carbon isotopes) along with the common form of carbon, C-12.

The mass number gives the total number of protons and neutrons in an atom of an element, but it does not convey the absolute mass of the atom. To work with the masses of elements, we use comparative masses. Initially, Dalton and the other pioneers of the atomic theory used the lightest element hydrogen and compared masses of other elements to hydrogen. The modern system uses C-12 as the standard and defines one atomic mass unit (amu) as 1/12 the mass of one C-12 atom. One amu is approximately 1.66×10^{-24}g. This standard means the masses of individual protons and neutrons are slightly more than 1 amu as shown in Table 4.6.

The atomic mass of an element is actually the average mass of atoms of that element in atomic mass units. Average mass must be used because atoms of any elements exist as different isotopes with different masses. Thus, the atomic mass of carbon is slightly higher than 12 with a value of 12.011. This is because most carbon (99%) exists as C-12, but heavier forms of carbon also exist.

Ions

Mass numbers and atomic masses focus on the nucleus of the atom. Little has been

Table 4.6
Masses of Fundamental Atomic Particles

Particle	Mass (amu)	Mass (grams)	Charge
Proton	1.007276	1.673×10^{-24}	+1
Neutron	1.008664	1.675×10^{-24}	0
Electron	5.485799×10^{-4}	9.109×10^{-28}	-1

said about electrons because the mass of electrons is negligible compared to the mass of protons and neutrons. Electrons have only about 1/2,000 the mass of protons and neutrons. For an atom to remain electrically neutral, the number of electrons must equal the number of protons. When a neutral atom gains or loses electrons, a charged particle called an **ion** results. This process is known as **ionization**. Positive ions are referred to as **cations**, and negative ions are call **anions**. Atoms gain or lose one or more electrons to become ions. Ions are written using the elements symbol and writing the charge using a superscript. A simple equation can be written to symbolize the ionization process. For example, when lithium loses an electron to form Li^+, the equation is

$$Li \rightarrow Li^+ + e^-$$

Several other equations representing the ionization process are

$$F + e^- \rightarrow F^-$$
$$Ca \rightarrow Ca^2 + 2e^-$$
$$O + 2e^- \rightarrow O^{2-}$$

Atoms do not gain or lose electrons in a haphazard fashion. Chemical reactions involve the loss and gain of electrons. In fact, the chemical behavior of all substances is dictated by how the substances' valence electrons interact when the substances are brought together. This subject is introduced in the next chapter in a preliminary discussion of compounds, and the concepts form the basis of our discussion on bonding.

Summary

The resurrection of Democritus' atom by Dalton at the start of the nineteenth-century was one of those seminal events in the history of science. Dalton's atom, while not immediately accepted, nonetheless pro-

vided an organizing framework to guide chemical thought for the next one hundred years. Dalton's atomic theory followed closely on the heels of Lavoisier's work, and the work of these two scientists lit the path that led to the development of modern chemistry. The atomic theory eventually became the touchstone by which new chemical theories were judged, and it played a fundamental role in the development of new branches of chemistry such as organic chemistry, biochemistry, and nuclear chemistry. The atomic theory and the problem of atomic weights was at the heart of the development of the periodic table in the mid-1800s.

As the twentieth-century unfolded, the internal structure of the atom began to unfold, and as it did, immediate practical uses were made of this knowledge. Roentgen's x-rays were applied to medical diagnosis almost immediately after their discovery in 1895. While humans have always pondered the ultimate nature of matter, theoretical knowledge also provided the hope of harnessing the atom's energy. This hope was realized in December of 1942 when Enrico Fermi (1901–1954) created the first controlled chain reaction beneath the football stadium at the University of Chicago. This experiment led three years later to the use of atomic weapons to end World War II. Nuclear fission was initially viewed as a new source of energy with vast potential. The promise of nuclear power has yet to be realized as nuclear accidents occurring at Three Mile Island, Pennsylvania, in 1979 and Chernobyl, Ukrainia, in 1986 raised serious questions in the public's mind about the safety of nuclear power. Current research on fusion may eventually lead to it being employed as an energy source. More will be said about this topic in the chapter on nuclear chemistry (Chapter 17).

While the hope of obtaining energy from atoms has diminished in recent years, there are numerous other applications based on the atomic nature of matter. With simple atoms and molecules as starting blocks, organic synthesis led to the development of the modern chemical industry. Over the years, a host of materials have been produced that we naturally accept as part of everyday life, for example, teflon, polyvinyl chloride (pvc), polyester, and nylon. This process continues as witnessed by the creation of the relatively new discipline of material science. Advances in medicines and pharmaceuticals are yet another aspect of organic synthesis that owes its existence to a fundamental knowledge of atomic theory.

The atomic theory is so rooted in today's science that we readily accept the existence of atoms. Even though few of us have observed them directly, we have faith that everything around us is made of them. Yet we continue to probe the structure of the atom. It is interesting that we continue to seek greater knowledge about the atom and its smallest constituents in order to understand the largest entity we can think of—the universe. Just as Democritus' quest was to build a philosophical system to explain nature, we continue to probe the atom in hope of understanding the universe and our place in it.

Atoms Combined

Introduction

Atoms rarely exist as individual units. Atoms combine with each other to produce the familiar substances of everyday life. Chemistry is largely the study of how atoms combine to form all the different forms of matter. The reason atoms combine involves the subject of chemical bonding, which is explored in Chapter 7. In this chapter, the grouping of atoms into different types of compounds is examined. In the first half of the chapter, chemical nomenclature is discussed. Some of the basic rules for naming compounds are presented. Atoms combine and are rearranged through chemical reactions. The last half of the chapter examines the basic process of chemical reactions and classifies several different types of reactions.

Mixtures, Compounds, and Molecules

An element is a pure substance that cannot be broken down chemically. The smallest unit of an element is an atom. When substances combine and retain their original identity, the combination is called a mixture. In a mixture, two or more substances are brought together and no chemical reaction takes place. Mixtures may consist of substances in the same or different physical states. For example, bronze is an alloy consisting of the two solid metals, copper and tin. The most common solutions are mixtures of a solid in a liquid. Mixtures may be classified as homogenous or heterogeneous. Homogenous mixtures consist of a uniform distribution of the substances throughout the mixture. Air is a homogenous mixture of nitrogen, oxygen, argon and other trace gases. Heterogeneous mixtures do not have a uniform consistency. A bottle of salad dressing consisting of separate layers of oil and vinegar is an example of a heterogeneous mixture.

Substances resulting from the chemical combination of two or more elements in fixed proportions are called compounds. The elements in compounds are chemically bonded to each other. Most elements do not exist in their free native state but are found in combination with other elements as chemical compounds. Compounds may exist as ionic compounds or molecular compounds. In ionic compounds, the constituent elements exist as ions. Ions are atoms or groups of atoms that carry a charge by virtue of losing or gaining one or more electrons.

Ionic compounds result from the combination of a positive ion known as a cation and a negative ion called an anion. Salt is an ionic compound in which sodium cations and chloride anions chemically combined. Molecular compounds contain discrete molecular units. Molecular units or molecules are the smallest unit of a molecular compound. Atoms in a compound are held together by covalent bonds. Bonds dictate how atoms are held together in a compound or molecule, but for now, just think of ionic compounds as compounds composed of ions, and molecular compounds as compounds composed of molecules. Sugar, water, and carbon dioxide are examples of molecular compounds.

While a compound consists of two or more elements, a molecule consists of two or more atoms. The water molecule is composed of two atoms of hydrogen and one atom of oxygen. Not all molecules will contain different elements. The oxygen molecule is made of two oxygen atoms, and the ozone molecule consists of three oxygen atoms.

Naming Elements

Just as the modern science of chemistry developed from the ancient arts and alchemy, so did the language of chemistry progress from ancient roots. During ancient times, humans knew of seven metals and each of these was associated with the seven known celestial bodies and the seven days of the week. The characteristics of the metal and celestial bodies were thought to be related, and astrological symbols represented an early form of naming chemicals. Gold was associated with the glowing sun, and a circle, considered to be the most perfect shape, was used to represent it. Silver was related to and represented by the moon. Saturn, the most distant planet known at the time, moved slowly across the sky and was associated with the heavy metal lead. The seven metals and their associated celestial body and ancient symbols are shown in Table 5.1.

As the ancients and alchemists discovered more substances, an increasing number of symbols were required to represent the substances. Because different civilizations used different symbols for the same substance, confusion resulted and there was no common language to transfer chemical knowledge.

This situation persisted up to the nineteenth century. Even Lavoisier, who put so much effort in constructing an unambiguous chemical language, used pictures to represent elements. Lavoisier and his colleagues used letters enclosed in a circle, short

Table 5.1
Metal, Celestial Bodies, and Ancient Symbols

Metal	Body	Symbol
Gold	Sun	○
Silver	Moon	☾
Mercury	Mercury	☿
Copper	Venus	♀
Iron	Mars	♂
Tin	Jupiter	♃
Lead	Saturn	♄

straight lines, and semicircles to represent substances. Hydrogen, sulfur, carbon, and phosphorus were represented as ⊂, ∪, ⊃, ∩, respectively.

Dalton too would not divorce himself from the use of pictures to represent elements and compounds. Dalton resorted to a series of circles containing various patterns and letters to represent substances. A sample of Dalton's nomenclature is shown in Table 5.2.

In 1814, the eminent Swedish chemist Jacob Berzelius (1779–1848) discarded the old sign language of chemical symbols and proposed a new system based on the initial letter of the element. Berzelius, in compiling the Swedish Pharmacopia, used the initial letter of its Latin name to symbolize an element. If two elements had the same first letter, Berzelius would include a second letter that the two elements did not have in common. Using Berzelius' system, hydrogen, oxygen, nitrogen, carbon, phosphorus, and sulfur became H, O, N, C, P, and S, respectively. Berzelius' system utilized traditional Latin names for existing chemicals, and this explains why some of our modern symbols seem unrelated to its English name. For example, gold comes from aurum (Au),

sodium from natrium (Na), and potassium from kalium (K).

During the nineteenth and twentieth centuries, as new elements were identified, the discoverer received the honor of naming the element. Different trends in assigning names developed at different times. Element names were based on mythological figures, celestial bodies, color, chemical properties, geographical areas, minerals, derived names, and people. Table 5.3 gives the derivation of names and symbols for the common elements.

Naming Compounds

Berzelius' method of assigning letters to represent elements was also applied to compounds. Berzelius used superscripts to denote the number of atoms in a compound. Thus, water would be H^2O and carbon dioxide CO^2. Later these superscripts were changed to the current practice of designating the number of atoms using subscripts. The absence of a subscript implies the subscript one. Using symbols we can write the chemical formula for ammonia:

$$N \qquad H_3$$
$$\uparrow \qquad \uparrow$$
1 atom of   3 atoms of
Nitrogen   Hydrogen

The rule for naming compounds depends on the type of compound. For ionic compounds consisting of two elements (binary compounds), we start by naming the cation element. After the cation element is named, the stem of the anion is used with the ending "ide" added to the stem.

Anion
oxygen   →   oxide
nitrogen   →   nitride
chlorine   →   chloride
sulfur   →   sulfide

Table 5.2
Dalton's Chemical Symbols

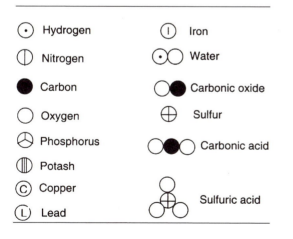

⊙ Hydrogen	① Iron
◐ Nitrogen	⊙◯ Water
● Carbon	◯● Carbonic oxide
◯ Oxygen	⊕ Sulfur
⊗ Phosphorus	◯●◯ Carbonic acid
⦀ Potash	
ⓒ Copper	Sulfuric acid
Ⓛ Lead	

Table 5.3
Symbols and Name Derivations for Common Elements

Element	Symbol	Derivation
Aluminum	Al	Latin, alum was the ore aluminum potassium sulfate
Argon	Ar	Greek, argos meaning inactive
Barium	Ba	Greek, barys meaning heavy
Beryllium	Be	Greek, berylios comes from stone beryl
Bromine	Br	Greek, bromos, to stink
Calcium	Ca	Latin, calx, word for chalk
Carbon	C	Latin, carbon, word for charcoal
Chlorine	Cl	Greek, khloros, yellow-green color
Copper	Cu	Latin, cuprum, from ore cyprium found on island Cypern
Fluorine	F	Latin, fluere, to flow, fluorspar used as flux in metallurgy
Gold	Au	Aurum, yellow, Aurora-goddess of dawn
Helium	He	Greek, helios, sun, first found in spectrum of sun
Hydrogen	H	Greek, hydros (water) + gen (producing), H produces H_2O when combusted
Iodine	I	Greek, iodes, violet color
Iron	Fe	Latin, ferrum, possibly meaning firm
Lead	Pb	Latin, plumbum, origin unknown
Magnesium	Mg	Greek, geographic region of Magnesia
Mercury	Hg	Greek, hyragyrum, liquid silver
Neon	Ne	Greek, neos, new
Nickel	Ni	German, word for devil
Nitrogen	N	Greek, niter, saltpeter (KNO_3)
Oxygen	O	Greek, oxys (acid) + gen (producing)
Phosphorus	P	Greek, light + pherus (bearing), light of white phosphorus
Potassium	K	German, kalium
Radium	R	Latin, ray
Silicon	Si	Latin, silex, flint, found in flint
Silver	Ag	Latin, argentums, shining
Sulfur	S	Latin, sulfur, enemy of copper
Tin	Sn	Latin, stannum, melts easily
Uranium	U	Greek god Uranos, named after planet
Zinc	Zn	German, zinke, spike

Examples of ionic compounds named using this system follow:

NaCl   sodium chloride
CaF_2 calcium fluoride
BaO   barium oxide
NaI   sodium iodide

Two or more atoms may combine to form a polyatomic ion. Common polyatomic ions are listed in Table 5.4.

The names of polyatomic ions may be used directly in compounds that contain

Table 5.4
Polyatomic Ions

Ion	Formula
Ammonium	NH_4^+
Carbonate	CO_3^{2-}
Bicarbonate	HCO_3^-
Hydroxide	OH^-
Nitrate	NO_3^-
Nitrite	NO_2^-
Phosphate	PO_4^{3-}
Sulfate	SO_4^{2-}
Sulfite	SO_3^{2-}

Table 5.5
Greek Prefixes

Prefix	Number
mono	1
di	2
tri	3
tetra	4
penta	5
hexa	6
hepta	7
octa	8
nona	9
deca	10

atoms of each element in the compound or molecule. Prefixes are given in Table 5.5.

Prefixes precede each element to indicate the number of atoms in the molecular compound. The stem of the second element is used with the "ide" suffix. The prefix "mon" is dropped for the initial element, that is, if no prefix is given, it is assumed the prefix is one. Examples of molecular compounds are:

CO_2	carbon dioxide
CO	carbon monoxide
CCl_4	carbon tetrachloride
N_2O_4	dinitrogen tetroxide

them. Hence, NaOH is sodium hydroxide, $CaCO_3$ is calcium bicarbonate, and $Ba(NO_3)_2$ is barium nitrate.

Naming molecular compounds uses Greek prefixes to indicate the number of

Some compounds are known by their common name rather than their systematic names. Many of these are common household chemicals. Table 5.6 summarizes some of these common chemicals.

Table 5.6
Common Chemicals

Common Name	Systematic Name	Formula
Baking soda	sodium hydrogen carbonate	$NaHCO_3$
Bleach	sodium hypochlorite	$NaClO$
Borax	sodium tetraborate decahydrate	$Na_2B_4O_7 \cdot 10H_2O$
Brimstone	sulfur	S_8
Chalk	calcium carbonate	$CaCO_3$
Epsom salt	magnesium sulfate heptahydrate	$MgSO_4 \cdot 7H_2O$
Gypsum	calcium sulfate dihydrate	$CaSO_4 \cdot 2H_2O$
Lime (quicklime)	calcium oxide	CaO
Lime (slaked)	calcium hydroxide	$Ca(OH)_2$
Limestone	calcium carbonate	$CaCO_3$
Lye	sodium hydroxide	$NaOH$
Magnesia	magnesium oxide	MgO
Marble	calcium carbonate	$CaCO_3$
Milk of magnesia	magnesium hydroxide	$Mg(OH)_2$
Muriatic acid	hydrochloric acid	$HCl_{(aq)}$
Plaster of Paris	calcium sulfate	$CaSO_4$
Potash	potassium carbonate	K_2CO_3
Saltpeter	potassium nitrate	KNO_3
Talc	magnesium silicate	$Mg_3S_4O_{10}(OH)_2$
Vinegar	acetic acid	CH_3COOH
Water	dihydogen oxide	H_2O

Writing Chemical Formulas

Knowing the names of the elements and a few basic rules allows us to name simple compounds given the chemical formula. We also can reverse the process. That is, if we know the name of the compound, we should be able to write the chemical formula. The process is straightforward for molecular compounds because prefixes are included in the names. Hence, the formula for sulfur dioxide is SO_2 and carbon monoxide is CO.

Writing the formula for ionic compounds requires us to know the charge of the cation and anion making up the ionic compound. Information on the charge of common ions can be obtained from the periodic table. More will be said about this in Chapter 7, but for now, a few basic rules will help us write the formulas for simple ionic compounds:

1. Elements in the first column (Group 1) of the periodic table form ions with a +1 charge.

2. Elements in the second column (Group 2) of the periodic table form ions with a +2 charge.

3. Generally speaking, elements in Group 13 (the column starting with the element B) form ions with a +3 charge.

4. Oxygen and sulfur generally form ions with a –2 charge.

5. The halogens, Group 17 (the column with F at the top), form ions with a –1 charge.

Using these rules and the fact that a compound carries no net charge allow us to write the compound's formula. This is illustrated for the compound magnesium chloride. Magnesium is a Group 2 element, so it forms Mg^{2+} ions. Chloride is a halogen and forms Cl^- ions. For the compound magnesium chloride to be neutral, two chloride ions must combine with one magnesium ion to give $MgCl_2$. An easy way of determining

the formula on an ionic compound is to write the elements with their charges over them and then crisscross the charges to make subscripts.

Notice that two choride ions, each with a –1 charge, are necessary to balance the +2 charge of the magnesium ion. If the subscripts have a common factor, then that common factor should be divided into both subscripts. Therefore, the formula for barium oxide would be BaO.

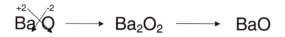

The same method can be applied when a polyatomic ion is one of the ions. For example, the formula for sodium sulfate would be Na_2SO_4.

Chemical Reactions

Chemical changes take place through chemical reactions. Chemical symbols are used to write chemical reactions in a chemical equation. For example, when a mixture of hydrogen gas (H_2) and oxygen gas (O_2) ignites, the hydrogen and oxygen combine to form water. A chemical equation expressing the combination of hydrogen and oxygen to form water is

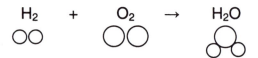

Let's examine the chemical equation above more closely. Hydrogen and oxygen are written as molecules rather than atoms. Hydrogen, oxygen, and a number of other gases (Cl_2, N_2, Br_2, I_2) exist in their natural state as

diatomic (two atoms) molecules. The reason for this diatomic form becomes apparent after we discuss electron configuration and bonding. The plus sign in the equation means "reacts with." The equation states a molecule of hydrogen reacts with a molecule of oxygen to yield a water molecule. Hydrogen and oxygen are called reactants. Reactants appear on the left side of a chemical equation. Water is the product in the reaction; products appear on the right side of chemical equations. The standard practice for writing a chemical equation is to write the individual reactants on the left side separated by a plus sign and use an arrow pointing toward the product(s) on the right side. The circles drawn below the equation are used to represent the atoms. The circles indicate this chemical equation is not balanced. One of the basic principles of chemistry that must be reflected in the chemical equation is conservation of mass. An accounting of the atoms in the formation of water shows an imbalance in the number of oxygen atoms:

Reactants	Products
2 atoms of hydrogen	2 atoms of hydrogen
2 atoms of oxygen	1 atom of oxygen

The process by which we make a chemical equation conform to conservation of mass is called balancing the equation. A chemical equation is balanced by placing whole number coefficients in front of reactants and/or products to make the number of atoms of each element equal on both sides of the equation. In our example, this is done by placing a 2 in front of both hydrogen and water

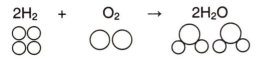

$$2H_2 \quad + \quad O_2 \quad \rightarrow \quad 2H_2O$$

By balancing the equation, we see that the number of hydrogen and oxygen atoms is equal on both sides of the equation:

Reactants	Products
4 atoms of hydrogen	4 atoms of hydrogen
2 atoms of oxygen	2 atoms of oxygen

The balanced equation now reads two molecules of hydrogen react with a molecule of oxygen to yield two molecules of water.

Additional information may be included when writing a chemical equation. One common practice is to indicate the physical state of substances by use of the subscripts (s), (l), (g), (aq) to indicate solid, liquid, gas, or aqueous (dissolved in water), respectively. Hence, the combustion of carbon to form carbon dioxide would be represented as:

$$C_{(s)} + O_{2(g)} \rightarrow CO_{2(g)}$$

A small symbol over the arrow may be used to indicate conditions necessary for the reaction to take place. For example, a Δ (Greek delta) over the arrow indicates that heat is required for the reaction, and an hν indicates ultraviolet radiation is needed for the reaction to take place.

Types of Reactions

The reactions of hydrogen and oxygen to form water and carbon and oxygen to yield carbon dioxide are examples of combination reactions. A **combination** or **synthesis reaction** results when two or more substances unite to form a compound. Many other types of reactions exist. Three other common types of reactions are **decomposition, single replacement**, and **double replacement**.

Decomposition is the opposite of combination. Decomposition takes place when a compound breaks down into two or more simpler substances. An electric current can be used to decompose water into hydrogen and oxygen:

$$\text{2H}_2\text{O}_{(l)} \quad \xrightarrow{\substack{\text{current} \\ \text{electric}}} \quad \text{2H}_{2(g)} + \text{O}_{2(g)}$$

Decomposition reactions are used to obtain pure metals from their ore.

In a single replacement reaction, one element replaces another element in a compound. An example of a single replacement is when sodium is added to water. The reaction of sodium and water is represented by the following chemical equation:

$$\text{2Na}_{(s)} + \text{2H}_2\text{O}_{(l)} \rightarrow \text{2NaOH}_{(aq)} + \text{H}_{2(g)}$$

Sodium displaces the hydrogen in the water molecule. The displaced hydrogen is liberated from the water resulting in the formation of hydrogen gas and sodium hydroxide.

Double replacement involves two elements exchanging places in a reaction. A common type of double replacement reaction occurs when a strong acid reacts with a strong base. The reaction of hydrochloric acid and sodium hydroxide would be an example of a double replacement reaction:

$$\text{HCl}_{(aq)} + \text{NaOH}_{(aq)} \rightarrow \text{NaCl}_{(aq)} + \text{H}_2\text{O}_{(l)}$$

In the above reaction, hydrogen and sodium exchange places. This reaction results in the acid and base neutralizing each other.

From Molecules to Moles

Our discussion of balancing chemical equations has focused on accounting for atoms and molecules in both reactants and products. Although a balanced equation tells us how many atoms of an element it may take to form so many molecules, it is impractical to speak in these terms for common applications. For example, if we want to know how much carbon dioxide is produced when some vinegar is added to bak-

ing soda, a practical unit for the carbon dioxide would be liters or milligrams. Industrial chemists often measure quantities in gallons and tons. Atoms and molecules are just too small to be convenient for practical applications. The unit used to define the amount of a substance in a chemical reaction is the **mole** (mol). A mole is the quantity of a substance that contains as many particles (atoms, molecules, ions) as there are atoms in exactly 12 grams of carbon-12.

The use of carbon-12 as a standard was adopted in 1961 by international agreement. Originally, Dalton had assigned an atomic mass of 1 for hydrogen, and the atomic masses of all other elements were based on this value. Using a value of 1 for hydrogen gave an atomic mass of 15.9 for oxygen. Because oxygen was typically used in determining how elements combined, it was decided it made more sense to base atomic masses using an oxygen standard. Oxygen was assigned a value of 16; this designation meant hydrogen had an atomic mass of slightly greater than 1. A complication developed, though, because oxygen consists of three isotopes. Different samples of oxygen containing slightly different ratios of the isotopes would have different masses, which presented problems because physicists were using oxygen samples to determine atomic masses. In 1961, physicists and chemists agreed to use the isotope carbon-12 as the standard for atomic mass measurements. Carbon-12 with six protons and six neutrons was assigned a value of 12.0000 for determining atomic masses.

What exactly does it mean to have as many atoms as there are in 12 grams of C-12? The number is known as **Avogadro's number**, and its accepted value is 6.022×10^{23}. Therefore, from the definition of a mole we can say there are 6.02×10^{23} atoms in a mole of C-12. The term Avogadro's number was coined by Jean Baptiste Perrin (1870–1942) who was the first to experimentally determine an accurate value of this constant in 1908. One mole of any substance contains approximately 6.02×10^{23} particles of the substance. One mole of hydrogen atoms is equal to 6.02×10^{23} hydrogen atoms, one mole of hydrogen molecules equals 6.02×10^{23} molecules of hydrogen, and one mole of toothpicks is 6.02×10^{23} toothpicks.

The fact that both the mole and the mass of an element are based on carbon-12 enables us to relate mole and mass. A molar mass is defined as the mass in grams of one mole of a substance, and it can be obtained directly from an element's atomic mass. We can use the elements hydrogen and nitrogen to illustrate this concept. Periodic table entries for both elements are shown below. The whole number above the element is the atomic number and gives the number of protons in the nucleus. The number below the element's symbol is the molar mass (as well as the atomic mass):

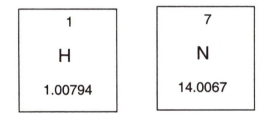

The entries indicate one mole of hydrogen atom equals 1.00794 gram and one mole of nitrogen atoms equal 14.0067 grams. Because each hydrogen molecule contains two hydrogen atoms, a mole of hydrogen molecules contains two moles of hydrogen atoms. Therefore, one mole of hydrogen molecules, H_2, would equal $2 \times 1.00794 = 2.01588$ grams. In a similar fashion, the molar mass of ammonia, NH_3 can be calculated by totaling the molar masses of constituent elements:

NH_3 N $1 \times 14.0067 =$ 14.0067
    H_3 $3 \times 1.00794 =$ $+\underline{3.0238}$
                        17.0305

Therefore, one mole of ammonia contains 6.022×10^{23} NH_3 molecules and has a mass of 17.0305 grams; conversely, 17.0305 grams of ammonia equals one mole of ammonia.

Coefficients in a balanced chemical equation are typically taken to represent moles. For the reaction of hydrogen and nitrogen to form ammonia, the equation

$$N_{2(g)} + 3H_{2(g)} \rightarrow 2NH_{3(g)}$$

can be read as one mole of nitrogen molecules reacts with three moles of hydrogen molecules to yield two moles of ammonia molecules. Knowing the moles of reactants and product(s) in the balanced equation allows us to express mass relationships. The ammonia reaction expressed in terms of mass tells us that approximately 28 grams of nitrogen reacts with 6 grams of hydrogen to yield 34 grams of ammonia.

Having a balanced chemical equation and knowing the relationship between mass and moles allows us to predict how much reactant is necessary to yield a certain amount of product. This knowledge has important applications in industrial chemistry, environmental chemistry, nutrition, and in any situation where reactions take place. The balanced equation is a recipe for a chemical reaction. Just as it is necessary to know the amount of eggs, flour, sugar, and salt to bake a cake, we need to know the amount of ingredients that go into a chemical reaction. The balanced chemical equation gives the quantities of different reactants that are required to produce a specific amount of product.

We can illustrate this with a practical problem having environmental implications. The greenhouse effect refers to the heating of the Earth due to the trapping of infrared radiation by certain gases. A principal greenhouse gas is carbon dioxide, CO_2.

International efforts in recent years have attempted to limit the amount of carbon dioxide produced by individual countries. Because carbon dioxide is a principal product when fossil fuel is burned, any limitation on carbon dioxide has far-reaching effects on a nation's economy. Automobiles are a chief producer of CO_2 in the United States. How much CO_2 is produced by automobiles? Let's calculate the amount of CO_2 produced when a gallon of gas is burned. We can simplify the problem a bit by assuming the gasoline is octane and it reacts completely with oxygen to produce carbon dioxide and water according to the equation:

$$2C_8H_{18} + 25O_2 \rightarrow 16CO_2 + 18H_2O$$

octane + oxygen → carbon dioxide + water

One gallon of octane has a mass of approximately 2,600 grams. The molar mass of octane is 114 grams, so 2,600 grams of octane is almost 23 moles of octane. Notice that the coefficients of CO_2 and C_8H_{18} in the equation are 16 and 2, respectively. Therefore, the balanced equation tells us that for every mole of octane burned, eight moles of carbon dioxide are produced. This means 182 moles of CO_2 result (23×8). Because carbon dioxide's molar mass is 44 grams, the amount of CO_2 produced is about 8,000 grams. It is interesting to note that the ratio of carbon dioxide produced to gasoline (octane) burned is more than three to one. Translated into English units, this means that for every gallon (a gallon is about six pounds) of gas burned, 18 pounds of carbon dioxide are produced. Given the number of cars in operation in this country, we see that a tremendous amount of carbon dioxide is emitted to the atmosphere every day. Industries and municipalities regularly use similar calculations to predict the amount of pollution they are emitting to the atmosphere. Such calculations help them conform to the Clean Air Act.

Balanced chemical reactions are critical to the chemical industry. Plant managers must know the amount of reactants necessary to yield a product to keep production lines moving. If reactants or products are not pure, this must be taken into account. Alternative reactions can be examined to find the most cost-effective process. For example, a fertilizer may be produced using one process with sulfuric acid as a reactant or another process that starts with ammonium carbonate. Working through the balanced equations for each process would help decide which process to adopt.

A final example of an application of balanced chemical equation involves the use of one form of the breath analyzer used by police to determine if someone was driving under the influence of alcohol. The chemical equation guiding the reaction when someone breaths into the instrument is

$$3C_2H_5OH + 2K_2Cr_2O_7 + 8H_2SO_4 \rightarrow 3CH_3COOH$$
$$+ 2Cr_2(SO_4)_3 + 2K_2SO_4 + 11H_2O$$

ethanol + potassium + sulfuric   acetic acid + chromic + potassium + water
      dichromate   acid →   sulfate   sulfate
       orange   **green**

The equation shows that alcohol in the form of ethanol reacts with an orange solution of potassium dichromate to produce a green solution of chromic sulfate. The source of alcohol is the suspect's breath. More alcohol produces a greater color change. The breath analyzer measures this color change and coverts this measurement into an amount of alcohol in the blood.

In many instances, the ratio of reactants available is different than that given by the balanced chemical equation. When this happens, the reactant in the smallest relative abundance is said to be limiting, while the other reactant is referred to as the excess reactant. Again, using the ammonia reaction, we see the ratio of hydrogen to nitrogen is 3 to 1. If three moles of both hydrogen and nitrogen are brought together, the amount of hydrogen will limit the reaction. Nitrogen will be in excess; two extra moles of nitrogen would be left over after the reaction is complete. An analogy may help to illustrate the idea of excess and limiting reactants. Suppose you wanted to make bicycles. For each bicycle we need two tires and one frame. Although other parts are required, we can simplify the analogy by focusing on tires and frames. If we had twenty tires and twenty frames, we could build ten bikes. The tires would limit the number of bikes, and the frames would represent the excess reactant. Even if we had one hundred frames, we could build only ten bikes with twenty tires. A chemical reaction is analogous to building bikes. One substance can be totally consumed while another remains when the reaction ceases.

One last point is that just because a chemical equation indicates a chemical reaction takes place, does not mean it will.

Many conditions are required for a chemical reaction to proceed. Conditions such as heat, light, and pressure must be just right for a reaction to take place. Furthermore, the reaction may proceed very slowly. Some reactions occur in a fraction of a second, while others occur very slowly. Consider the difference in the reaction times of gasoline igniting in a car's cylinder versus the oxidation of iron to form rust. The area of chemistry that deals with how fast reactions occur is known as kinetics (Chapter 12). Finally, not all reactions go to completion. The amount of product produced based on the chemical equation is known as the theoretical yield. The amount actually obtained expressed as a percent of the theoretical is the actual yield. In summary, it's best to think of a chemical equation as an ideal representation of a reaction. The equation provides a general picture of the reaction and enables us to do theoretical calculations, but in reality reactions deviate in many ways from that predicted by the equation.

Elements and the Periodic Table

Introduction

If there were a flag that represented the science of chemistry, it would be the periodic table. The periodic table is a concise organizational chart of the elements. The periodic table not only summarizes important facts about the elements, but it also incorporates a theoretical framework for understanding the relationships between elements. The modern periodic table attests to human's search for order and patterns in nature. As such, the periodic table is a dynamic blueprint for the basic building blocks of our universe. This chapter examines the development of the modern periodic table and presents information on how the modern periodic table is organized.

Discovery of the Elements

The organization of the elements into the first modern periodic table in 1869 depended on two important developments. One was that enough elements had to be identified to organize the elements into a meaningful pattern. Nine elements were known in ancient times. These were gold, silver, iron, mercury, tin, copper, lead, carbon, and sulfur. Bismuth, zinc, antimony, and arsenic were isolated in the alchemical period and Middle Ages. By the start of the nineteenth century, approximately thirty elements had been identified, and this number doubled over the next sixty years. Currently, 116 elements have been identified, but scientists at research labs throughout the world are attempting to create more using high speed particle accelerators. Figure 6.1 gives the approximate dates of the discovery of elements during different time periods.

As the number of elements increased, so did attempts to organize them into meaningful relationships. Johann Döbereiner (1780–1849) discovered in 1829 that certain elements had atomic masses and properties that fell approximately mid-way between the masses and properties of two other elements. Döbereiner termed a set of three elements a triad. Thus, chlorine, bromine, and iodine form a triad; Döbereiner proposed several other triads (lithium-sodium-potassium, calcium-strontium-barium). Döbereiner recognized that there was some sort of relationship between elements, but many elements did not fit in any triad group, and even those triads proposed displayed numerous inconsistencies.

Figure 6.1
Discovery Dates of the Elements

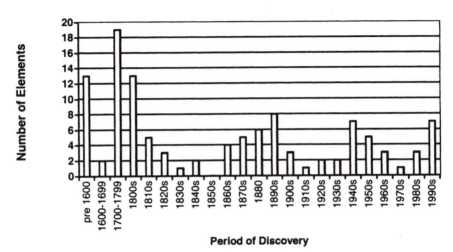

John Newlands (1837–1898) attempted to organize the elements by listing them in order of increasing atomic mass. Newlands prepared a table with elements grouped in columns of seven elements. Several rows in Newlands' table contained related elements and also included Döbereiner's triads. Because there was some evidence of characteristics repeating with every eighth element, Newlands proclaimed a law of octaves for the elements. Newlands' table listed the elements in a continuous manner and did not leave any blank spaces. His adherence to a rigid organizational scheme resulted in a number of inconsistencies. Other chemists ridiculed Newlands when he presented his table at conferences and meetings between 1863 and 1865. In fact, there was so much skepticism regarding his work that he was unable to get it published. Years later Newlands was vindicated when others recognized his insight of recognizing periodic patterns among the elements.

The second development that led to the modern periodic table was the acceptance of specific atomic masses for the elements. While today we readily accept the masses given in a periodic table, there was much

disagreement regarding masses in the mid-nineteenth century. A major problem that was not recognized until that time is that some elements exist in their natural state as atoms, while others exist as molecules. The element hydrogen (H) has an atomic mass of approximately 1, but hydrogen exists as a diatomic molecule (H_2) with a mass of approximately 2. The existence of molecules and **Avogadro's law** was the key for determining universally accepted atomic masses. Although Avogadro put forth his ideas in 1811, the concept of molecules as a basic unit of matter was not widely accepted in the mid-nineteenth century. The situation changed when Stanislao Cannizzaro (1826–1910), a disciple of Avogadro's ideas, resurrected Avogadro's work at the First International Chemical Congress convened in Karlsruhe, Germany, in 1860. Cannizzaro vigorously defended the existence of molecules using Avogadro's law as a basis for determining the correct atomic masses of the elements. Cannizzaro circulated a paper he had written in 1858 outlining his ideas on the matter. This paper helped persuade several prominent chemists on the utility of incorporating Avogadro's law into

Figure 6.2
Periodic Trend of Melting Point

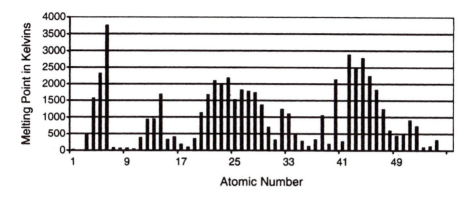

their thinking. Cannizzaro's crusade and the Karlsruhe Conference ultimately resulted in universally accepted values for atomic masses.

Two chemists in attendance at the Karlsruhe were Julius Lothar Meyer (1830–1895) and Dmitri Mendeleev (1834–1907). These two independently developed the periodic law and constructed their own versions of the periodic table. Meyer based his table primarily on the physical properties of the elements. Meyer plotted atomic volume against the atomic mass and noticed the periodicity in volumes of the elements. Other physical properties also showed periodic trends. Figure 6.2 shows how the melting point of the first fifty-five elements rises and falls in a roughly periodic fashion as atomic number increases. Based on his analysis, Meyer published his periodic table in 1870.

Despite Meyer's efforts, Mendeleev receives the majority of credit for the modern periodic table. Mendeleev's table resulted from years of work examining the properties of the elements. Mendeleev's method involved writing information about the elements on individual cards and then trying to organize the cards in a logical order. His genius resulted from modifications he introduced into his table that others had not included. He produced his table in

1869 as part of a text he had written for chemistry. Mendeleev did not always arrange the elements in strict order of atomic masses, and he claimed that several of the accepted atomic masses were wrong. More important, he left blank spaces in his table where he claimed undiscovered elements existed. Two of the empty spaces fell right after aluminum and silicon, respectively. Mendeleev named these undiscovered elements eka-aluminum and eka-silicon ("eka" is the Sanskrit word for first, so eka-aluminum was first after aluminum). Using his knowledge about the periodic properties of the elements, he predicted the chemical and physical properties of eka-aluminum and eka-silicon. In 1875, the element gallium (named after ancient France, Gaul) was discovered, which turned out to be Mendeleev's eka-aluminum. The properties of gallium were almost identical to those Mendeleev had predicted as seen in Table 6.1. Germanium (named after Germany) discovered in 1886 turned out to be Mendeleev's eka-silicon. Scandium (named for Scandanavia) discovered in 1879 was another element Mendeleev had predicted with his table.

Mendeleev's periodic table (Figure 6.3) was used with little modification until well into the twentieth century. The top of each

Table 6.1
Comparison of Mendeleev's Prediction and Actual Properties for Eka-Aluminum (Gallium)

	Mendeleev's Prediction	Actual Properties
Atomic Mass	68	69.7
Melting Point	low	29.8°C
Density	5.9 g/cm³	5.9 g/cm³
Oxide	X_2O_3	Ga_2O_3

group gives the general formula for how the group's elements combine with hydrogen and/or oxygen. Blank spaces appear at atomic masses of 44, 68, 72, and 100 for undiscovered elements. At the turn of the century, the discovery of the noble gases led to the addition of another column. Since the time of Mendeleev's original table, many different versions of the periodic table have appeared. These include three-dimensional, circular, cylindrical, and triangular shapes. Yet, Mendeleev's basic arrangement has stood the test of time, and the periodic table is the universally recognized icon of chemical knowledge.

The Modern Periodic Table

Elements in the modern periodic table are arranged sequentially by atomic number in rows and columns. Mendeleev and his contemporaries arranged elements according to atomic mass. In 1913, Henry Moseley's (1887–1915) studies on the x-ray diffraction patterns for metals showed a relationship between the spectral lines and the atom's nuclear charge. Moseley's work established the concept of atomic number, the number of protons in the nucleus, as the key for determining an element's position in the periodic table. Rows in the periodic table

are referred to as **periods**, and columns are called **groups**. The periods run left to right, and the groups from top to bottom. A group may also comprise a **chemical family**. The first and second groups are the **alkali metals** and alkaline earth metals, respectively. The group starting with fluorine (F) is the **halogen** family (halogen means salt forming). The last group on the far right of the table is the noble gas family.

Groups are numbered across the top of the table. Unfortunately, Americans and Europeans have different group numbering systems that at times cause confusion. The American system uses Roman numerals followed by either the letter A or B. The A groups are known as main group or representative elements. Those groups with the letter B comprise the transition elements. Main groups fall at both ends of the periodic table with transition elements in the center. The European system also uses Roman numerals and the letters A and B, but the A is used for groups on the left half of the table and B for groups on the right half. In recent years, scientists have suggested eliminating the letters and using the Arabic numerals 1–18 to number each group starting on the left. We use the latter method to identify groups.

The Main Group Elements

To understand how elements are organized into different groups, we need to consider the electron configuration of elements. Specifically, it is the electron configuration of the valence electrons that characterizes groups. Remember, the valence electrons are those electrons in the outer shell. We can use Group 1 to illustrate how the valence electron configuration defines the group. The electron configurations of Group 1 elements are listed on page 66.

Figure 6.3
Mendeleev's 1869 Periodic Table (Top) and 1870 Table (bottom). Image from Edgar Fahs Smith Collection, University of Pennsylvania Library.

Ueber die Beziehungen der Eigenschaften zu den Atomgewichten der Elemente. Von D. Mendelejeff. — Ordnet man Elemente nach zunehmenden Atomgewichten in verticale Reihen so, dass die Horizontalreihen analoge Elemente enthalten, wieder nach zunehmendem Atomgewicht geordnet, so erhält man folgende Zusammenstellung, aus der sich einige allgemeinere Folgerungen ableiten lassen.

			Ti = 50	Zr = 90	? = 180
			V = 51	Nb = 94	Ta = 182
			Cr = 52	Mo = 96	W = 186
			Mn = 55	Rh = 104,4	Pt = 197,4
			Fe = 56	Ru = 104,4	Ir = 198
			Ni = Co = 59	Pd = 106,6	Os = 199
H = 1			Cu = 63,4	Ag = 108	Hg = 200
	Be = 9,4	Mg = 24	Zn = 65,2	Cd = 112	
	B = 11	Al = 27,4	? = 68	Ur = 116	Au = 197?
	C = 12	Si = 28	? = 70	Sn = 118	
	N = 14	P = 31	As = 75	Sb = 122	Bi = 210?
	O = 16	S = 32	Se = 79,4	Te = 128?	
	F = 19	Cl = 35,5	Br = 80	J = 127	
Li = 7	Na = 23	K = 39	Rb = 85,4	Cs = 133	Tl = 204
		Ca = 40	Sr = 87,6	Ba = 137	Pb = 207
		? = 45	Ce = 92		
		?Er = 56	La = 94		
		?Yt = 60	Di = 95		
		?In = 75,6	Th = 118?		

1. Die nach der Grösse des Atomgewichts geordneten Elemente zeigen eine stufenweise Abänderung in den Eigenschaften.

2. Chemisch-analoge Elemente haben entweder übereinstimmende Atomgewichte (Pt, Ir, Os), oder letztere nehmen gleichviel zu (K, Rb, Cs).

3. Das Anordnen nach den Atomgewichten entspricht der *Werthigkeit* der Elemente und bis zu einem gewissen Grade der Verschiedenheit im chemischen Verhalten, z. B. Li, Be, B, C, N, O, F.

4. Die in der Natur verbreitetsten Elemente haben *kleine* Atomgewichte

	Gruppe I.	Gruppe II.	Gruppe III.	Gruppe IV.	Gruppe V.	Gruppe VI.	Gruppe VII.	Gruppe VIII.
Typische Elemente	H 1 Li 7	Be 9,4	Bo 11	C 12	N 14	O 16	F 19	
1. Periode: Reihe 1	Na 23	Mg 24	Al 27,3	Si 28	P 31	S 32	Cl 35,5	
1. Periode: - 2	Ka 39	Ca 40	—44	Ti 50(?)	V 51	Cr 52	Mn 55	Fe56, Co59, Ni56, Cu [63
2. Periode: Reihe 3	(Cu 63)	Zn 65	—68	—72	As 75	Se 78	Br 80	
2. Periode: - 4	Rb 85	Sr 87	(Yt 88) (?)	Zr 90	Nb 94	Mo 96	—100	Ru104, Rh104, Pl106, [Ag108
3. Periode: Reihe 5	(Ag 108)	Cd 112	In 113	Sn 118	Sb 122	Te 125	J 127	
3. Periode: - 6	Cs 133	Ba 137	—137	Ce 138 (?)	—	—	—	
4. Periode: Reihe 7	—	—	—	—	—	—	—	
4. Periode: - 8	—	—	—	—	Ta 183	W 184	—	Os 199 (?), Jr 198, Pt [197, Au 197
5. Periode: Reihe 9	(Au 197)	Hg 200	Tl 204	Pb 207	Bi 208	—	—	
5. Periode: - 10	—	—	—	Th 232	—	Ur 240	—	
Höchste salzbild. Oxyde	R^2O	R^2O^2 od. RO	R^2O^3	R^2O^4 o. RO^2	R^2O^5	R^2O^6 o. RO^3	R^2O^7	R^2O^8 od. RO^4
Höchste H-Verbindung				RH^4	RH^3	RH^2	RH	(R^2H) (?)

992

H $1s^1$

Li $1s^2\ 2s^1$

Na $1s^2\ 2s^2\ 2p^6\ 3s^1$

K $1s^2\ 2s^2\ 2p^6\ 3s^2\ 3p^6\ 3d^{10}\ 4s^1$

Rb $1s^2\ 2s^2\ 2p^6\ 3s^2\ 3p^6\ 3d^{10}\ 4s^2\ 4p^6\ 4d^{10}\ 4f^{14}\ 5s^1$

Cs $1s^2\ 2s^2\ 2p^6\ 3s^2\ 3p^6\ 3d^{10}\ 4s^2\ 4p^6\ 4d^{10}\ 4f^{14}\ 5s^2$
$5p^6\ 5d^{10}\ 6s^1$

Fr $1s^2\ 2s^2\ 2p^6\ 3s^2\ 3p^6\ 3d^{10}\ 4s^2\ 4p^6\ 4d^{10}\ 4f^{14}\ 5s^2$
$5p^6\ 5d^{10}\ 6s^2\ 6p^6\ 7s^1$

We see from the electron configurations that all Group 1 elements have a valence configuration of ns^1, where n is the period of the element. Electron configurations for elements in Groups 2 and 13–18 show a regular pattern similar to that demonstrated above. Each group would have its own distinctive **valence electron configuration**:

Group	Valence Electron Configuration
1	ns^1
2	ns^2
13	ns^2np^1
14	ns^2np^2
15	ns^2np^3
16	ns^2np^4
17	ns^2np^5
18	ns^2np^6

Groups 1 and 2 together are often referred to as the s-block, and Groups 13–18 are the p-block elements. These names come from the valence orbital being filled in each block. In the s-block, the s orbital is filled, while in the p-block, the p orbital is progressively filled across a period.

The valence electron configuration explains why groups define chemical families that exhibit similar chemical properties. Because it is the valence electrons that determine an element's chemical behavior, and groups have similar valence electron configurations, then elements in the same group display similar chemical behavior. The electron configuration also explains the

periodic law. We can demonstrate this by examining how one property, the atomic radius, changes moving across a period. Lithium is the first element in the 2nd period with an atomic number of 3. It has an atomic radius of 152 picometers (pm). As we move across the 2nd period, the atomic number increases in increments of one. Each time the atomic number increases one more proton is added to the nucleus and one additional electron is added to the 2nd electron shell. Because each electron goes into a shell where the principal quantum number is 2, and the principal quantum number gives the average distance from the nucleus, we would expect the radii of atoms across the period to stay constant. In reality, moving across the period there is a tendency for the radii to decrease. This happens because as protons and electrons are added, the positive nucleus exhibits a relatively stronger pull on the cloud of electrons in the 2p orbital. This trend continues across the period until we reach sodium. At this point, an additional electron must go into the 3rd shell, and the radius displays a "jump" from neon. After this abrupt increase in sodium's radius occurs, a trend similar to that for period 2 repeats itself in period 3. This trend is demonstrated in Figure 6.4, which is a graph of atomic radii versus atomic number for periods 2 and 3.

The periodic trend of a decrease in atomic radii across a period is readily seen in the Figure 6.4. Other properties related to atomic radii show a similar pattern. Knowing that the elements exhibit a general periodic trend allows us to predict unknown properties for elements and aided in the discovery of unknown elements. The periodic nature of the elements supported the development of the quantum theory. The elements show a periodic pattern in both their properties and electron configurations. The periodic trend in properties of the elements

Figure 6.4
Change in Atomic Radii for Periods 2 and 3

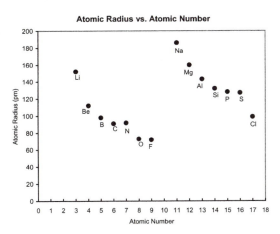

Moving across each period the f orbital is progressively filled; the 4f orbital is filled for the lanthanides, and the 5f orbital is filled for the actinides. These elements are sometimes referred to as the rare earths, because it was originally difficult to separate and identify these elements. Rare earths are actually not scarce, but the term "rare earths" is still used for the lanthanides and actinides. A more accurate modern term for these two periods are the inner transition elements.

should be interpreted as a general pattern that elements in the periodic table follow. There are many examples of deviations from the periodic law. It should not be viewed as a strict rule, but as pattern that presents general trends among the elements.

d and f Block Elements

The elements in Groups 3 to 12 are called the transition elements or transition metals. Transition elements are characterized by a filling of the d orbitals from left to right across a period. Collectively these elements form the d-block of the periodic table. The d-block groups do not demonstrate the consistent pattern displayed by main group elements. In general, though, the pattern of filling the d orbital gives transition elements a nd^{1-10} configuration. For example Group 3 elements have nd^1 electrons, and Group 12 nd^{10} electrons, where n represents the period.

The two periods at the bottom of the periodic table represent two special groups that fall between the s and d blocks. The period beginning with lanthanum (La) is called the lanthanides. The next period beginning with actinium (Ac) is the actinides. Elements in these two periods form the f-block.

Metals, Metalloids, and Nonmetals

Figure 6.6 summarizes different blocks, families, and areas of the periodic table. Most elements can be classified as metals. Metals are solid at room temperature, are good conductors of heat and electricity, and form positive ions. Moving across the table from left to right elements lose their metallic characteristics. The metalloids, also known as the semi-metals, have properties intermediate between metals and nonmetals. Because they display characteristics of both conductors and nonconductors, elements such as silicon and germanium find wide use in the semi-conductor industry. Nonmetals are found on the far right of the periodic table. Nonmetals are poor conductors and are gases at room temperature.

Periodic Problems

Don't be surprised if you see a periodic table that looks different than the one presented in this book. As already mentioned, slightly different versions exist depending on the conventions used to prepare the table. Some tables start the lathanides and actinides in the f-block with cesium and thorium. Another problem is the placement of hydrogen. Hydrogen is positioned above lithium in Figure 6.5, which indicates it can act like

Figure 6.5
Blocks, Families, and Areas of the Periodic Table

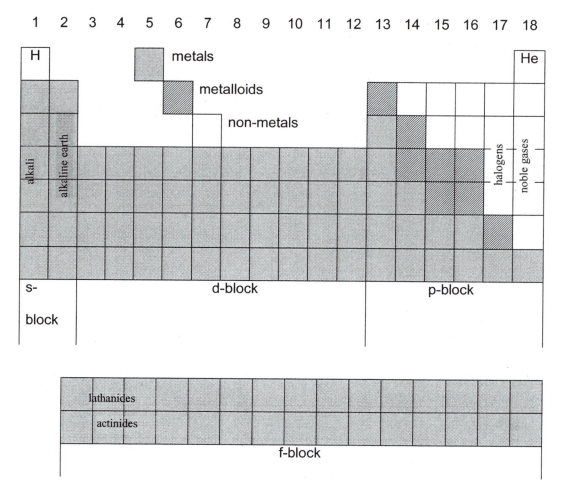

an alkali metal, but hydrogen should not be considered an alkali metal. It often acts as a nonmetal like a halogen because, like other halogens, it needs one more electron to achieve the electron configuration of a noble gas. Hydrogen is a gaseous nonmetal that at time behaves like a halogen. Some tables list hydrogen separately to indicate that it is not associated with any family.

The problem concerning the use of letters to identify groups has already been mentioned. A similar but more serious problem involves the naming of newly discovered ele-

ments. The elements beyond uranium, with atomic numbers greater than 92, are not natural. These heavy elements are synthesized in particle accelerators. New elements are created by bombarding an existing heavy element with neutrons or the nucleus of another element such as hydrogen, carbon, or oxygen. In the latter case, fusion of the nuclei of the heavy element and the bombarding element creates a new element. The newly created heavy element is unstable and may last for only a fraction of a second before it decays. The fact that only a few atoms of a new element are created for only a fraction of a second makes detection difficult. It is common for separate research labs to claim priority of the discovery of new elements. Because elements are named by their discoverers, it is important to establish priority in order to assign a name to the new element. During the past twenty years, American, Russian, and German scientists have had several disputes over discovery claims. This confusion has led to some elements not having a universally accepted name. The problem involves the naming of the transferium elements, that is, those elements whose atomic number is greater than 100.

IUPAC (International Union of Pure and Applied Chemistry) has attempted to settle the controversies and determine priority with mixed success. A temporary system proposed to name newly discovered elements is based on using a symbol for each of the digits zero through nine:

0	nil	5	pent
1	un	6	hex
2	bi	7	sept
3	tri	8	oct
4	quad	9	enn

Using this system, element 104 is named unnilquadium (un + nil + quad + ium). The ium is added to indicate the element is a metal. Unnilquadium's symbol is Unq. Combining terms to give an element's name is just a temporary solution until an accepted name is determined. Stringing the terms together, while convenient, results in rather awkward names. One problem IUPAC faced when trying to decide an element's name was their insistence that an element could not be named for a living person. In the mid-1990s, Americans called element 106 seaborgium, named after Glenn Seaborg (1912–1997) who at the time was still alive. In 1997, after Seaborg's death, the names of elements 101 to 109 were established:

101	medelevium	106	seaborgium
102	nobelium	107	bohrium
103	lawrencium	108	hassium
104	rutherfordium	109	meitnerium
105	dubnium		

While this solved the dilemma for elements up to number 109, new elements continue to be discovered. In 1994, elements 110 and 111 were identified, and in 1996, element 112 was discovered. In recent years, elements 113, 114, 116, and 118 have been discovered, and very soon all elements between 104 and 118 should be identified. Elements 104–118 will fill another row below the actinides, and are called the transactinides. Beyond element 118 lie the superactinides waiting for more powerful accelerators and better detectors to be created so they can be identified.

Summary

From ancient times, humans have pondered what the universe is made of. Early philosophers proposed fire, earth, water, and air either individually or in combination as the building blocks of nature. Lavoisier defined an element operationally as a substance that cannot be broken down chemically. Using this definition, the number of elements has increased from around 30 in Lavoisier's time to over 115 today. The initial search for elements involved classical methods such as replacement reactions, electrochemical separation, and chemical analysis. New methods such as spectroscopy greatly advanced the discovery of new elements during the twentieth century. The last half century has been marked by the synthesis of elements by humans.

Though the quest to produce more elements continues, amazingly our observable universe is composed of a relatively few common elements. Hydrogen and helium makes up over 90% of the universe. Only six elements make up over 96% of the Earth as shown in Table 6.2.

Over 99% of the human body is composed of only seven elements (Table 6.3), yet trace amounts of a number of elements are essential for vital functions, for example, iron in hemoglobin or zinc in enzymes.

Although our world is dominated by a few common chemicals, all the natural chemical elements are essential to both our survival and our society. We have come a long way in learning how to combine less than one hundred elements into millions of different substances that enhance our lives. The list of products is endless, but consider what life would be like without using the elements to produce fertilizers, medicines, clothing,

Table 6.2
Elemental Composition of Earth by Mass

Element	Percentage
Iron	34.6
Oxygen	29.5
Silicon	15.2
Magnesium	12.7
Nickel	2.4
Sulfur	1.9

Table 6.3
Elemental Composition of Human Body by Mass

Element	Percentage
Oxygen	65.0
Carbon	18.5
Hydrogen	9.5
Nitrogen	3.2
Calcium	1.5
Phosphorus	1.0
Potassium	0.4

building materials, paper, and electronics. We often lose sight of the fact that elements are the building blocks of nature, and all substances are essentially elements packaged in different bundles. Just consider that we breath oxygen, drink hydrogen and oxygen, eat carbon (and host of other elements), calculate with silicon and oxygen computer chips, light our homes with tungsten, advertise with neon, drink out of aluminum, store foods in tin, disinfect with iodine and chlorine, and the list goes on. We live in a sea of elements, and we ourselves are part of that sea.

7

Chemical Bonding

Introduction and History

Only a small portion of matter consists of elements in uncombined atomic states. Hydrogen, which comprises over 90% of the universe, exists as molecular hydrogen, H_2. Likewise, the nitrogen and oxygen in the air we breath exist as the diatomic molecules N_2 and O_2. Practically all substances consist of aggregates of atoms in the form of compounds and molecules. The fact that matter exists as combinations of atoms leads naturally to several basic questions:

- What forces hold atoms together?
- Why do certain elements combine to form compounds?
- How are atoms connected to form compounds?

The concepts of chemical bonding provide the answers to these questions and are fundamental to the science of chemistry.

The atomists led by Leucippus and Democritus explained chemical combination by proposing that atoms consisted of various shapes, which contained projections. Different-shaped atoms interlocked like pieces of a jigsaw puzzle or atoms with hook-like projections latched onto other atoms. By proposing atoms of various sizes and shapes, the atomist Greeks explained the existence of the three states of matter. Solids were composed of atoms with jagged edges, while liquid and gases were made of particles with smooth, rounded edges. Aristotle rejected the idea of atoms and considered matter to be continuous in nature. Adopting this viewpoint eliminated the need to address chemical combination.

Traditional theories on chemical combination were based on projecting observable phenomenon at the macroscopic level to the microscopic level. Hence, the mechanical explanations using size and shape persisted well into the 1600s. Alchemical explanations projected human qualities on matter to account for specific reactions and combinations. Substances that combined had a natural attraction or "love" for one another. Conversely, other substances "hated" each other and did not combine.

Isaac Newton (1642–1727) was one of the first to speculate on the true nature of chemical attraction. Newton believed an attractive force similar to that of gravity, magnetism, and electricity also applied to matter at the molecular level. Other scientists based their ideas of chemical attraction on Newton's law of universal gravitation. Comte de Buffon (1701–1788) proposed chemical combination resulted from the gravitational attraction between particles of various shapes. Buffon believed the attractive force depended on the inverse square of the distance between particles, while the specific shapes of particles explained while specific substances combined.

While several natural philosophers sought a theoretical explanation for chemical reactions and combinations, many individuals were content to merely catalog the combining properties of substances. The study of how substances combined became known as **chemical affinity**. Tables of chemical affinity were developed to summarize different reactions. Etienne Francois Geoffroy (1672–1731) presented his ideas on chemical affinity in 1718 (Figure 7.1). Others such as Tobern Bergman (1735–1784) and Berthollet advanced Geoffroy's ideas. Bergmann realized that chemical affinity depended on conditions such as temperature and quantities of reactants.

Experiments on electricity at the start of the nineteenth century shed light on the nature of chemical affinity. Humphrey Davy (1778–1829) used the voltaic cell invented by Alessandro Volta (1745–1827) to isolate pure metals from their oxides. Davy used

Figure 7.1
Diagram of Geoffroy's Table of Affinities

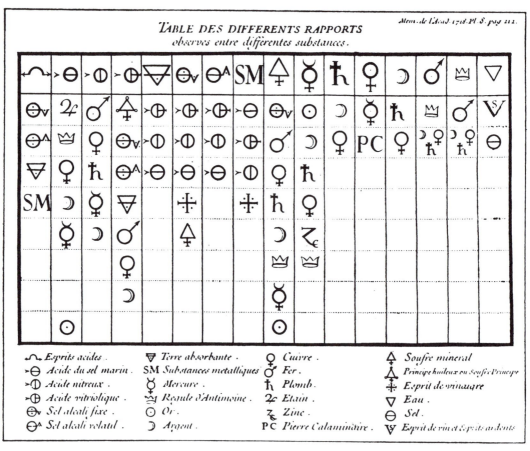

Source: Leicester and Klickstein, 1952.

electrolysis to discover a number of metals including potassium, sodium, calcium, barium, and magnesium. Davy's work provided evidence that chemical affinity was somehow related to electricity. Michael Faraday (1791–1867), Davy's assistant and successor, expanded upon Davy's studies and believed electrical attraction and chemical affinity not only were related, but also were actually a manifestation of the same force. Berzelius proposed a model in which a compound was held together because it contained a positive part and negative part. Berzelius' model explained affinity for simple inorganic compounds, but it fell short when applied to more complex compounds, especially organic compounds. Berzelius' polar model consisting of an attraction between positive and negative components of a compound introduced the concept of the ionic bond.

To explain the structure and composition of organic molecules, Berzelius proposed radicals as basic building blocks. Berzelius believed radicals had opposite electrical charges and were held together by electrical attraction. According to Berzelius, organic radicals consisted of indestructible combinations of positive hydrogen and negative carbon. Auguste Laurent (1808–1853) disputed Berzelius' ideas on organic structure and proposed that organic molecules had a nucleus to which radicals, single atoms, or groups of atoms were attached. The nucleus of the organic molecule dictated its nature. Laurent carried out reactions in which chlorine was substituted for hydrogen in organic compounds. Because both carbon and chlorine were considered negative, Laurent's work refuted Berzelius' radical theory. Laurent believed a process of substitution was a better explanation for how organic molecules were built.

Berzelius' idea of charged radicals lost favor to Laurent's substitution model after Berzelius died in the mid-1800s. Using Laurent's model as a starting point, it became apparent that a central atom serving as a nucleus would only accept a certain number of atoms or groups when it combined with other atoms. For example, oxygen would only accept two atoms or radicals, nitrogen would combine with three, and sodium would only combine with one. Edward Frankland's (1825–1899) work on organometallic compounds led him to propose the theory of combining power. Frankland used the term "atomicity" in a paper he delivered in 1852 to indicate the tendency of an atom to combine with a specific number of other atoms. Frankland's ideas on combination were advanced by Friedrich Kekulé (1829–1896) in the 1860s and led directly to the concept of valence. The term "valence" (Latin for power) started to be used about this time.

By the start of the twentieth century, the stage had been set for advancing the idea of chemical affinity. Thomson's discovery of the electron provided a theoretical foundation upon which new ideas of affinity could be based. Chemical affinity and valence had been based on empirical evidence; the discovery of the electron and a clearer picture of atomic structure provided the framework for new ideas on chemical bonding. The ionic bond involving the transfer of electrons from one atom to another was firmly established at the start of the twentieth century, but it had numerous shortcomings that made it difficult to explain certain observations. Specifically, halogen molecules, such as Cl_2 and I_2 could not be explained using a polar model based of ionic bonds. The problem was how could two halogens, which were considered to be negatively charged, be attracted to each other. A simple polar explanation also fell short when trying to explain bonding in organic molecules.

Figure 7.2
Lewis's Representation of Valence Electrons in Cubic Arrangements

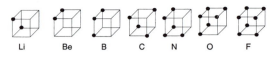

Figure 7.3
Lewis's Representation of Single Bond

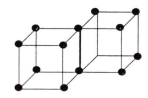

Figure 7.4
Lewis's Representation of Double Bond

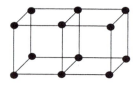

Richard Abegg (1869–1910) noted that the noble gases were extremely stable and suggested other atoms may gain or lose electrons to achieve a stable electron configuration similar to the noble gases. One of the pioneers in developing new concepts on affinity was the American chemist Gilbert Newton Lewis (1875–1946). Lewis expanded Abegg's idea into the **octet rule**. Lewis had proposed as early as 1902 that electrons were arranged in an atom around a nucleus in a cubic configuration as represented in Figure 7.2.

In 1913, Lewis first hinted at a nonpolar bond in which electrons were not transferred between atoms, but shared by atoms. It should be noted that at about the same time as Lewis was developing his theories similar ideas were proposed by the German Walther Kossel (1888–1956). Lewis published his ideas on bonding in a 1916 paper. In this paper, Lewis proposed that electrons at the corners of cubic atomic structures were shared. According to Lewis, two cubes sharing an edge represented the affinity of halogens. This shared edge would comprise the single bond holding the atoms together. In this arrangement, each atom would be surrounded by eight electrons (the octet rule) and obtain the stable electron configuration of a noble gas (Figure 7.3).

A double bond formed when atoms shared a common face with four electrons being shared (Figure 7.4).

Lewis's model established the idea of the nonpolar **covalent bond**, although his idea of the cubic arrangement of electrons had several major flaws, for example, how to represent a triple bond in which six electrons must be shared. Irving Langmuir (1875–1946) further developed Lewis's idea on the nonpolar bond from 1919 to 1921. The nonpolar model became known as the Lewis-Langmuir model. The Lewis-Langmuir model established the covalent bond to complement the ionic bond to explain chemical affinity.

Dot Formulas, the Octet Rule, and Ionic Bonds

Lewis and Langmuir's ideas were being developed as the modern view of the atom was unfolding. Knowledge of atomic structure enables us to develop a more accurate picture of chemical bonding. A chemical bond is simply a force that holds atoms together. Chemical bonding involves the valence electrons, those electrons in the outer shells of atoms. One method of representing an atom and its valence electrons is by using **Lewis electron dot formulas**. A Lewis electron dot formula consists of an element's symbol and one dot for each valence electron. The representative elements and noble gases contain between one

Figure 7.5
Lewis's Dot Structure of Representative
Elements and Noble Gas

Li •	• Be •	• B •	•C•	•N•	•O•	: F •	:Ne:

and eight valence electrons according to their position in the periodic table. The Lewis structure for the first element in each family of the representative elements and noble gases are shown in Figure 7.5.

As mentioned previously, in 1916, Lewis noted that noble gases were particularly stable and did not form compounds. Lewis used these facts to formulate the octet rule. The noble gases have their outer electron shell filled with eight electrons. (Helium is an exception with only two electrons in its outer shell.) The octet rule says that the most stable electron configuration of an atom occurs when that atom acquires the valence electron configuration of a noble gas. That is, when an atom can acquire eight (octet) electrons in its valence shell (or two for hydrogen to become like helium).

Using the octet rule and knowing the valence configuration of the elements explains why ionic compounds form. Strong metals such as the alkalis and alkaline earths have a strong tendency to lose electrons. When an alkali metal loses an electron, it obtains an octet valence electron configuration and acquires a +1 charge. Alkaline earth metals lose two electrons and become cations with a +2 charge. Conversely, nonmetals tend to accept electrons. The halogens have high electron affinities and readily accept a single electron to become anions with a –1 charge. By accepting a single electron, halogens obtain an octet electron configuration. Similarly, elements such as oxygen and sulfur accept two electrons and acquire a –2 charge. When atoms of a strong metal and nonmetal react, electrons are transferred from the metal to nonmetal. Because the metal and nonmetal

ions produced in this process have opposite charges, the ions are attracted to each other. This electrical attraction is the ionic bond.

We can use Lewis dot formulas to represent the transfer of electrons in the formation of ionic compounds. For example, the formation of the ionic compound sodium fluoride, NaF, can be represented using Lewis dot formulas and valence electron configurations:

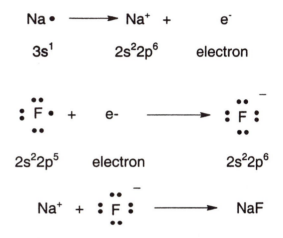

Sodium donates an electron to fluorine, and in the process, sodium becomes the sodium ion, Na^+, and fluorine becomes the fluoride ion, F^-. The net outcome of the transfer of an electron in this case results in both sodium and fluorine obtaining a valence electron configuration similar to the noble gas neon. The Na^+ and F^- are held together by the electrostatic attraction between the oppositely charged ions.

More than a single electron may be transferred when an ionic bond is formed. Two examples of ionic compound formation involving the transfer of more than one electron are the formation of calcium chloride and magnesium oxide:

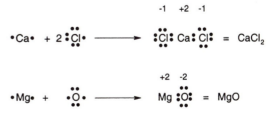

In the first reaction, a single calcium atom donates two electrons to two chlorine atoms. This results in the formation of a Ca^{2+} ion and two Cl^- ions. The positive ion and two negative ions are attracted to each other. In the formation of magnesium oxide, magnesium donates two electrons to oxygen resulting in Mg^{2+} and O^{2-}

In summary, ionic bonds form when there is a transfer of electrons between atoms of different elements. The result of this transfer produces oppositely charged ions. The ions produced generally obtain the valence electron configuration of noble gases, that is, conform to the octet rule. The oppositely charged ions produced are held together by electrostatic attraction. This attractive force is the ionic bond.

Covalent Bonds

Lewis and many other chemists had recognized the shortcomings of the ionic bond. When diatomic molecules, such as H_2 or Cl_2, were considered, there was no reason why one atom should lose an electron and an identical atom should gain an electron. There had to be another explanation for how diatomic molecules formed. We have seen how the octet rule applies to the formation of ionic compounds by the transfer of electrons. This rule also helps explain the formation of covalent bonds when molecules (covalent compounds) form. Covalent bonds result when atoms share electrons. Using fluorine, F, as a representative halogen, we can see how the octet rule applies to the formation of the F_2 molecule. Each fluorine atom has seven valence electrons and needs one more electron to achieve the stable octet valence configuration. If two fluorines share a pair of electrons, then the stable octet configuration is achieved:

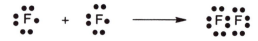

A shared pair of electrons creates a single covalent bond in F_2. Often the double dots forming the bond are replaced with a short dash to represent the bond (F-F). The other halogens form diatomic molecules with single bonds in a similar fashion.

Other diatomic molecules, such as oxygen and nitrogen, are also held together by covalent bonds, but they share more than a single pair of electrons. Oxygen, with its six valence electrons, needs two additional electrons to achieve an octet. If two oxygens share four electrons, each oxygen atom obtains the stable valence configuration of neon. The four shared electrons create a double bond.

$$\ddot{\cdot O\cdot} + \cdot \ddot{O}\cdot \longrightarrow \ddot{\cdot O}::\ddot{O}\cdot = \ddot{\cdot O}=\ddot{O}\cdot = O_2$$

Similarly, sharing six electrons unites two nitrogen atoms. Each nitrogen obtains an octet, and the two nitrogen atoms are united by a triple bond.

$$\cdot\ddot{N}\cdot + \cdot\ddot{N}\cdot \longrightarrow :N :: N: = :N\equiv N: = N_2$$

Electronegativity and the Polar Covalent Bond

It is easy to see from the examples in the previous section how two identical atoms can share electrons to achieve an octet and form diatomic molecules. Because each of our examples dealt with identical atoms, the electrons can be considered to be shared equally by each atom. The bond formed when the atoms are equally shared can be thought of as a pure covalent bond. But what happens in covalent compounds? Remember, a compound contains two different elements. When atoms of two different elements are held together by covalent bonds, there is an unequal sharing of the electrons. The sharing of electrons in a covalent bond may be compared to you and a friend sharing a flashlight while walking down a dark street. If you and your friend both held the light exactly between you, there would be an equal sharing of the light. It's much more likely, though, that one of you holds the flashlight and this person receives more of the light. The light, while shared, is not shared equally. When atoms of two different elements share electrons, one element will have a greater attraction for the electrons. A measure of an element's attraction for a shared pair of electron is given by the element's **electronegativity**. The basic concept of electronegativity dates back to the early nineteenth century, but it was not until 1932 that Linus Pauling (1901–1994) was able to calculate electronegativity. Pauling used bond energies to derive a number to represent an element's attraction for a shared pair of electrons in a covalent bond. Table 7.1 gives the electronegativities of the common elements. The higher the element's electronegativity, the greater the element's attraction for the shared electrons.

Table 7.1 demonstrates that the elements with the highest electronegativities are located in the upper right corner of the periodic table and electronegativities decrease moving down and to the left. Fluorine has the highest electronegativity. Using the electronegativities, we can represent the sharing of electrons in a more realistic manner. For example, for HCl we see the electronegativity of hydrogen is 2.1 and chlorine is 3.0. Therefore, chlorine has a greater attraction for the bonding pair of electrons in the HCl molecule. Because the shared pair of electrons forming the covalent bond is more attracted to the chlorine in the molecule, the bond is referred to as a **polar covalent bond**. Because of the unequal sharing of the electrons in HCl, the H part of the molecule tends to have a positive charge, and the chlorine part a partial negative charge. Hydrogen chloride is a polar molecule. Whenever we have diatomic molecules consisting of two different elements, the molecule is generally polar. It should be remembered that the electrons surrounding atoms should not be viewed as stationary discrete points. Electrons are in constant motion in regions best described by quantum theory. For HCl, it is more appropriate to say that the electron pair forming the covalent bond spends more time near the chlorine atom, or the probability of finding the bonding pair nearer the chlorine is greater. Figure 7.6 represents this situation.

Table 7.1
Electronegativities of Elements

H 2.1																
Li 1.0	Be 1.5											B 2.0	C 2.5	N 3.0	O 3.5	F 4.0
Na 0.9	Mg 1.2											Al 1.5	Si 1.8	P 2.1	S 2.5	Cl 3.0
K 0.8	Ca 1.0	Sc 1.3	Ti 1.5	V 1.6	Cr 1.6	Mn 1.5	Fe 1.8	Co 1.9	Ni 1.9	Cu 1.9	Zn 1.6	Ga 1.6	Ge 1.8	As 2.0	Se 2.4	Br 2.8
Rb 0.8	Sr 1.0	Y 1.2	Zr 1.4	Nb 1.6	Mo 1.8	Tc 1.9	Ru 2.2	Rh 2.2	Pd 2.2	Ag 1.9	Cd 1.7	In 1.7	Sn 1.8	Sb 1.9	Te 2.1	I 2.5
Cs 0.7	Ba 0.9	La- 1.0-1.2	Hf 1.3	Ta 1.5	W 1.7	Re 1.9	Os 2.2	Ir 2.2	Pt 2.2	Au 2.4	Hg 1.9	Tl 1.8	Pb 1.9	Bi 1.9	Po 2.0	At 2.2
Fr 0.7	Ra 0.9															

Figure 7.6
The electron density is greater nearer the chlorine atom in the HCl molecule. Dots represent the nuclei.

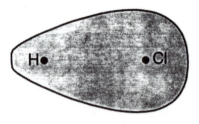

The point to emphasize with polar covalent bonds is that atoms do not equally share the bonding electrons.

Electronegativities provide a convenient tool for classifying the type of bonding present in a compound. The closer the electronegativities of two elements, the greater the degree of sharing of electrons. In the extreme case for homonuclear diatomic molecules (H_2, Cl_2, O_2, etc.) the electronegativities are equal, and the bond is a pure covalent bond. As the difference in electronegativities increases, the bond becomes less covalent and more ionic. While we can characterized bonds as covalent, polar covalent, and ionic, these terms should be interpreted as describing bonds in a general sense. The true nature of a bond should be viewed more as a continuum moving from ionic when the difference in electronegativities is large to pure covalent when the electronegativities are the same. Table 7.2 is useful when using electronegativity to characterize the nature of a bond.

Polar Molecules and Hydrogen Bonds

In the previous section, HCl was described as a polar molecule containing a polar covalent bond. The unequal sharing of electrons in a bond leads to what is referred to as a **dipole**. A dipole is symbolized using a modified arrow pointing toward the negative end of the dipole. For HCl, this can be represented as

$$\overset{\longmapsto}{H \;\vdots\; \overset{\bullet\bullet}{\underset{\bullet\bullet}{Cl}}\vdots}$$

The dipole in HCl causes the H end of the molecule to possess a partial positive charge and the Cl end to possess a partial negative charge. This is denoted as

$$\overset{\delta+\;\delta-}{HCl}$$

with the δ character indicating a partial charge.

The dipole created in HCl creates a polar molecule. The presence of polar covalent bonds does not necessarily mean the molecule will be polar. Some molecules contain two or more dipoles that cancel to give a nonpolar molecule. For example, in CO_2 the two oxygens are attached to carbon by polar covalent bonds.

$$\overset{\longleftarrow\!\mapsto}{O=C=O}$$

The more electronegative oxygens attract the shared electrons that form the double covalent bonds. Thus, two polar double

Table 7.2
Percent Ionic Character Based on Difference in Electronegativity

Difference in electronegativity	0.2	0.4	0.6	0.8	1.0	1.2	1.4	1.6	1.8	2.0	2.2	2.4	2.6	2.8	3.0	3.2
% ionic character	1	4	9	15	22	30	39	47	55	63	70	72	82	86	89	92

covalent bonds are present in CO_2, but because of the symmetry of the dipoles the molecule itself is nonpolar.

When determining whether a molecule is polar or nonpolar, it is important to consider the geometry of the molecule. Carbon dioxide is nonpolar because it is a straight molecule in which the dipoles balance each other so that the center of negative charge coincides with the center of positive charge. Nonpolar CO_2 can be contrasted with H_2O:

Water is a bent molecule. In water, the polar covalent bonds lead to dipoles in which the centers of positive and negative charge do not coincide. This makes water a polar molecule.

In water, the hydrogen atoms are bonded to an oxygen atom. Oxygen has a high electronegativity, which means the bond has a high polarity. The fact that hydrogen is a small atom also means that its positive nucleus can closely approach an oxygen atom on a neighboring water molecule. The partial positive charge of the small hydrogen atoms and partial negative charge of oxygen create a highly polar molecule, which creates a strong attraction between water molecules. The strong attraction between hydrogen in one water molecule and oxygen in another water molecule is called the hydrogen bond. Hydrogen bonding is not unique to water, but occurs when hydrogen is bonded to atoms of oxygen, nitrogen, or fluorine. The fact that hydrogen is small coupled with the high electronegativities of oxygen, nitrogen, and fluorine lead to hydrogen bonding. This is why compounds of these elements have unique physical properties, for example, high boiling points, high heat capacity.

Bond Energies

A bond is a force holding atoms together. When atoms come together to form compounds and molecules, the atoms acquire a more stable electron configuration. The more stable electron configuration means that the chemical potential energy of the system is lowered, and energy is released. To decompose a compound, energy must be supplied to break the chemical bond(s) holding the atoms together. The energy necessary to break a chemical bond is referred to as the **bond energy**. Bond energy is measured in kilojoules per mole of bonds. Average bond energies for some common covalent bonds are given in Table 7.3.

From Table 7.3, it is seen that double bonds are stronger than single bonds, and triple bonds are stronger than double bonds. Table 7.3 shows that the bond energies for covalent bonds are about several hundred kilojoules per mole. For comparison, the amount of energy associated with ionic bonds is given by the lattice energy. The lattice energy is the force that holds ions together in an ionic compound. Lattice energies range from approximately 1,000 kilojoules per mole to several thousand kilojoules per mole. This indicates that ionic bonds are roughly a magnitude stronger compared to covalent bonds. The strength of hydrogen bonds ranges between 10 and 40 kilojoules per mole, a magnitude smaller than covalent bonds.

Table 7.3
Bond Energies in kJ per Mol

H-H	436	C-N	276
C-H	412	Cl-Cl	243
C-C	347	N-O	176
O-O	142	N-N	193
O-H	463	N≡N	418
C-O	350	C=C	615
H-F	570	O=O	498
H-Cl	432	C=O	745
N-H	390	C≡C	830
C-Cl	326	N≡N	940

The Metallic Bond

Several properties characteristic of metals include high thermal and electrical conductivity, ductibility, and malleability. The properties of metals can be attributed to the special forces holding metal atoms together. In a metal, the atoms are arranged in a closely packed structure. Rather than valence electrons being shared between two adjacent atoms, the electrons move freely through the crystalline structure. For example, in solid sodium eight other sodium atoms surround each sodium atom, as shown in Figure 7.7.

Each sodium atom contains a $3s^1$ valence electron. Metals have a tendency to lose electrons and form positive ions. The valence electrons loosely held in metals are pooled and belong to the crystal structure as a whole. The positive metal ions are continually being formed as they lose their valence electrons, and these electrons are shared among the metal atoms. In this situation, we

Figure 7.7
Arrangement of Sodium Atoms in Metallic Sodium

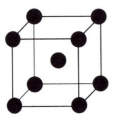

Figure 7.8
Sodium Ions in a Pool of Delocalized Valence Electrons

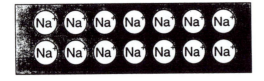

say the electrons are **delocalized** and are free to move throughout the crystal. Figure 7.8 represents the sodium ions embedded in a sea of the pooled electrons.

The Na^+ ions are surrounded by a pool of delocalized electrons, which acts as the "glue" that holds the metallic atoms together. This arrangement is referred to as the metallic bond. This mobile pool of electrons accounts for the characteristic properties of metals. For example, because the electrons are loosely attached, rigid bonds are not formed and atoms can easily be shaped because the electrons move freely throughout the structure.

Bonding and Molecular Geometry

While the basic principles of bonding provide insights on how and why atoms combine to form compounds and molecules, the information does not give us a picture of the structure of substances. Molecular structure depends on the interaction between the valence electrons of atoms making up the molecule. We know that electrons carry a negative charge and like charges repel each other. In molecules containing two or more bonds, the shape or molecule geometry of the molecule can be predicted using a model known as the valence shell electron pair repulsion **(VSEPR) model**. As the name of the model implies, the shape of the molecule is derived from the repulsion of electron pairs in the molecule.

A few examples will illustrate how VSEPR is used to predict molecular geometry. Beryllium chloride, $BeCl_2$, has the Lewis structure

$$Cl—Be—Cl$$

The central beryllium atom has two bonding pairs of electrons. These bonding pairs repel each other and attempt to remain as far apart as possible. The arrangement that satisfies this condition is linear, with each chloride atom separated by an angle of 180°.

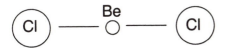

Now consider boron trifluoride, BF_3 in which boron is covalently bonded to three fluorine atoms. For the electron pairs to remain as far apart as possible, the arrangement is triangular with a 120° angle between fluorines:

The Lewis dot structure for BF_3 shows the central boron atom being surrounded by only six electrons, which violates the octet rule. This illustrates that the octet rule, while providing general guidelines, has exceptions.

For a final example, let's consider methane, CH_4 with four bonding electrons surrounding the central carbon atom. VSEPR predicts a tetrahedral arrangement with bond angles of 109.5°.

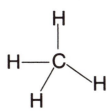

· In each of our examples, only bonding pairs of electrons surrounded the central atom. Many times the central atom contains a single lone pair or lone pairs of electrons in addition to bonding pairs. The presence of lone pairs, which occupy more space than bonding pairs, affects the repulsive forces between the valence electrons in molecules.

When lone pairs are present, several different repulsive forces exist: between bonding pairs, between bonding pair and a lone pair, and between lone pairs. The VSEPR model is based on different repulsive forces for the three situations. These are shown here:

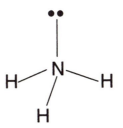

The difference in repulsive forces between electron pairs means that when lone pairs are present the geometries change. Let's examine two common substances to see how the presence of lone pairs affects the geometries of molecules. Ammonia, NH_3, contains three bonding pairs and one lone pair surrounding the nitrogen atom:

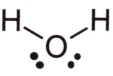

The four pairs of electrons surrounding nitrogen are similar to the electron structure in methane. Methane, though, has four bonding pairs, and ammonia has three bonding pairs and one lone pair. The repulsive force between the single lone pair and three bonding pairs is greater than the repulsive force between the bonding pairs. This results in the hydrogens being "squeezed" together so that the H–N–H bond angles are 107° rather than the 109.5° found in methane.

Water contains two bonding pairs and two lone pairs:

Table 7.4
Simple Molecular Geometries

Geometry	Number of lone pairs	Number of bonding pairs	Lewis dot structure	Shape
Linear	0	2	B∶A∶B	
Trigonal planar	0	3		
Tetrahedral	0	4		
Trigonal pyramidal	1	3		
Bent	2	2		

Like ammonia, the structure is similar to the tetrahedral structure of methane. The two lone pairs repel each other in order to be as far apart as possible. The squeezing of the hydrogens in water is even greater than that in ammonia. The H–O–H bond angle in water is 104.5°.

The VSEPR model lets us predict molecular geometry based on the number of bonding pairs and lone pairs on the central atom. Table 7.4 summarizes several of the simpler geometries based on the number of bonding pairs and lone pairs possessed by the central atom.

Modern Bonding Theory

The use of Lewis dot structures and VSEPR to explain bonding and predict molecular geometry portrays electrons as discrete, fixed units of charge. It is important to remember that in the modern quantum view of the atom electrons occupy a probabilistic region of space surrounding the nucleus. Electrons are described in terms of waves using a probabilistic model. While quantum theory is not required to understand the basic concepts of bonding, the actual structure and properties of many substances can be explained only using quantum mechanical bonding theories. Two modern theories used to describe bonding are the valence bond theory and the **molecular orbital theory**. These two models can be compared by applying them to the hydrogen molecule, H_2.

Valence bond theory assumes that covalent bonds form by atoms sharing valence electrons in overlapping valence orbitals. In valence bond theory, the individual atoms possess valence orbitals that take part in bonding. The overlap of valence orbitals is used to explain bonding. For the hydrogen molecule, each hydrogen atom has one valence electron in a 1s orbital. As two hydrogen atoms approach each other, each electron is attracted to the other atom's nucleus, while the electrons repel each other. The overlapping 1s orbitals that result when the hydrogen atoms combine is the covalent bond:

In valence bond theory, the strength of the bond depends on the degree of overlap. The greater the overlap, the stronger the bond. One feature of the bond valence theory is that the orbitals may combine to produce hybrid orbitals. For example, an s and three p orbital may combine to form four sp^3 hybrid orbitals.

A second quantum mechanical bonding theory is molecular orbital theory. This theory is based on a wave description of electrons. The molecular orbital theory assumes that electrons are not associated with an individual atom but are associated with the entire molecule. Delocalized molecular electrons are not shared by two atoms as in the traditional covalent bond. For the hydrogen molecule, the molecular orbitals are formed by the addition of wave functions for each 1s electron in each hydrogen atom. The addition leads to a bonding molecular

Figure 7.9
Bonding Molecular Orbital Diagram for H_2

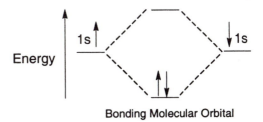

Bonding Molecular Orbital

orbital when the waves reinforce each other and to an antibonding orbital when the waves cancel each other. A bonding molecular orbital diagram for the hydrogen molecule is shown in Figure 7.9.

The two electrons in the hydrogen molecule occupy the bonding orbital with the lowest energy. The energy of this orbital is lower than the energy of the electrons in individual hydrogen atoms resulting in a more stable configuration.

One way of understanding molecular orbitals is to relate them to atomic orbitals. We know electrons fill atomic orbitals starting at the lowest energy levels. As more electrons are added, the atomic orbitals fill. Certain electron configurations are more stable than others, for example, the noble gas configuration. In molecular orbital theory, electrons fill molecular orbitals. Electrons in certain molecular orbitals are more stable than when they exist as electrons in individual atoms. The stability of molecular orbitals is useful in predicting whether a molecule will form. The more stable the formation of molecular orbitals is, the more likely a molecule will form. Molecular orbitals in which the arrangement of electrons produces a higher-energy, less-stable configuration predict nonbonding situations, that is, a molecule will not form. For example, the molecular orbital diagram for He_2 results in two electrons in a bonding orbital and two electrons in a nonbonding orbital.

There is no decrease in energy in this arrangement; therefore, we can conclude that He molecules do not form.

There are many aspects to valence bond theory and molecular orbital theory. Each theory has its strengths and weaknesses, and chemists must use both theories individually and collectively to explain chemical behavior. This section has provided just a brief glimpse into modern bonding theory. More detailed introductions can be found in many college texts.

8

Intermolecular Forces and the Solid and Liquid States

Introduction

The previous chapter dealt with chemical bonding and the forces present between the atoms in molecules. Forces between atoms within a molecule are termed **intramolecular** forces and are responsible for chemical bonding. The interaction of valence electrons between atoms creates intramolecular forces, and this interaction dictates the chemical behavior of substances. Forces also exist between the molecules themselves, and these are collectively referred to as **intermolecular** forces. Intermolecular forces are mainly responsible for the physical characteristics of substances. One of the most obvious physical characteristics related to intermolecular force is the phase or physical state of matter. Solid, liquid, and gas are the three common states of matter. In addition to these three, two other states of matter exist—plasma and **Bose-Einstein condensate**.

Plasma is a gas in which most of the atoms or molecules have been stripped of some or all of their electrons. The electrons removed still stay within the vicinity of the atom or molecule. This state generally occurs when a gas is subjected to tremendous temperatures, is electrically excited, or is bombarded by radiation. Plasma is the principal component of interstellar space and comprises over 99% of the universe. The plasma state is also present in gas-emitting light sources such as neon lights. The state of matter known as the Bose-Einstein condensate was theoretically predicted by Satyendra Nath Bose (1894–1974) and Albert Einstein (1879–1955) in 1924. A Bose-Einstein condensate forms when gaseous matter is cooled to temperatures millionths of a degree above absolute zero. At this temperature atoms, which normally have many different energy levels, coalesce into the lowest energy level. At this level, atoms are indistinguishable from one another and form what has been termed a "super atom." The first experimentally produced Bose-Einstein condensate was reported 1995, and currently, a number of research groups are investigating this new state of matter.

In this chapter, we focus on solids and liquids. In the next chapter, we turn our attention to gases. Except where noted, the discussion focuses on the states of matter under normal atmospheric conditions, that is, 1 atmosphere pressure and room temperature.

Characteristics of Solids, Liquids, and Gases

Most of the naturally occurring elements exist as solids under normal conditions (Figure 8.1). Eleven elements exist as gases, and only two exist as liquids.

Liquids and gases share some common characteristics, such as the ability to flow and together are classified as fluids. Likewise, liquids also share some characteristics with solids, and these two states are called condensed states. Solids, liquids, and gases can be differentiated by their properties. Solids and liquids have relatively strong intermolecular attraction compared to gases, in which the intermolecular attraction can be considered to be nonexistent under normal conditions. Solids and liquids contain molecules and ions arranged close together with little empty space between particles. Therefore, compared to gases, solids and liquids have high densities and are nearly incompressible. Gases are characterized by predominantly void space through which the molecules move. In solids, molecules and ions are held rigidly in place preventing any appreciable movement of particles with respect to one another. Solid have a fixed shape and volume. In liquids, the intermolecular forces are weaker than in solids, and so liquids have the ability to flow and take the shape of the object containing the liquid. Table 8.1 summarizes some of the characteristics of the solid, liquid, and gas states.

Intermolecular Forces

Intermolecular forces are responsible for the condensed states of matter. The particles making up solids and liquids are held together by intermolecular forces, and these forces affect a number of the physical properties of matter in these two states. Intermolecular forces are quite a bit weaker than the covalent and ionic bonds discussed in Chapter 7. The latter requires several hundred to several thousand kilojoules per mole to break. The strength of intermolecular forces are a few to tens of kilojoules per

Figure 8.1
States of elements under normal conditions. Light gray indicates gas, dark gray indicates liquids, all the rest are solids.

H																	He
Li	Be											B	C	N	O	F	Ne
Na	Mg											Al	Si	P	S	Cl	Ar
K	Ca	Sc	Ti	V	Cr	Mn	Fe	Co	Ni	Cu	Zn	Ga	Ge	As	Se	Br	Kr
Rb	Sr	Y	Zr	Nb	Mo	Tc	Ru	Rh	Pd	Ag	Cd	In	Sn	Sb	Te	I	Xe
Cs	Ba	La	Hf	Ta	W	Re	Os	Ir	Pt	Au	Hg	Tl	Pb	Bi	Po	At	Rn
Fr	Ra	Ac	Rh	Db	Sg	Bh	Hs	Mt	Uun	Uuu	Uub						

Ce	Pr	Nd	Pm	Sm	Eu	Gd	Tb	Dy	Ho	Er	Tm	Yb	Lu
Th	Pa	U	Np	Pu	Am	Cm	Bk	Cf	Es	Fm	Md	No	Lr

Table 8.1
Characteristics of Solids, Liquids, and Gases

State	Intermolecular Force	Compressibility	Shape	Relative Density	Volume
Solid	strong	incompressible	definite shape	high	definite
Liquid	strong	nearly incompressible	assumes that of container	high	definite
Gas	practically zero	compressible	fills container	low	assumes volume of container

mole. One of the strongest and most important types of intermolecular forces is the hydrogen bond. Its strength can be as high as 40 kilojoules per mole. The hydrogen bond was introduced in Chapter 7. Let's examine the hydrogen bond before looking at several other types of intermolecular attractions.

Hydrogen bonds are formed between molecules that contain hydrogen covalently bonded to atoms of oxygen, fluorine, or nitrogen. Hydrogen bonded to any one of these elements produces a highly polar molecule. The hydrogen takes on a partial positive charge, and the electronegative element has a partial negative charge. The hydrogen on one molecule has a high attraction for the electronegative atom on an adjacent molecule. It is important to remember that the hydrogen bond is an intermolecular force.

Water is the most common substance displaying hydrogen bonding. The hydrogen on one molecule is attracted to an unshared pair of electrons held by oxygen on an adjacent water molecule (Figure 8.2). Because each oxygen in water has two unshared pairs of electrons and there are also two hydrogen atoms in water, a network of water molecules forms, held together by hydrogen bonds. Hydrogen bonding has a pronounced effect on the physical properties of water. For example, the boiling points of compounds

Figure 8.2
Hydrogen bonding in water. Hydrogen bonds are represented by dashed lines. Only several water molecules are shown and not all bonds are displayed. The pattern for the central oxygen atom continues to form a network of hydrogen-bonded water molecules.

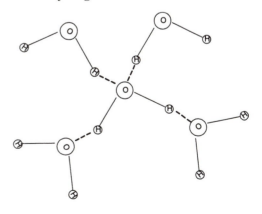

Figure 8.3
Trend in Boiling of Several Group 16 Compounds

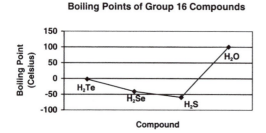

Boiling Points of Group 16 Compounds

generally decrease up a group in the periodic table. The reason for this decrease involves intermolecular forces and is explained shortly. In the oxygen group, the boiling points of H_2Te, H_2Se, H_2S, and H_2O are graphed in Figure 8.3.

The trend follows the expected trend for the first three compounds with the boiling point decreasing from the largest compound H_2Te to H_2S. According to the normal pattern, water should have the lowest boiling point. If the trend displayed for the first three compounds continued, then water's boiling point should be lower than that of H_2S. The reason that water has an abnormally high boiling point compared to similar compounds is that water molecules exhibit hydrogen bonding. Because the water molecules are hydrogen bonded, it takes more energy to cause water to leave the liquid state and enter the gas state. It's just as though you were with a group of friends, and someone tried to pulled you away from the group. It would be a lot harder to pull you, or any of your friends, away from the group if you were all holding on to each other. In essence, hydrogen bonding is how water molecules hold on to each other. A pattern similar to that shown in Figure 8.3 exists for Groups 15 and 17 hydrogen compounds with ammonia, NH_3 and hydrogen fluoride, HF, having the highest boiling points in these two groups, respectively. This

is not surprising because both ammonia and hydrogen fluoride exhibit hydrogen bonding.

Another interesting property of water is that its maximum density is at 4°C. In general, almost all substances have a greater density in their solid state than in their liquid state. Hydrogen bonding is also responsible for this unique property of water. Because each water molecule has two hydrogen atoms and two lone pairs of electrons, the water molecules can form a three dimensional network of approximately tetrahedrally bonded atoms. Each oxygen is covalently bonded to two hydrogen atoms and also hydrogen bonded to two oxygen atoms (Figure 8.2). When water exists as ice, the molecules form a rigid three-dimensional crystal. As the temperature of ice increases to the melting point of 0°C, hydrogen bonding provides enough attractive force to maintain the approximate structure of ice. The increase in temperature above water's melting point causes thermal expansion, and this would normally lead to an increase in volume and a corresponding decrease in density. Remember, density is mass divided by volume so a larger volume results in a smaller density. The reason water actually becomes more dense is that at the melting point and up to 4°C there is enough energy for some of the water molecules to overcome the intermolecular attraction provided by the hydrogen bonding. These water molecules occupy void space in the still-present approximate crystalline structure. Because more molecules of water are occupying the same volume, the density increases. This phenomenon continues until the maximum density is reached at 4°C. At this point, the trapping of additional water molecules in the void space is not great enough to overcome the thermal expansion effect that lowers the density; therefore, solid ice is less dense than liquid water.

The fact that the maximum density of water is at 4°C has significant environmental impacts. Consider the freezing of a freshwater lake in winter. As the temperature decreases, the surface water becomes progressively denser and sinks to the bottom. This process helps carry oxygen from the surface to deeper water. The cycling of a lake's waters also helps to bring nutrients to the surface. This process is sometimes referred to as the fall overturn. Once the lake reaches a temperature of 4°C, the water is at its maximum density. Further cooling results in less dense water, which doesn't sink, eventually forming a layer of ice if temperatures are cold enough. The ice layer effectively insulates the rest of the lake. Because the water was replenished with oxygen during the fall overturn, the lake generally does not become **anoxic**. If water behaved like most substances, the entire lake would cool down to its freezing point, and then the entire lake would freeze solid.

Several other unique properties of water can be attributed to hydrogen bonding. We discuss these after we have had a chance to examine some of the other intermolecular forces. The hydrogen bond is actually an unusually strong form of what is known as a **dipole-dipole force**. Polar molecules give rise to dipole-dipole forces. A dipole-dipole force results from the electrostatic attraction between the partial positive and negative charges present in polar molecules. The strength of the dipole-dipole force is directly proportional to the strength of the dipole moment of the molecule.

Just as two polar molecules, like opposite ends of a magnet, are attracted to each other, a polar molecule may be attracted to an ion. This gives rise to an **ion-dipole force**. The negative ends of polar molecules are attracted to cations and the positive end to anions. The charge on the ion and the strength of the dipole moment determine the

Figure 8.4
Ion-Dipole Forces. M^+ represents a cation and X^- an anion.

strength of the ion-dipole force. When an ionic compound such as salt, NaCl, dissolves in water, ion-dipole forces come into play. The positive hydrogen end of water and negative chloride ion are attracted to each other, while the negative oxygen end and positive sodium ion are attracted to each other (Figure 8.4). Polar molecules tend to be soluble in water, while nonpolar molecules are insoluble. This fact is illustrated by a mixture of oil and water. The nonpolar oil does not dissolve in water, and two separate layers result. A similar effect occurs with the polar vinegar and nonpolar oil portions of a salad dressing.

Hydrogen bonding, dipole-dipole, and ion-dipole forces all involve polar molecules, and all involve electrostatic attractions between opposite charges. A force that is present in both polar and nonpolar molecules is known as a **dispersion** or **London force**. The London force is named for Fritz London (1900–1954) who described this force in 1931. Dispersion forces are a consequence of the quantum nature of the atom. Electrons move around atomic nuclei in a probabilistic fashion, but they occupy specific energy levels. In nonpolar molecules, on average, the movement of the electrons are such that the centers of positive and negative charge are the same. It is important to realize, though, that this is an average state and does not represent what is occurring at any one instance. Because the electrons are in constant motion, the electrons assume an asymmetrical distribution producing an temporary dipole. An instant later, this situation

Figure 8.5
On average, the electron distribution in helium is symmetrical, and the centers of positive and negative charge are the same as shown on the left. At any instant, though, the electrons may assume an asymmetrical distribution as shown on the right. In this situation, a temporary dipole is formed.

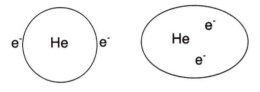

Table 8.2
Freezing Points of Noble Gases

Noble Gas	Freezing Point
Neon	-250°C
Argon	-189°C
Krypton	-157°C
Xenon	-112°C

changes. Because electrons are constantly moving and changing position, temporary dipoles are continually being formed and dissolved (Figure 8.5). Temporary dipoles induce other temporary dipoles in adjacent atoms. That is, the negative and positive ends of the dipole will attract or repel electrons in adjacent atoms. This continual process leads to an ever-present attraction between nonpolar (as well as polar) molecules.

Dispersion forces depend on temporary dipoles inducing dipoles in adjacent atoms or molecules. The ease with which electrons in an atom or molecule can be distorted to form a temporary dipole is known as **polarizability**. The larger the atom or molecule is, the greater its polarizability. This is because larger atoms have more electrons, and these electrons are located farther from their respective nuclei. Electrons farther from the nucleus are held less tightly, and this results in their greater polarizability. This is why larger molecules in a series tend to have higher boiling points. For example, the boiling points of methane (CH_4), propane (C_3H_8), and butane (C_4H_{10}) are $-161°C$, $-42°C$, and $0°C$, respectively.

An example of how dispersion forces and polarizability affect physical properties is seen in the halogens. Moving down the halogen group fluorine and chlorine are gases, bromine is a liquid, and iodine a solid. This is just what would be expected. Dispersion forces become stronger moving down the halogen group as the atoms increase in size and are more polarizable. Another illustration is seen in Table 8.2 which lists the freezing point of the Nobel gases. The freezing point increases down the group. If the temperature of a mixture of the gases listed in the table was lowered, Xenon would freeze first because of the presence of greater dispersion forces.

Dipole-dipole, ion-dipole, and dispersion forces are collectively known as **van der Waals forces**. Johannes Diederick van der Waals (1837–1923) received the 1910 Nobel Prize in physics for his work on fluids. We have seen how hydrogen bonding and van der Waals forces affect the physical properties of substances, and more is said about these forces as we examine the different states of matter.

Crystalline Solids

Solids are the most obvious state of matter for us; they are ubiquitous and come in all shapes and sizes. Solids are classified as either **crystalline solids** or **amorphous solids**. A crystalline solids displays a regular, repeating pattern of its constituent particles throughout the solid. At the microscopic

Figure 8.6

A crystalline solid is composed of a repeating unit known as the unit cell. Lattice points are occupied by atoms, molecules, or ions.

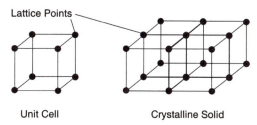

Lattice Points

Unit Cell   Crystalline Solid

level, crystalline solids appear as crystals. An amorphous solid does not display a regular, repeating pattern of its constituent particles. Crystalline solids are composed of atoms, molecules, or ions that occupy specific positions in a repeating pattern. The position occupied by the particles are referred to as **lattice points**. The most basic repeating unit making up the crystalline solid is known as the **unit cell**. The unit cell repeats itself throughout the crystalline solid (Figure 8.6).

Crystalline solids can be classified according to the type of particles occupying the lattice points and the type of intermolecular forces present in the solid. Ions occupy the lattice points in an **ionic solid**, and the crystal is held together by the electrostatic attraction between cations and anions. Common table salt is an example of a crystalline solid. In **molecular solids**, molecules occupy the lattice points, and van der Waals forces and/or hydrogen bonding predominate in this type of crystalline solid. Common table sugar, sucrose, is an example of a common molecular solid. **Covalent crystals** are generally composed of atoms linked together by covalent bonds in a three-dimensional array. Carbon in the form of graphite or diamond is an example of a covalent solid. The final type of solid is a **metallic solid** in which **metallic bonding** is characterized by a delocalized sea of electrons holding together metal cations.

Much of our knowledge on the structure of crystalline solids comes from **x-ray crystallography** or x-ray diffraction. **Diffraction** is the scattering or bending of a wave as it passes an obstacle. X-rays exist as electromagnetic waves. They are particularly useful for probing crystalline solids, because their wavelengths are roughly on the same scale as that of atoms, 10^{-10}m. Visible light cannot be used for probing crystalline solids, because the wavelength of visible light is far too large to produce an image. A simple analogy may help to put this concept in perspective. Say you wanted to pick up a very tiny grain of rice with your fingers. It would be impossible to use your thumb and index finger in the usual manner to do this; your fingers are just too large. You could pick up the grain with a pair of fine tweezers, though. Just as your fingers are too large to perform certain tasks, visible light is too large to probe at the atomic level. X-rays are the right size to "observe" crystalline solids.

Max Theodor Felix von Laue (1879–1960) first predicted x-rays could be diffracted by a crystal in 1912. He received the Nobel Prize in 1914 for his discovery of x-ray crystallography. One year later, William Henry Bragg (1862–1942) and his son William Lawrence Bragg (1890–1951) shared the Physics Nobel Prize for expanding on von Laue's work. In x-ray crystallography, an x-ray beam is focused through a crystal. The particles making up the crystal scatter the x-rays resulting in a pattern of **constructive interference** and **destructive interference**. Constructive interference results when waves are in phase and reinforce each other; in destructive interference, the waves are out of phase and cancel each other (Figure 8.7). Because the particles (atoms, molecules, or ions) making the crystalline

Figure 8.7
In constructive interference, waves reinforce each other. In destructive interference, waves cancel each other.

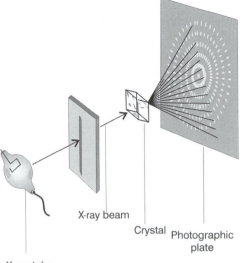

Figure 8.9
Three Simple Unit Cells for Crystalline Solids

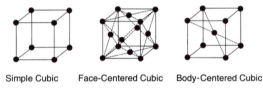

ence (Figure 8.8). The diffraction pattern can then be used to help decipher the structure of the crystal.

The x-ray diffraction pattern reveals how the particles are arranged in a crystalline structure. A number of common arrangements exist among crystalline solids. These can be classified according to the arrangement of particles making up the unit cell. Three common arrangements are simple cubic, face-centered cubic, and body-centered cubic (Figure 8.9). In sodium chloride, the Na^+ and Cl^- ions are positioned in a face-centered cubic arrangement. The alkali metals follow a body-centered cubic arrangement. In addition to the arrangement of particles in the crystalline solid, interpretation of the x-ray diffraction pattern allows chemists to determine the distance between particles and how much void space is present in the solid.

solid have an ordered arrangement, the x-rays scattered by the particles produce an ordered pattern. The pattern can be captured on a photographic plate or film with dark areas representing constructive interference and light areas showing destructive interfer-

Figure 8.8
Picture of X-ray Diffraction Pattern (Rae Déjur)

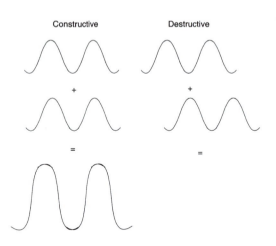

Amorphous Solids

The term "amorphous" comes from the Greek word for shape, "morph," and means disordered or without shape (a + morph). Amorphous solids do not display a regular three-dimensional arrangement of particles. The most common amorphous solid is glass. We normally use the term "glass" in association with silica-based materials, although the term is sometimes used with any amorphous solid including plastics and metals. The technical definition for glass is an optically transparent fusion product of inorganic

Figure 8.10
Both quartz and silica glass are primarily composed of silica, SiO$_2$.

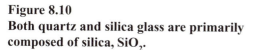

● Silicon
○ Oxygen

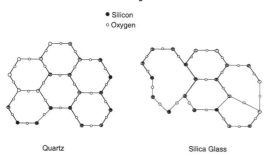

Quartz   Silica Glass

material that has cooled to a rigid state without crystallizing. Our discussion in this section will primarily focus on silica-based glass. Both crystalline quartz and the silica glass are primarily composed of silica, SiO$_2$. The difference between quartz and glass is that quartz displays **long-range order** (Figure 8.10). In quartz, the silicon is covalently bonded to the oxygen in a tetrahedral arrangement. When glass is produced, SiO$_2$ is heated to an elevated temperature and then rapidly cooled. The rapid cooling does not allow the SiO$_2$ to form a regular crystalline structure. The result is a solid that behaves like a very viscous liquid when it is heated. Glass is sometimes called a solid solution and does flow at a very slow rate. This flow can sometimes be seen in old window glass where the bottom is slightly thicker than the top.

Quartz is a crystalline solid and has an orderly structure, but glass is an amorphous solid that is disordered. The structures are actually arranged in a three-dimensional tetrahedral pattern. They are shown in Figure 8.10 as two-dimensional representations.

The first silica glass was produced around 3500 B.C. in Mesopotamia (present day Syria and Iraq), although there is evidence of early production in Egypt and Phoenicia (Lebanon). The melting point of pure SiO$_2$ is 1713°C, but the mixing of other substances with the SiO$_2$ lowers its melting point. Egyptians added natron (sodium carbonate) to SiO$_2$. Some glasses are produced at temperatures as low as 600°C. As the art of glass making developed, individuals discovered how to produce different glasses by adding various substances to the silica melt. The addition of calcium strengthened the glass. Other substances imparted color to the glass. Iron and sulfur gave glass a brown color, copper produced a light blue, and cobalt a dark blue. Manganese was added to produce a transparent glass, and antimony to clear the glass of bubbles. Most modern glass produced is soda-lime glass and consists of approximately 70% SiO$_2$, 15% Na$_2$O (soda), and 5% CaO (lime). Borosilicate glass is produced by adding about 13% B$_2$O$_3$. Borosilicate glass has a low **coefficient of thermal expansion**, and therefore, is very heat resistant. It is used extensively in laboratory glassware and in cooking where it is sold under the brand name Pyrex®.

Liquids

Liquids share certain properties with both solids and gases. Like solids, liquids are almost incompressible and occupy a specific volume; like gases liquids flow. The intermolecular forces present in liquids are weaker than in solids. Liquid particles are not held rigidly in place and are able to slide past one another. The ability to flow varies greatly among liquids. Anyone familiar with the saying "slow as molasses" can appreciate the difference in the ability of molasses to flow compared to other liquids such as water. **Viscosity** is a measure of a liquid's resistance to flow. The ability of a liquid to flow is related to several factors including intermolecular forces, size of particles, and structure. The stronger the intermolecular forces in liquids are, the higher the viscosity. Small particles can slide past neighboring particles easier, and for this reason, substances composed of smaller particles have lower viscosities. Another factor affecting viscosity is the size and shape of molecules. Structures composed of long chains have higher molecular masses, and a greater probability of molecules entangling. The dependence of size and shape on viscosity can be pictured by contrasting the difference in dumping out a box of marbles versus dumping out a box of rubber bands. The marbles flow readily out of the box, while the rubber bands become entangled with each other and do not flow smoothly.

Anyone who has heated honey or syrup is aware that viscosity decreases with increasing temperature. Viscosity is a primary concern in lubricants, and a practical example of viscosity deals with the labeling of motor oils. Motor oils have an arbitrary viscosity rating given by their SAE (Society of Automotive Engineers) numbers. The higher the number is, the higher the viscosity. This means that oils with higher SAE numbers are thicker. Most common motor oils are produced to have multiviscous characteristics to meet the needs of both cold start-up conditions and the elevated temperatures during operation. A typical SAE number would be 5W-30 or 10W-40. The first number can be considered the base number appropriate for cold conditions (W stands for winter). Cold conditions are considered to be 0°F. The second number is the equivalent viscosity at an elevated temperature, say 200°F. Multiviscous oils are used to meet the typical driving cycle. When an engine is cold, it needs a lower viscosity (thinner) oil to lubricate the engine effectively. As the engine warms up, the oil inside also warms and becomes even less viscous. To compensate for the lowering of the viscosity with increased temperature, multiviscous oils contain carbon **polymer** additives. These polymers are coil shaped at low tem-

peratures and uncoil as the engine temperature increases. The uncoiled polymers increase the viscosity, thereby compensating for the heating effect.

Another characteristic dependent on the intermolecular forces is the **surface tension** of the liquid. Surface tension results from the unbalanced forces on molecules at the surface of a liquid. Figure 8.11 shows how surface tension results from these unbalanced forces. Consider water as the liquid in Figure 8.11. A water molecule in the interior of the liquid is surrounded on all sides by other water molecules. Attractive intermolecular forces pull the molecule equally in all directions and these forces balance out. A water molecule on the surface experiences an unbalanced force toward the interior of the fluid. This unbalanced force pulls on the surface of the water putting it under tension. This situation is similar to the tightening of the head of drum. The tension causes the surface of the water to act like a thin film. If you carefully use tweezers to place a clean needle on the surface of water, surface tension will allow the needle to float even though the needle is denser than water.

Figure 8.11
Surface tension results from the fact that surface molecules are pulled toward the interior of the liquid as compared to interior molecules where the forces are balanced.

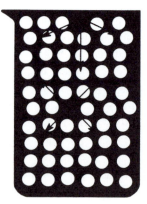

Surface tension is directly related to the magnitude of intermolecular forces in a liquid. The greater the intermolecular force is, the greater the surface tension. For this reason, water has a high surface tension. Because of the surface tension, water drops will assume a spherical shape. This shape minimizes the surface area of the drop. Related to surface tension are the properties of **cohesion** and **adhesion**. Cohesion refers to the attraction between particles of the same substance. That is, it is the intermolecular force between molecules of the liquid. Adhesion is the attraction between particles of different substances. Adhesion is seen in the ability of water to rise in thin tubes by capillary action. The adhesion between water and glass is greater than the cohesion between water molecules. The glass pulls water up the interior of the tube, and the cohesive force between water molecules creates a concave **meniscus** (Figure 8.12). The water rises until the cohesive force balances the weight of the water pulled up on the side of the tube. Because smaller diameter tubes contain less volume of water, water rises higher as the tube diameter decreases.

Certain chemicals have the ability to lower the surface tension of water. This allows water to spread out over a surface rather than bead up. Wetting agents decrease the cohesive forces between water molecules, and this helps water to spread over the surface of an object by adhesive force.

Material Science

The material in this chapter has focused on the solid and liquid states. Up until recent times, there has been a clear distinction between these states. Recent advances in modern science have led to materials that blur the distinction between states of matter. Material science is a relatively new interdisciplinary area of study that applies knowledge from chemistry, physics, and engineering. Material scientists study the properties, structure, and behavior of materials. Material science can further be divided into branches that deal in areas such as ceramics, alloys, polymers, superconductors, semi-conductors, corrosion, and surface films. Material scientists continue to modify our traditional view of the different states of matter. A good example of a material that straddles two states of matter is **liquid crystals**. The term "liquid crystals" sounds contradictory, but it is an appropriate name for molecules classified as liquid crystals. There are close to 100,000 different organic compounds identified as liquid crystals. They are slender, rod-like molecules that have mobility like a liquid but display order by having their axes line up parallel to each other. An analogy might be that a liquid is like the people walking in one direction down a busy sidewalk, while a

Figure 8.12
In capillary action, the adhesion between the walls of the tube and the liquid is greater than the cohesion between liquid molecules. The liquid rises up the tube and forms a meniscus that is concave upward. The smaller the tube is, the greater the height the liquid rises.

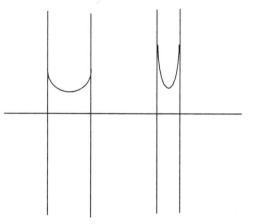

liquid crystal is like a marching band. On the sidewalk, people flow in a definite direction, but they are not aligned. People in the marching band are aligned in columns. There is a definite structure to the marching band, but members of the band still flow.

Liquid crystals are all around us. All you have to do is look at a calculator, a digital clock or watch, or a laptop computer and you see liquid crystals. Devices with liquid crystals are so commonplace that it is hard to believe that the first digital watch was not marketed until 1973. If you examine a liquid crystal display closely, you will often notice a figure 8 arrangement consisting of seven bars. The liquid crystals in these displays are controlled by small electric fields; the crystals are very sensitive to these electric fields and orient themselves according to the electric field. A liquid crystal display works by rotating **polarized light** so that it either passes through or is blocked by a second polarizer. Think of unpolarized light as light waves vibrating in all directions. In this condition, light is said to be incoherent, or in simple terms it is all scrabbled up. A polarizer lines the light waves up in one direction making the waves coherent. In short, it unscrabbles the light waves. A polarizer is like an optical grating, allowing light that has the right orientation to pass through it. This is like a narrow passage allowing only those people who are turned sideways to squeeze through. When liquid crystals cause the light to be blocked by the second polarizer, the display registers a dark image (Figure 8.13). These dark images are displayed as numbers or letters in the liquid crystal display. By mixing other chemicals with liquid crystal molecules, color displays can be made.

Liquid crystals are just one of the many advances made in material science during the last thirty years. Many other new substances with interesting properties have

Figure 8.13
In a liquid crystal display, unpolarized light passes through a polarizer. The polarized light is either unaffected (1) or rotated (2) by the liquid crystals depending on the voltage applied to the liquid crystals. Light either passes through the second polarizer and is reflected by the mirror producing a light image or is blocked by the second polarizer in which case a dark image appears.

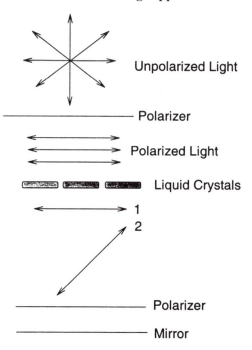

been discovered. One that captured widespread attention starting in 1985 was the discovery of **buckminsterfullerenes** or "bucky balls." This naturally occurring form of carbon was named after the American architect F. Buckminster Fuller (1895–1983) who designed the geodesic dome. Buckminsterfullerene is said to be the most spherical molecule known. The first buckminsterfullerene identified was C-60 by Richard Smalley (1943–), Robert Curl (1933–), and Harold Kroto (1939–). These three shared the 1996 Nobel Prize in chemistry for their pioneering discovery. C-60 has the shape of a soccer ball with the sixty carbons making

Figure 8.14
Buckminsterfullerene

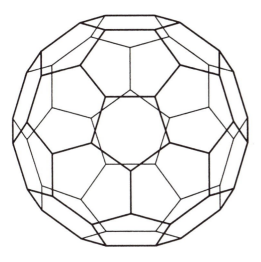

up twelve pentagonal and twenty hexagonal faces (Figure 8.14). Since the original discovery of C-60, numerous other fullerenes of various shapes have been produced ranging from C-28 to C-240. Research into their potential use is currently underway, but their expense, which is several times that of gold, currently limits widespread use.

Other advances in material science have helped humans mimic nature in the production of certain materials. For example, in the last half of the twentieth century we have learned how to produce synthetic diamonds. Diamonds were first produced commercially in the 1950s by General Electric by subjecting graphite to temperatures of 2,500°C and pressures approaching 100,000 atmospheres. Currently, well over a hundred companies produce synthetic diamonds.

Other advances in material science are being made in a number of areas, including materials collectively known as "smart" or "intelligent" materials. These materials have the ability to adapt to their surroundings. A common example is lenses used in eyeglasses made from material that darken in bright light. Memory metals have the ability to return to a preformed shape at certain temperature. These metals are used in orthodontic devices and guidewires. A popular memory metal is the nickel-titanium alloy called Nitinol. Memory metals assume a "parent" shape above a specific transition temperature. Below the transition temperature, the metal can be formed into various shapes. When the memory metal is heated above the transition temperature, the metal goes back to its parent shape as though it has remembered it. Shape memory polymers that work similar to metals have recently been introduced. Advances in memory material may make it possible to repair bent objects by simply heating the object. Another group of materials is known as self-healing materials. To understand how self-healing materials are designed, think of how a cut heals. When you cut yourself, a number of signals muster a chemical response to the wound. Over several days the wound heals. Self-healing plastics are currently being studied. Materials such as plastics develop numerous micro-cracks through everyday use. As time goes on, these micro-cracks propagate and grow, eventually leading to failure of the product. One concept is to strategically position liquid plastic-filled capsules in self-healing plastics. When the plastic object is dropped or handled in a certain way, the capsules would break open and flow into the micro-cracks where it would harden. This self-healing process would lengthen the normal life of materials. Stealth weaponry is yet another example of how materials can be modified to behave in a totally new manner. Stealth airplanes are built with materials designed to absorb radar waves rather than reflect a signal back to the receiver.

Research in material science continues to modify the substances that make up our world. As the twenty-first century unfolds, new materials will improve our lives. Just consider biomaterials used for bone replacement, drug delivery microchips, and skin grafts; composites used in cars, planes, and other transportation vehicles; building materials that are lighter, provide insulation, and have superior strength. All we have to do is take a good look around us to realize that many of the common materials present today were not available even fifty years ago, for example, teflon, fiberglass, and semi-conductors. Material scientists continue to improve old materials and create new ones. In this manner, they are the modern version of the ancient alchemists. Like alchemists, material scientists strive to perfect the basic elements into something more perfect, although without the mystical connection. As these new alchemists continue to perfect materials, chemistry will play a central role in their quest.

Gases

Introduction

This chapter continues to explore states of matter by focusing on the gas state. Table 8.1 summarized the main features of gases. In the gaseous state, molecules are much farther apart than in either solid or liquids. Because of this distance molecules in the gaseous state have virtually no influence on each other. The independent nature of gas molecules means intermolecular forces in this state are minimal. A gas expands to fill the volume of its container. Most of the volume occupied by the gas consists of empty space. This characteristic allows gases to be compressible, and gases have only about 1/1,000 the density of solids and liquids.

Eleven elements (hydrogen, helium, oxygen, nitrogen, fluorine, chlorine, argon, neon, krypton, xenon, and radon) exist as gases under normal conditions. Additionally, many molecular compounds exist as gases. The most common gases are those found in our atmosphere. Nitrogen and oxygen comprise most of the atmosphere (Table 9.1). The third most abundant gas in the atmosphere is argon. Only trace amounts of carbon dioxide, helium, neon, methane, krypton, and hydrogen are present in the atmosphere.

Pressure

One of the most important properties characterizing a gas is its **pressure**. Pressure is defined as the force exerted per unit area. Atmospheric pressure is the force exerted by the atmosphere on the Earth's surface. A common device used to measure atmospheric pressure is the barometer. The barometer was invented in 1643 by Evangelista Torricelli (1608–1647). Torricelli, who was a student of Galileo, was presented with the problem of why water could not raise more than 32 feet with a suction pump. Galileo explained the rise of water up a pipe by invoking the statement "nature abhors a vacuum." When a pump created a vacuum, Galileo believed water rose to fill the void created, but somehow there was a limit of approximately 32 feet. Torricelli reasoned correctly that the height to which water could be pumped was due to atmospheric pressure. In his study of the pump problem, Torricelli took a sealed tube approximately four feet in length and filled it with mercury.

Table 9.1

Composition of Earth's Atmosphere by Volume

Gas	Formula	Percent
Nitrogen	N_2	78.08
Oxygen	O_2	20.95
Argon	Ar	0.90
Carbon Dioxide	CO_2	0.037
Neon	Ne	0.0018
Helium	He	0.00053
Methane	CH_4	0.00028
Krypton	Kr	0.00011
Hydrogen	H_2	0.000040

Figure 9.1

In a mercury thermometer, atmospheric pressure, P, forces mercury to rise in the tube to a height of approximately 30 inches

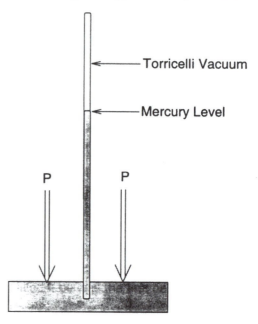

He inverted the tube in a pool of mercury. Torricelli noticed that the mercury in the tube fell, creating a vacuum at the top of the tube. The final height of the mercury in the tube was roughly 28 inches. Because mercury has 13.5 times the density of water, Torricelli proposed that water should rise about 13.5 × 28 inches or about 32 feet. Torricelli reasoned that atmospheric pressure placed a limit on how high a liquid could raise in the tube. His inverted mercury-filled tube was the first barometer (Figure 9.1).

Many units are used to express pressure. Because pressure is defined as force per unit area, a common unit used in the United States is pounds per square inch. This unit is commonly used for tire inflation pressure. The atmospheric pressure at sea level is about 14.7 pounds per square inch. In the metric system, the basic unit for force is the **newton**, abbreviated N, and area is mea-

sured in squared meters, m^2. Therefore, the metric unit for pressure is newtons per squared meter, N/m^2. A N/m^2 is also known as a **pascal**, abbreviated Pa. As we have seen, atmospheric pressure can also be expressed by measuring the height of mercury in a barometer. This measurement is given in inches in the United States and is how atmospheric pressure is reported in daily weather reports. In the metric system, the height of mercury is given in millimeters of mercury. A common unit used by chemists to express pressure is atmospheres. One atmosphere is equal to normal atmospheric pressure at sea level. The various units for pressure and the value of standard atmospheric pressure are summarized in Table 9.2.

The values in column three of Table 9.2 are atmospheric pressure at sea level. At the top of Mt. Everest, the atmospheric pressure is only one-fourth of that at sea level.

Table 9.2
Pressure Units and Values for Standard Atmospheric Pressure

Pressure Units	Abbreviation	Standard
atmosphere	atm	1.00
millimeters of mercury	mm Hg	760
inches of mercury	in Hg	29.92
pounds per square inch	psi	14.70
newtons per square meter = pascal	N/m^2 = Pa	101,300

The pressure in Denver is about three-fourths of that at sea level. While changes in elevation have pronounced effects on atmospheric pressure, temperature and moisture also affect atmospheric pressure. Changes in the moisture and temperature of air masses produce high and low pressure systems that move across the continent dictating our weather. Pressure increases underwater because the water pressure adds to the atmospheric pressure. Ten meters (33 feet) of water is equivalent to 1 atmosphere of pressure.

We can understand how atmospheric pressure is just the weight of the atmosphere pushing down on the Earth's surface, but how can we apply the basic definition of pressure to a confined gas? To expand our concept of pressure and provide a basic framework for understanding the behavior of gases, we use a simple model for a confined gas. This model is known as the **kinetic molecular theory**. The kinetic molecular theory states:

1. A gas consists of small particles, either individual atoms or molecules, moving around randomly.
2. The total volume of the gas particles is so small compared to the total volume the gas occupies that we can consider the total particle volume to be zero. This means that a gas consists almost entirely of empty space.

3. The gas particles act independently of one another. A particle is not attracted to nor repelled from any other particle.
4. Collision between gas particles and between gas particles and the walls of the container are elastic. This means that the total **kinetic energy** of the gas particles is constant as long as the temperature is constant.
5. The average kinetic energy of the gas particles is directly proportional to the absolute temperature of the gas.

The kinetic molecular theory is used throughout the discussion of gases and should become clearer as examples are used that illustrate this theory.

First, the kinetic molecular theory is used to explain pressure. We can use a sealed syringe as our container and assume it contains a volume of air. To simplify our discussion, we will also assume air consists of 80% nitrogen and 20% oxygen (Figure 9.2). The nitrogen and oxygen molecules move randomly in the barrel of the syringe. The molecules collide with the walls of the syringe barrel and the face of the syringe's plunger (as well as with each other). The collisions exert a constant force on the inside surface area of the walls of the syringe barrel. The force exerted per unit area of the syringe's internal surface is the pressure of the gas. Because the molecules move randomly throughout the syringe barrel, the pressure is the same throughout the syringe. If we assume the pressure outside the syringe is 1 atmosphere and that the plunger

Figure 9.2
Syringe filled with air that is assumed to be 80% nitrogen and 20% oxygen

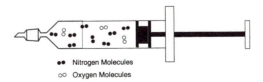

•• Nitrogen Molecules
○○ Oxygen Molecules

is not moving, then the pressure inside the syringe must also be 1 atmosphere. So the pressure of a confined gas is nothing more than the force caused by the constant bombardment of the gas particles on the sides of the container.

The Gas Laws

Continuing to use a syringe as a container, the basic gas laws can be explained. These laws apply to what is referred to as an **ideal** or **perfect gas**. An ideal or perfect gas can be thought of as a gas that conforms to the kinetic molecular theory. In reality, gas molecules do have volume and exert forces on each other. Under normal conditions of temperature and pressure, though, the kinetic molecular theory explains the behavior of gases quite well. It is only when a gas is at very low temperatures and/or under extremely high pressure that a gas no longer behaves ideally.

The definition of pressure assumed the volume of the air contained in the syringe was constant. What would happen if the plunger is pushed while making sure that the opening remained sealed (Figure 9.3)? Pushing in on the plunger obviously decreases the volume of the syringe's barrel. Because the volume has decreased, the inside surface area has also decreased. This means that the frequency of collisions per

unit area has increased, which translates into an increase in pressure.

Remember, because pressure is force divided by area, if the area decreases the pressure increases. When we push in on the plunger, the pressure increases to some value above 1 atmosphere. The smaller the syringe's barrel volume becomes, the higher the pressure exerted by the air in syringe. The relationship between volume and pressure is known as Boyle's Law. Boyle's Law was first mentioned in Chapter 3 and is named after Robert Boyle. Simply stated, Boyle's Law says that the pressure and volume of an ideal gas are inversely related, as one goes up, the other goes down. Boyle's Law can be stated mathematically as

$$\text{Pressure} \times \text{Volume} = \text{constant}$$
$$\text{Or}$$
$$PV = k$$

So if the gas in the syringe was originally at 1 atmosphere and the volume was 50 mL, then PV would equal 50 atm-mL. If we pushed in on the plunger to decrease the volume to 25 mL, then the pressure would have to increase to 2 atmospheres for PV to remain constant. Remember, when we apply Boyle's Law, the only two variables that change are pressure and volume. We are assuming the temperature of the gas and the number of molecules of gas in the syringe remain constant.

What happens when the pressure of a gas remains constant and the temperature and volume change? Before looking at the relationship between temperature and volume, though, the concept of temperature must be understood. Temperature is one of those terms that is continually used, but rarely given much thought. As long as we can remember, we have had our temperature taken, seen and heard the daily temperature reported, and baked foods at various temperatures. Intuitively, we think of tempera-

Figure 9.3
Pushing in on the plunger decreases the volume, which causes an increase in pressure due to the collision frequency increasing

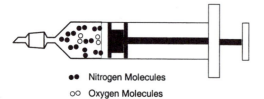

●● Nitrogen Molecules
○○ Oxygen Molecules

ture in terms of hot and cold, but what does temperature actually measure? Temperature is a measure of the random motion of the particles making up a substance. Specifically, temperature is a measure of the kinetic energy of the particles in a substance. Kinetic energy is energy of motion. A body possesses kinetic energy when it moves. The formula for kinetic energy is

$$\text{Kinetic Energy} = \frac{1}{2} \times \text{mass} \times \text{velocity}^2$$
$$\text{K.E.} = \frac{1}{2} mv^2$$

Kinetic energy depends on the mass and velocity of an object. Each of the nitrogen and oxygen molecules in the syringe contain a certain amount of kinetic energy by virtue of the fact that they have mass and are moving. If we could somehow measure the speed of particles in the syringe, we would discover that they move at various velocities. While most move at near some average velocity, some are moving much slower and others much faster than average. A graph of the percent of molecules at a certain velocity versus the velocity is called a Maxwell-Boltzmann distribution, named for Ludwig Boltzmann (1844–1906) and James Clerk Maxwell (1831–1879) (Figure 9.4). A velocity of 400 meters per second is roughly 800 miles per hour and represents the typi-

Figure 9.5
Boltzmann Curves at Higher and Lower Temperatures (Rae Déjur)

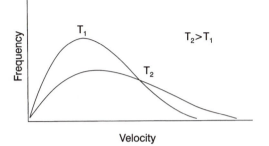

cal velocity gas molecules would have at room temperature. The shape of Figure 9.5 is dependent on temperature. At a higher temperature, a greater percentage of the molecules would be moving with greater velocities, and at a lower temperature, a greater percentage with slower velocities (Figure 9.5).

Higher and lower velocities translate into higher and lower kinetic energy of the molecules, respectively. We can now consider what a change in temperature means at the molecular level. An increase in temperature implies more energetic particles and a decrease in temperature less energetic particles. When a mercury thermometer is placed in a hot oven, it is exposed to more collisions with greater kinetic energies. The kinetic energy of the gas particles is transferred to the kinetic energy of the mercury, which causes it to expand and register a higher temperature. The expansion of the mercury can be compared to a rack of billiard balls being struck by the cue ball. At the lower temperature, the column of mercury in the thermometer is like the racked balls. An energetic cue ball is like the energetic gas molecules in the oven. The cue ball transfers its kinetic energy to the racked balls causing them to spread out. Similarly, the energetic gas molecules in the oven impart their energy to the mercury atoms

Figure 9.4
Distribution of Speeds in Gas—Maxwell-Boltzmann Distribution (Rae Déjur)

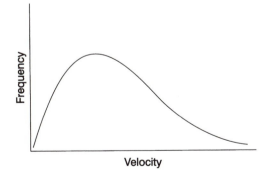

causing them to expand up the thermometer column. Cooling would cause the mercury's volume to decrease. If the mercury was cooled to its freezing point of –39°C, the formation of solid mercury might be considered analogous to racking up the billiard balls.

The two most common scales used to measure temperature are the Fahrenheit and Celsius scales. The German Daniel Fahrenheit (1686–1736) proposed his temperature scale in 1714 in Holland. Fahrenheit invented the modern mercury thermometer and calibrated his thermometer using three different temperature standards. He assigned a value of 0 to the lowest temperature he could obtain using a mixture of ice, salt, and liquid water. A value of 30 was used for a mixture of ice and freshwater. The third point was set at 96 based on the oral temperature of a healthy person. Using his scale Fahrenheit determined the boiling point of water would be 212. He later changed the standard of the ice-freshwater mixture from 30 to 32 in order to have an even 180 temperature divisions between the freezing and boiling points of water. The latter change resulted in 98.6°F being the accepted normal body temperature of a healthy person.

Anders Celsius (1701–1744) from Sweden devised his temperature scale in 1742. Celsius assigned a value of 0 to the boiling point of water and 100 to the temperature of thawing ice. Instrument makers soon reversed the 0 and 100 to give us the modern freezing and boiling points of water as 0°C and 100°C, respectively. The relationship between the Fahrenheit and Celsius temperature scales are given by the two equations:

Degrees Celsius =
$$\frac{5}{9} \times (\text{Degrees Fahrenheit} - 32)$$
Degrees Fahrenheit =
$$(\frac{9}{5} \times \text{Degrees Celsius}) + 32$$

Both the Fahrenheit and Celsius temperature scales are based on the physical characteristics of water. The use of water to establish a zero point means that both 0°F and 0°C are arbitrary and not true zero points. Because temperature is a measure of the random motion of the particles making up a substance, true zero on either the Fahrenheit or Celsius scale would imply that there is no motion at 0°F and 0°C. We know that substances at these temperatures particles possess an ample amount of motion and kinetic energy. A temperature scale that has a true zero is known as an absolute temperature scale. One problem that arises when using the Fahrenheit and Celsius temperature scales is because they are not absolute scales, the mathematical comparisons do not make sense. For example, if we compare equal quantities of water at 10°C and 20°C, we might expect the water at 20°C to be twice as hot or have twice as much energy as the water at 10°C. This is not the case. As an analogy, consider a child who is 3 feet tall and an adult who is 6 feet tall. Our absolute scale for measuring a person's height is the vertical distance from the bottom of our feet, and using this absolute scale the adult is twice the height of the child. But what if both the child and adult stood on top of 5-foot ladders and we measured their heights from the ground? The child would now measure 8 feet and the adult 11. The adult on this revised scale is no longer twice as tall as the child. The Fahrenheit and Celsius temperature scales are like measuring heights of people standing on ladders. It might sound ridiculous to measure people standing on ladders, but the heights would seem normal if that's how we have always measured height.

The most common absolute temperature scale used by scientists is the Kelvin temperature scale. The Kelvin temperature scale was proposed by William Thompson,

Lord Kelvin (1824–1907). The Kelvin temperature scale has an **absolute zero**. True comparisons can be made using the Kelvin scale. A substance at a temperature of 400 Kelvins contains particles with twice as much kinetic energy as a substance at 200 Kelvins. Absolute zero is the temperature where the random motion of particles in a substance stops. It is the absence of temperature. Absolute zero is equivalent to −273.16°C. How this value is determined is discussed shortly after we discuss our next gas law. The relationship between Kelvin and Celsius temperature is

Kelvins = Degrees Celsius + 273.16

For most work, it is generally sufficient to round off 273.16 to 273.

Now that the physical meaning of temperature has been explored, the relationship between temperature and volume of an ideal gas can be examined. Let's consider a syringe containing a specific volume of air at a specific temperature. The syringe is originally at room temperature, about 20°C, and placed in a pot of boiling water at 100°C. In the boiling water, energy is transferred to the nitrogen and oxygen molecules in the syringe barrel, and they move faster. The number of collisions with the interior walls of the syringe's barrel increases. If the pressure within the syringe is to remain constant, the volume must increase. Another way of putting this is that an increase in temperature causes the pressure within the syringe to increase. Because the pressure outside the syringe is 1 atmosphere, the gas will push out on the plunger until the pressure in the syringe returns to 1 atmosphere. There is a direct relationship between temperature and volume of an ideal gas. As temperature increases, volume increases. The direct relationship between temperature and volume in an ideal gas is known as **Charles' law**. Jacques Alexandre Charles (1746–

1823) was an avid balloonist. Charles made the first hydrogen-filled balloon flight in 1783 and formulated the law that bears his name in conjunction with his ballooning research.

Charles' law can be stated mathematically as

$$\frac{Volume}{Temperature} = Constant$$

$$\frac{V}{T} = k$$

When applying Charles' Law it is important to remember that the absolute temperature must be used. Let's use the syringe example to determine how much the volume would increase if the temperature was raised from 20°C to 100°C. Let's assume the original volume at 20°C is 25 mL. First we'll convert the temperatures to Kelvins, and then apply Charles' Law. The calculations are

20°C + 273 = 293 Kelvins

100°C + 273 = 373 Kelvins

$$\frac{25 \ mL}{293 \ Kelvins} = \frac{V}{373 \ Kelvins}$$

Solving for V, we find that the volume at 100°C is approximately 32 mL.

Louis Gay-Lussac continued the ballooning exploits initiated by Charles, ascending to over 20,000 feet in a hydrogen balloon in the early 1800s. **Gay-Lussac's law** defines the relationship between the pressure and temperature of an ideal gas. If the temperature of the air in the syringe increases while keeping the volume constant, the gas particles speed up and make more collisions with the inside walls of the syringe barrel. As we have seen, an increase frequency in the number of collisions of the gas particles with a container's wall translates into an increase in pressure. Gay-Lussac's law says that pressure is directly proportional to temperature. Gay-Lussac's law expressed mathematically is

$$\frac{Pressure}{Temperature} = Constant$$

$$\frac{P}{T} = k$$

Again, the temperature must be expressed in Kelvins. If the pressure and temperature of the air in the syringe are originally 1.0 atmosphere and 293 Kelvins and the syringe is placed in boiling water, the pressure will increase to approximately 1.3 atmospheres. The calculations are

$$\frac{1.0 \ atmosphere}{293 \ Kelvins} = \frac{Pressure \ Final}{373 \ Kelvins}$$

Pressure Final = 1.0 atmospheres

$$\times \frac{373 \ Kelvins}{293 \ Kelvins} = 1.3 \ atmospheres$$

The relationship between temperature and pressure provides a method for determining the value for absolute zero. By measuring the pressure of a gas sealed in a constant volume container at different temperatures and extrapolating to a pressure of 0 atmospheres gives the value for absolute zero (Figure 9.6).

Figure 9.6
The value of absolute zero may be determined by extrapolating a plot of pressure versus temperature to a value of 0 atmosphere. This gives a value for absolute zero of −273.16°C

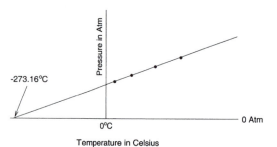

Boyle's, Charles', and Gay-Lussac's laws explain the relationships between pressure, volume, and temperature of an ideal gas. In the examples thus far, the amount of gas in the syringe was considered to be constant. The amount of gas is measured in moles (the standard symbol for moles is n). In Chapter 4, we learned that Avogadro's hypothesis stated that equal volumes of gases contain an equal number of molecules. This means that the volume of a gas is directly proportional to the number of molecules present, and therefore, the number of moles of gas. This relationship is known as **Avogadro's law**. If temperature and pressure are held constant and somehow more molecules are added to the syringe barrel, the volume would increase. A better example to illustrate Avogadro's Law is to think about blowing up a balloon. As you blow up the balloon, you are adding more gas molecules (principally carbon dioxide) and the volume increases. It is assumed the pressure and temperature stay constant as the balloon inflates, and that the increase in volume is due to the increase in the number of moles of gas entering the ballon. Because the relationship between pressure and moles is direct, the relationship is expressed mathematically as

$$\frac{Volume}{moles} = Constant$$

$$\frac{V}{n} = k$$

The four fundamental gas laws are summarized in Table 9.3.

The Ideal Gas Law

The gas laws discussed in the previous section are limited, because they only allow us to examine the relationship between two variables at a time. Fortunately, all four laws can be combined into one general law called

Table 9.3
Summary of Four Laws for an Ideal Gas

Gas Law	Variables	Relationship	Equation
Boyle	Pressure, Volume	Inverse	$P_1V_1 = P_2V_2$
Charles	Volume, Temperature	Direct	$\dfrac{V_1}{T_1} = \dfrac{V_2}{T_2}$
Gay-Lussac	Temperature, Pressure	Direct	$\dfrac{P_1}{T_1} = \dfrac{P_2}{T_2}$
Avogadro	Volume, Moles	Direct	$\dfrac{V_1}{n_1} = \dfrac{V_2}{n_2}$

the **ideal gas law**. The ideal gas law relates the four quantities pressure, volume, moles, and temperature. The ideal gas law is given by the equation

$$\text{Pressure} \times \text{Volume} = \text{Moles} \times R \times \text{Temperature}$$
$$PV = nRT$$

In this equation, R is called the ideal gas law constant. Its value depends on the units used, but assuming pressure is measured in atmospheres, volume in liters, and temperature in Kelvins its value is 0.082 atm-L/mol-K. Other forms of the ideal gas law are

$$\frac{PV}{nT} = R \quad \text{or} \quad \frac{P_1V_1}{n_1T_1} = \frac{P_2V_2}{n_2T_2}$$

The four gas laws in the previous section are all special cases of the ideal gas law. We can use the ideal gas law to calculate the volume one mole of gas occupies at standard conditions. Standard conditions are 0°C (273 K) and 1 atm pressure. The volume at these conditions is known as a standard molar volume. Plugging the numbers into the ideal gas law equation gives a value of 22.4 liters for the standard molar volume.

Partial Pressure and Vapor Pressure

John Dalton was a contemporary of Charles and Gay-Lussac. As a meteorologist, he had a keen interest in the atmosphere and devoted much of his work to a study of the behavior of gases. Dalton's study of gases led him to formulate the law of partial pressure and the concept of vapor pressure. **Dalton's Law of Partial Pressure** states that in a mixture of gases each gas exerts a pressure independent of the pressure exerted by other gases in the mixture. The pressure of the individual gases in the mixture is the **partial pressure** of that gas. The sum of the partial pressures is the total pressure. Figure 9.7 depicts the concept of partial pressure.

Partial pressure is directly related to the moles of gas present. The partial pressures of gases in our atmosphere are approximately 0.78 atm for nitrogen, 0.21 atm for oxygen, and 0.01 atm for all the other gases combined.

At the same time that Dalton proposed his ideas on partial pressure, he developed the concept of vapor pressure. A **vapor** is the gaseous form of a substance that normally exists as a solid or liquid. A gas is a substance that exists in the gaseous states under normal conditions of temperature and pressure. The **vapor pressure** of a liquid is the partial pressure of the liquid's vapor at equilibrium. Liquids with strong intermolecular forces exert lower vapor pressures than those with weak intermolecular forces. In liquids with strong intermolecular forces, it is more difficult for the molecules to leave the liquid state and enter the gaseous state.

Figure 9.7
Dalton's Laws of Partial Pressure states that in a mixture of gases each gas exerts a pressure independent of the other gases. The total pressure is the sum of the partial pressures

Liquids with high vapor pressures are known as volatile liquids.

To understand vapor pressure, let's consider an empty jar that is partially filled with water and then covered with a lid. We will assume the space above the water in the jar contains only air when we screw on the jar's lid. After the lid is place on the jar, water molecules leave the liquid and enter the air above the liquid's surface. This process is known as **vaporization**. As time goes by, more water molecules fill the air space above the liquid, but at the same time, some gaseous water molecules condense back into the liquid state. Eventually, a point is reached where the amount of water vapor above the liquid remains constant. At this point, the rates of vaporization and **condensation** are equal, and equilibrium is reached. The partial pressure exerted by the water at this point is known as the equilibrium vapor pressure or just vapor pressure. Vapor pressure is directly related to the temperature, that is, the higher the temperature, the higher the vapor pressure. Table 9.4 gives

Table 9.4
Vapor Pressure of Water at Different Temperatures

Temperature Celsius	Vapor Pressure mm Hg
0	5.00
20	17.5
40	31.8
60	149
80	355
100	760

the vapor pressure of water at several temperatures. The amount of water vapor in the air is known as **humidity. Relative humidity** is the ratio of the actual vapor pressure compared to the equilibrium vapor pressure at a particular temperature. If the vapor pressure was 14.0 at a temperature of 20°C, then the relative humidity would be 14.0/17.5 = 80%.

The vapor pressure of a liquid dictates when a substance will boil. In fact, the boiling point of a substance is defined as the temperature at which the vapor pressure equals the external pressure. Typically, the external pressure is equal to atmospheric pressure, and we define the normal boiling point as the temperature when the vapor pressure equals 1 atmosphere. If we consider water heated on a stove, the bubbles that develop in the liquid contain water vapor that exerts a pressure at the specific vapor pressure of water at that temperature. For example, when water reaches 60°C, any bubbles that form will contain vapor at 149 mm Hg (see Table 9.4). At this pressure, and any other pressure below 760 mm Hg (1 atmosphere), the external pressure of 1 atmosphere causes the bubbles to immediately collapse. As the temperature of the water rises, the vapor pressure continually increases. At 100°C, the vapor pressure inside the bubbles finally reaches 760 mm Hg. The vapor pressure is now sufficient to allow the bubbles to rise to the surface without collapsing. At higher elevations where the external pressure is lower, liquids boil at a lower temperature. At the top of a 15,000-foot peak, water boils at approximately 85°C rather than 100°C. This increases the cooking time for items, as noted in the directions of many packaged food. If the external pressure is increased, the boiling temperature also increases. This is the concept behind a pressure cooker. The sealed cooker allows pressure to build up inside it and increases the boiling point allowing a substantial increase in cooking temperature and decrease in cooking time.

Applications of the Gas Laws

The pressure cooker is just one practical example of the behavior of gases in everyday life. In this section, we look at a few more relevant examples involving gases. A process analogous to the pressure cooker which we all can relate to is popping corn. Each kernel of corn contains about 15% water. The hard kernel acts as a constant volume container. When the kernel is heated, the temperature increases, and according to Gay-Lussac's law, the pressure increases. A small amount of the water inside the kernel vaporizes into superheated steam, but most of the water remains in liquid form because of the increased pressure. At some point when the pressure increases to several atmospheres, the kernel pops as the steam transforms the starchy kernel into gelatinous globules of popped corn.

One national news story in the year 2000 involved the defect in certain brands of tires and how tread separation caused vehicle accidents leading to personal injury and in some cases deaths. One prominent aspect of the tire story involved whether the tires were properly inflated to the correct pressure. The air inside a tire can be considered an ideal gas, and we can apply the gas laws to the air inside. The inflation pressure recommendations stamped on the sides of tires call for tires to be inflated under cold conditions, or before the vehicle is driven. Because the most important aspect of tire performance is correct inflation pressure, it is important to adhere to the recommended tire pressure. Many individuals disregard this recommendation. It is not uncommon for someone to pull into a gas station and fill a warm tire to the recommended pressure. How much can this affect the pressure in the tire? In this example, the important variables are pressure and temperature. We will assume that the amount of air in the tire and volume of the tire are constant, and will see how much of an error can be introduced by ignoring the inflation temperature. Let's assume the recommended inflation pressure is 30 pounds per square inch (psi), and we fill the tire on a hot day after pulling off the freeway and the tire temperature is 90°F (32°C). We will assume the cold pressure applies to a temperature of 70°F (21°C). Gay-Lussac's law can be applied to determine the actual cold inflation pressure.

We can write Gay-Lussac's Law as

$$\frac{P_{cold}}{T_{cold}} = \frac{P_{hot}}{T_{hot}}$$

$$P_{cold} = 30 \text{ psi} \times \frac{294 \text{ K}}{305 \text{ K}} = 28.9 \text{ psi}$$

This is a relatively small change, but consider the change if you inflate the tire in a heated garage at 70°F and then drive in winter when the temperature is 10°F. In this case, the pressure drop would be approximately 5 psi. A pressure change of this much can lead to as much as a 25% reduction in fuel efficiency. Additionally, under-inflated tires result in overloading and possible tire failure. Reduced tire pressures of as little as 4 psi were cited in some of the national stories involving tire failure and vehicle accidents.

Another area in which the gas laws play a key role is in scuba diving. At the surface, we breath air at a pressure of approximately 1 atmosphere. The partial pressures of nitrogen and oxygen are 0.78 and 0.20 atmosphere, respectively. A scuba diver breaths compressed air that is delivered at a pressure that corresponds to the pressure at the depth of the diver. Because 33 feet of water corresponds to 1 atmosphere of pressure, a diver at 33 feet would breath air at a total pressure of 2 atmospheres; one atmosphere of this pressure is due to the air pressure at sea level and 1 atmosphere is due to the pressure of the water. The partial pressures of the nitrogen and oxygen would be twice as great as that at the surface. Most serious problems in diving arise when divers descend to depths in excess of 100 feet. The total pressure at depths greater than 100 feet can cause problems such as nitrogen narcosis and oxygen toxicity. Nitrogen narcosis, also known as rapture of the deep, is due to breathing nitrogen under pressure. At a depth of 100 feet, the total pressure of air delivered to the diver is about 4 atmospheres. This results from a water pressure of slightly greater than 3 atmospheres plus the surface air pressure of 1 atmosphere. The partial pressure of nitrogen at this depth would be about 3.2 atmospheres. Nitrogen at this pressure produces an anesthetizing effect similar to nitrous oxide (laughing gas). The result is a state of nitrogen intoxication with loss of judgment, a state of euphoria, drowsiness, and impaired judgment. If nitrogen narcosis develops, it can be quickly reversed by moving to a shallower depth. To prevent nitrogen narcosis, deep divers use a mixture of helium, oxygen, and nitrogen. Helium is much less soluble in body tissues and like nitrogen is inert and does not play a role in metabolism.

Another example of a diving problem that is a direct consequence of Dalton's Law of partial pressure concerns oxygen toxicity. The deeper a diver descends, the greater the partial pressure of oxygen. At a depth of 130 feet, the total pressure is close to 5 atmospheres and the partial pressure of oxygen will be close to 1 atmosphere ($21\% \times 5$ atmospheres). What this means is that breathing compressed air at 130 feet is like breathing pure oxygen at the surface. Breathing pure oxygen or breathing compressed air at 130 feet is only a problem if done for several hours. Oxygen toxicity can damage the lungs, affect the nervous system, and in extreme cases, result in seizures. Recreational divers do not stay down long enough at great depths for this to become a problem, but it can be a problem for commercial divers. When the partial pressure of oxygen approaches 2 atmospheres, it may take only several minutes for symptoms of oxygen toxicity to appear. To prevent oxygen toxicity, divers use a lower percentage of oxygen for deep dives. For example, a diver at 200 feet might breath from a tank that contains only 4% oxygen.

Both nitrogen narcosis and oxygen toxicity involve deep diving. Other problems result when the diver ascends too quickly. We know from Boyle's Law that as pressure decreases, volume increases. Decompression sickness develops when a diver ascends too fast and nitrogen bubbles develop in the tissue. (This is also related to **Henry's law** and the solubility of gases in the blood; this topic is covered in Chapter 11.) Symptoms may include dizziness, paralysis, shock, and joint and limb pain. The term "the bends" comes from the pain that appears in body joints such as the elbows, knees, ankles, and shoulders. During a normal ascent from moderately deep depths (greater than 60 feet), a diver must be sure to take scheduled stops and allow gases such as nitrogen to come out of the body tissue slowly. Another serious problem with surfacing too quickly is an air embolism. An air embolism develops when air expands in the lungs and bubbles are forced into the blood vessels, possibly cutting off blood flow to the brain and resulting in a loss of consciousness.

Both decompression sickness and air embolism can be treated by placing the victim in a **hyperbaric** chamber. "Hyper" means greater or above normal and "baric" means weight or pressure. In a hyperbaric chamber, the victim is placed in an environment of increased pressure and enriched oxygen. By decreasing the pressure slowly, the gases can come out slowly. The enriched oxygen in the hyperbaric chamber helps to purge and dilute the nitrogen in the tissues.

A final example of a practical application of the ideal gas law involves snow making. Many ski resorts and Hollywood often employ snow guns to produce their own snow (Figure 9.8). A snow gun essentially uses compressed air to atomize water into droplets. The droplets are sprayed into the air where they condense into ice crystals. As the compressed air and water droplets leave the snow gun, the decrease in pressure causes the air to expand resulting in a drop in temperature. The cooling of air as it expands is also demonstrated when moist air masses from the Pacific move inland over the West Coast. The air masses will ride up the western edge of coastal mountain ranges such as the Cascades. As this moisture-laden air moves to higher elevations, the pressure decreases causing the air to expand and cool. When the relative humidity reaches 100%, precipitation results in the form of rain or snow. On the eastern side of the mountains, dry air masses descend and become compressed as they are subjected to increased pressure at lower elevations. This results in dry, desert-

**Figure 9.8
Snow Gun (Rae Déjur)**

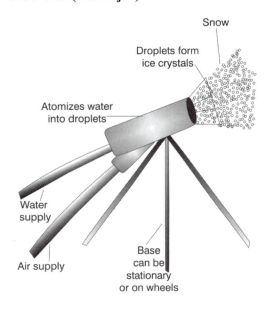

like conditions on the eastern side of the coastal mountain ranges.

The few examples illustrated above demonstrate how gases and the gas laws affect our lives on a daily basis. We are surrounded by gas and our lives are dependent on a steady supply of gas. Humans can survive a few weeks without food, a few days without water, but only a few minutes without a steady supply of air. In this chapter, we have focused on some of the unique properties of gases and the gas laws. We continue our exploration of gases when we examine phase changes in Chapter 10 and solubility in Chapter 11.

10

Phase Changes and Thermochemistry

Introduction

Chapters 8 and 9 explored the three common states (phases) of matter—solid, liquid, and gas. Substances are generally associated with one of these three phases under normal conditions of 1 atmosphere pressure and room temperature; yet phase changes constantly occur all around us. Take an ice cube out of the freezer, and in a few seconds it will start to melt changing from the solid to the liquid phase. The familiar phase changes of water can be used to explore changes of state and also introduce the topic of **thermochemistry**. Thermochemistry is the study of the energy changes associated with chemical processes. The term "chemical process" will be used throughout this chapter, and it includes both physical processes and chemical reactions. After looking at the thermochemistry of phase changes, chemical processes in general will be examined.

Changes of State

The phase changes of water result in the transitions between the solid, liquid, and gas states:

$$H_2O_{(s)} \leftrightarrow H_2O_{(l)} \leftrightarrow H_2O_{(g)} \qquad (1)$$

The double arrows indicate that two processes are possible at each stage. The terms freezing and melting indicate the general direction of the phase change. The former from liquid to solid, and the latter from solid to liquid. Similarly, condensation and vaporization characterize a phase change from gas to liquid and from liquid to gas, respectively. Equation 1 is somewhat misleading because it implies the liquid phase is always intermediate between the solid and gas phases. A solid can pass directly into the gas phase without passing through the liquid phase. This process is known as **sublimation**. Solid carbon dioxide, also known as dry ice, is a substance that sublimes under normal conditions. The change from a gas directly to a solid is called deposition. Table 10.1 summarizes the different phase changes.

Energy and Phase Changes

Before looking at the energy associated with phase changes, some basic terms related to thermochemistry need to be

Table 10.1
Phase Changes

Melting	Freezing
Solid → Liquid	Liquid → Solid
Sublimation	Deposition
Solid → Gas	Gas → Solid
Vaporization	Condensation
Liquid → Gas	Gas → Liquid

defined. Terms, such as "work," "heat," and "energy" are used every day, but they have very specific meaning in the context of chemistry. Whenever a chemical process such as a phase change or chemical reaction is examined, it is important to specify exactly what is being examined. This statement might seem obvious, but it is especially important when considering the energy aspects of a chemical process. The chemical system or simply **system** is that part of the universe a person is interested in and wants to examine. The concept of a system is a somewhat abstract concept. At times, it is very easy to define a system, while in other cases it might not be obvious what the system includes. For example, when you take an ice cube out of the freezer and observe how it changes, you can define your system as the ice cube. If you are looking at the reaction between vinegar and baking soda, the system might be a beaker containing these two substances. A system might be microscopic, or it might be as large as the planet. For example, a biochemist looking at the effect of toxins on humans might define a system as the human cell; whereas an atmospheric chemist looking at how the chemical composition of the atmosphere affects global temperatures might define the system as the entire planet and its

atmosphere. Everything outside the system is considered the **surroundings**. This means the entire universe excluding the defined system comprises the surroundings. A useful conceptual formula is

System + Surroundings = Universe

Energy is defined as the ability to do work or produce heat. In Chapter 9, kinetic energy was defined as the energy of motion. The motion of molecules and the electrons within the atoms make up the kinetic energy of a chemical system. **Potential energy** can be considered stored energy. Potential energy results from the position or configuration of a system. Just as the gravitational potential energy of an object is due to its position with respect to the Earth's surface, chemical potential energy results from the position of electrons with respect to nuclei. Chemical potential energy is stored in the bonds of compounds. During a chemical reaction, atoms are rearranged, and it is this reconfiguration of the atoms that causes the energy changes accompanying the reaction. The sum total of the kinetic and potential energy of all particles in the chemical system is termed the **internal energy**.

Inherent in the definition of energy are the terms "work" and "heat." Work is defined as force times distance. In physics, work is generally calculated by multiplying the resultant force applied to an object by how far the object moves. The physical definition of work can be modified into a chemical definition for our purposes. Chemical work can be considered expansion work. To understand chemical work, consider a chemical reaction that takes place in a container equipped with a frictionless piston (Figure 10.1). Let's assume a gas is generated during the chemical reaction, for example alka-seltzer plus water. The gas generated in the reaction expands in the container and forces the plunger out.

Figure 10.1

A reaction generates a gas, which causes the plunger to expand. The volume increases by an amount equal to ΔV. The external pressure is P. Because pressure = force(f)/area(A), then the force is equal to PA. The distance the plunger moves is d. Work = force $\times$ distance gives work = PAd. The figure shows Ad is equal to ΔV. This means work is equal to $P\Delta V$ (Rae Déjur)

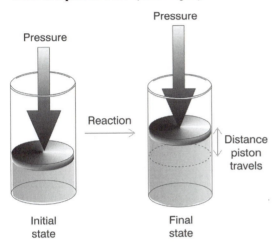

Pressure

Pressure

Reaction

Distance piston travels

Initial state

Final state

The external pressure outside the container can be assumed to be 1 atmosphere. It can be shown that the work done during the chemical reaction is equal to the external pressure times change in volume. This work is called **PV work**, and it is what is meant by chemical work.

Energy may also produce heat. Heat is defined as a transfer of energy from a body of higher temperature to a body of lower temperature. Energy and heat are often considered as synonymous terms, but heat involves the flow of energy. When examining thermochemistry, the transfer of energy between the chemical system and its surroundings is of primary concern.

Work, heat, and energy can be expressed using various units. The metric unit for energy is the **joule**, abbreviated J. The joule was named in honor of the English physicist James Prescott Joule (1818–1889).

The joule is a rather small unit of energy. It takes approximately 10,000 joules to raise the temperature of a coffee cup of water by 10°C. Often kilojoules (kJ) are used with 1 kJ equal to 1,000 J. Another popular unit is the **calorie**, abbreviated with a c. A **calorie** is the amount of heat necessary to raise one gram of water by 1°C. One calorie is equal to 4.18 joules. The physical unit of a calorie should not be confused with a food calorie. A food calorie, symbolized with a capital C, is equivalent to 1,000 calories. When you eat a piece of cake with 400 Calories, you have actually consumed 400,000 calories.

Heating Curves

Phase changes are most often the result of a heat added or released from a substance. Phase changes are also pressure related, but the thermal aspects of phase changes are the primary focus in this section. The overall transition between phases of a substance can be graphically displayed using a **heating curve**. A heating curved is a graph showing the temperature of the substance versus the amount of heat absorbed over time. Note that it could have been called a cooling curve and the transitions examined in terms of heat released over time. Whether called a heating curve or cooling curve, the graph would be exactly the same.

Figure 10.2 shows the heating curve for water. To understand Figure 10.2, the path of an ice cube on the curve is followed as heat is supplied to it. To make our example realistic, assume the ice cube has a mass of 10 grams and is taken out of a freezer where its temperature is -10°C. Consider what happens when the cube is placed in a pan on a stove and heat applied. Initially, the ice cube is at -10°C when it is removed from the freezer. The temperature of the ice cube

Figure 10.2
The Heating Curve for Water

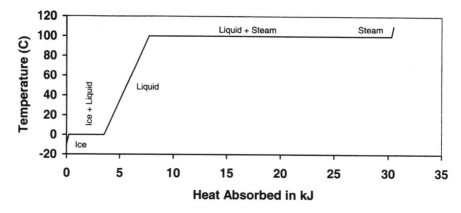

immediately starts to raise, but it does not start to melt until its temperature reaches 0°C. The amount of heat absorbed between −10°C and the melting point of 0°C can be determined using the **specific heat capacity**, or just specific heat, of ice. The specific heat of a substance is a measure of the amount of heat necessary to raise the temperature of one gram of a substance by 1°C. The specific heats of several common substances are listed in Table 10.2.

Table 10.2 demonstrates that the specific heat of a substance depends on its phase. The specific heat of liquid water is approximately twice that of ice and steam. Water has one of the highest specific heats compared to other liquids. The high specific heat of liquid water is directly related to the presence of hydrogen bonds. Because specific heat is a measure of the heat needed to change the temperature of a substance, and temperature is a measure of the random kinetic energy of the particles, substances with hydrogen bonds have a greater tendency to resist temperature change. The high specific heat of water explains why coastal environments have more moderate weather than areas at similar latitudes located inland. Water's high heat capacity

means coastal regions will not heat up or cool down as much as areas not adjacent to a body of water.

The relationship between heat, specific heat, and temperature change is given by the equation $Q = mc\Delta T$. In this equation, Q is the amount of heat absorbed or released in joules, m is the mass in grams, c is the specific heat of the substance, and ΔT is the

Table 10.2
Specific Heats of Some Common Substances

Substance	Specific Heat J/g°C
Steel	0.45
Wood	1.7
Ice	2.1
Liquid Water	4.2
Steam	2.0
Air	1.0
Alcohol	2.5

change in temperature equal to the final temperature minus the initial temperature. For our ice cube, the amount of heat needed to increase the temperature from $-10°C$ to $0°C$ is equal to $(10 \text{ g})(2.1 \text{ J/g°C})(10°C) = 210 \text{ J}$. If the temperature of a substance decreases, heat must be released from the substance. In this case, the change in temperature is negative resulting in a negative value for Q. A negative sign for Q always indicates that heat is being removed from the substance rather than being added to it.

Once the ice cube reaches $0°C$, the cube starts to melt. The heating curve plateaus at the melting point even though heat is still being added to the ice-liquid mixture. It may seem strange that even though heat is added the temperature does not increase, but this is because the added heat is used to convert ice to liquid water. The heat supplied at the melting point is not being used to increase the random kinetic energy of the water molecules, but to overcome intermolecular attractions. The energy supplied at the melting point goes into breaking the water molecules free from the crystalline structure resulting in the less-structured liquid state. As long as ice is present, any heat added goes into the phase change. The heat necessary to melt the ice is called the **heat of fusion** for water, and its value is 6.0 kJ per mole of water. Our 10-gram ice cube is equal to approximately 0.55 moles of water. This means it will require about 3.3 kJ or 3,300 joules to melt the ice cube. The heat of fusion of a substance is a measure of how much energy is required to convert a solid into a liquid. It would be expected that solids, which are tightly held together, would have high heats of fusion. Sodium chloride with its strong ionic bonds has a heat of fusion of 30.0 kJ per mole; aluminum has a heat of fusion 10.7 kJ per mole. Substances held together by weak dispersion forces have low heats of fusion. The heat of fusion of oxygen is 0.45 kJ per mole.

Once all the ice has melted, the heat added to the liquid water can now go to increasing the kinetic energy of the water molecules and the temperature begins to rise again. The rate at which the temperature rises is governed by the specific heat of liquid water, 4.2 J/g°C. The specific heat of liquid water is twice that of ice; therefore, the rate at which the temperature increases for liquid water is only half of what it was for ice. The heat necessary to increase liquid water from $0°C$ to $100°C$ can be found by multiplying 10 g by 4.2 J/g°C by $100°C$. This is equal to 4,200 joules. The heating curve reaches a second plateau at water's boiling point of $100°C$. At this temperature, any heat added to the water goes into breaking the hydrogen bonds as liquid water is converted to steam. Just as with the first plateau at the melting point, the temperature remains constant as long as the liquid and gas phases coexist. The wide plateau at the boiling point indicates that much more energy is needed for the vaporization process than for the melting process. The heat needed to convert liquid water to steam is called the **heat of vaporization** of water. The heat of vaporization for water is 41 kJ per mole. This value is roughly seven times the heat of fusion and indicates it takes only 1/7 of the energy to melt ice as compared to vaporizing water. Vaporization should take significantly more energy than melting because to convert liquid water to steam requires completely breaking the hydrogen bonds. In the gas phase, the water molecules can be thought of as independent molecules with minimal intermolecular attraction. The heat of vaporization of a substance is generally several times that of its heat of fusion. This is because the intermolecular forces present in the condensed phases must be overcome before a substance can be converted to gas.

In closing the discussion of the heating curve for water, a couple of points should be made. The last section of the heating curve represents the situation when all the liquid water has been converted to steam. At this point as the temperature of the steam begins to rise, superheated steam would be obtained. In our ice cube example, we would not observe this temperature rise because the water we heated in our pot (our system) would escape to the surroundings. It should also be remembered that phase changes can be considered in terms of a cooling curve. In this case, we would follow the curve from right to left and 41 kJ per mole of heat would have to be released to condense steam to liquid water and 6.0 kJ per mole would have to be released by liquid water for it to freeze (Figure 10.2).

Many everyday examples illustrate the concepts embodied in the heating curve. Perspiration is a physiological response for cooling the body. For sweat to evaporate from our skin, heat must be supplied. This heat comes from our body. If we consider our body as a chemical system, sweating is simply a means of transferring heat from our body to the surroundings. The same evaporative cooling effect occurs when stepping out of a shower or getting out of water after a swim. While the vaporization requires energy, a heating effect occurs during condensation and freezing. One practice used in agriculture to protect plants from frost damage is to spray the plants with water. Typically, growers using this method spray their crops with water the evening before a hard freeze is forecasted. When temperatures fall during the night and early morning, the water freezes releasing energy to the surrounding plants. This energy may be sufficient to keep the plants themselves from suffering frost damage. Constantly

occurring phase changes in the atmosphere figure prominently in our weather. Clouds, rain, snow, hail, and humidity all demonstrate phase changes on a grand scale. As noted in Chapter 9, ski resorts and the entertainment industry try to duplicate the snowmaking process by manufacturing snow. Snow guns use a source of water and compressed air to spray water droplets into the air. The expansion of the air as it leaves the snow gun creates a drop in the temperature of the surrounding air helping the droplets cool and fall as snow. Depending on the ambient temperature, the snow guns may have to be elevated and/or the water cooled to get water to undergo the phase change from liquid to solid.

Calorimetry

In the previous section, the thermochemistry of water's phase changes was examined. The heat of fusion and heat of vaporization were used to determine the energy flow that accompanies phase changes. All chemical processes involve energy. To characterize the energy changes involved in these processes, chemists use the term "heat of" followed by the process. For example, heat of vaporization characterizes the energy changes involved in the vaporization process. Table 10.3 defines a number of "heat of" processes chemists use.

A basic method for determining the energy change involved with many chemical processes is **calorimetry**. A calorimeter is an insulated container used to carry out a chemical process. A thermometer is used to measure temperature changes that take place during the process. A simple constant-pressure calorimeter is shown in Figure 10.3. This type of calorimeter derives its name from the fact that it is open to the atmosphere and the pressure remains constant during the process. Constant-pressure

Table 10.3
Energy Changes of Some Chemical Processes

Heat of fusion	Energy change involved in freezing or melting
Heat of vaporization	Energy change involved in condensation or vaporization
Heat of solution	Energy change involved when solute dissolves
Heat of dilution	Energy change involved in dilution
Heat of hydration	Energy change involved in hydration
Heat of combustion	Energy change involved during combustion
Heat of reaction	Energy change involved in chemical reaction
Heat of formation	Energy required to form compound from its elements

calorimeters are used to determine all the processes listed in Table 10.3 except combustion. The general procedure in using a constant-pressure calorimeter is to carry out the process in the calorimeter, observe the temperature change that takes place, and then use the equation $Q = mc\Delta T$ to determine the heat involved in the process. The processes involved using constant-pressure calorimeters generally involve solutions or employ water. A description of a simple experiment that can be used to determine the heat of fusion of water can illustrate the procedure. To determine the heat of fusion of water, a measured volume of water is placed in the calorimeter. The temperature of the water is measured before placing a measured mass of ice into the calorimeter. After the ice is placed in the calorimeter, the mixture is gently stirred until all the ice has melted. At this point, the final temperature of the water in the calorimeter is measured. The basic principle used to determine the heat of fusion assumes that heat required to melt the ice comes from the water in the calorimeter. The amount of heat that flows from the water to the ice is found by using the change in temperature of the water, the specific heat of water, and the mass of water. This heat is divided by the moles of ice (mass of ice/18 g per mole) to give an estimate of water's heat of fusion.

Figure 10.3
Simple Calorimeter (Rae Déjur)

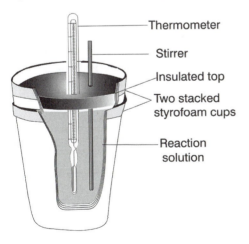

Thermometer

Stirrer

Insulated top

Two stacked styrofoam cups

Reaction solution

Figure 10.4
Bomb Calorimeter (Rae Déjur)

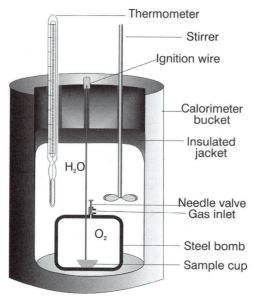

A second type of calorimeter is a constant-volume or bomb calorimeter (Figure 10.4). This type of calorimeter is used to determine the energy content of materials. As the name implies, the process takes place in a heavy walled metal "bomb" of constant volume. The sample is placed in the bomb with ample oxygen to sustain the combustion. The bomb is surrounded by water. The sample is ignited electrically, and the temperature change of the water is monitored. The energy content of the sample is calculated in a manner similar to that used in constant-pressure calorimetry. Bomb calorimetry is used extensively to determine the energy content of fuels and foods. The energy of food is expressed in food calories. The relationship between food calories and joules is 1 food calorie equals 4,180 joules. The conversions mean that a candy bar containing 200 food calories has an energy equivalent of 837,000 joules.

Heat of Reaction

Thus far our discussion on thermochemistry has focused on physical processes

such as phase changes. Bomb calorimetry involves a chemical reaction. In a chemical reaction, substances are rearranged, bonds are broken, and bonds are formed. The reconfiguration of atoms during a chemical reaction involves changes in the potential energy of the electrons. A lower chemical potential energy and a release of energy results when bonds are formed. Conversely, an energy input is required to break bonds. It is not surprising that practically all chemical reactions produce or absorb energy.

The combustion of natural gas, methane, to produce heat can be summarized in the following reaction:

$$CH_{4(g)} + 2O_{2(g)} \rightarrow CO_{2(g)} + 2H_2O_{(l)} + 890 \text{ kJ}$$

The energy released in this reaction is used to heat homes, generate electricity, and cook food. Figure 10.5 is a schematic diagram of the process. When heat is released during a chemical reaction, the reaction is said to be **exothermic**. Another way of saying this is that there is a heat transfer from the system to the surroundings. Examples of exothermic processes include combustion reactions, condensation, and freezing. Reactions in which the potential energy of the products

Figure 10.5
Energy involved in the combustion of methane, CH_4. In the combustion process, the chemical potential energy of the reactants is higher than the products. The difference in potential energy is the heat involved in the reaction

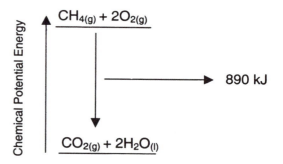

is greater than the potential energy of the reactants are called **endothermic** reactions. In an endothermic reaction, the heat flows from the surroundings into the system. Melting and vaporization are endothermic processes.

Another means to describe the transfer of energy into or out of a chemical system is in terms of internal energy. It was previously stated that internal energy was equal to the sum total of the potential and kinetic energy of all particles making up a chemical system. It is impossible to measure directly the internal energy of a system, but the **first law of thermodynamics** states that the change in internal energy of a system is equal to the heat plus work:

$$\Delta E = q + w \qquad (3)$$

The first law of thermodynamics simply says that energy cannot be created or destroyed. With respect to a chemical system, the internal energy changes if energy flows into or out of the system as heat is applied and/or if work is done on or by the system. The work referred to in this case is the PV work defined earlier, and it simply means that the system expands or contracts. The first law of thermodynamics can be modified for processes that take place under constant pressure conditions. Because reactions are generally carried out in open systems in which the pressure is constant, these conditions are of greater interest than constant volume processes. Under constant pressure conditions Equation 3 can be rewritten as

$$\Delta E = q_p - P\Delta V \qquad (4)$$

Written as Equation 4, q_p is the heat exchange at constant pressure and $P\Delta V$ is the work done. It is important to understand the significance of the mathematical signs in Equation 4, and how they relate to the change in internal energy. When q_p is posi-

tive, heat is transferred from the surroundings into the system causing the internal energy to increase. Pressure is always a positive value, but ΔV (Δ means "change in," so ΔV means change in volume) can be positive or negative. When ΔV is positive, work is being done by the system on its surroundings. Think of a syringe in which a reaction results in the production of gas that expands and forces the plunger out. In this case, ΔV is positive and this causes the internal energy of the system to decrease. When the gas expands, some of the internal energy of the system is being used to do work on its surroundings. There are actual four cases that describe how heat and work can affect the internal energy of the system. These are summarized in Table 10.4.

Reactions under constant pressure conditions are so prevalent in chemistry that the quantity q_p is defined as the change in enthalpy of a chemical system and is symbolized as ΔH. Substituting ΔH for q_p in Equation 4 and rearranging gives $\Delta H = \Delta E + P\Delta V$. The change in enthalpy is typically used to characterize a reaction as exothermic or endothermic. When q_p is positive, heat flows into the system increasing the potential energy of the system, and this results in an endothermic reaction. A negative q_p signifies a heat flow out of the system decreasing the potential energy and giving an exothermic reaction.

When writing a chemical reaction, ΔH should be included with the reactants and products; it is assumed that the reaction takes place under the standard conditions of 25°C and 1 atmosphere. Several simple reactions with their ΔHs are

$$2H_{2(g)} + O_{2(g)} \rightarrow 2H_2O_{(g)} \; \Delta H = -484 \text{ kJ}$$
$$N_{2(g)} + 3H_{2(g)} \rightarrow 2NH_{3(g)} \; \Delta H = -93 \text{ kJ}$$
$$6CO_{2(g)} + 6H_2O_{(l)} \rightarrow C_6H_{12}O_6 +$$
$$6O_{2(g)} \; \Delta H = +2,816 \text{ kJ}$$

The first two reactions given are for the formation of water and ammonia, respectively.

Table 10.4

Four Cases Describing How Heat and Work Affect Internal Energy of a System

q_p	ΔV	ΔE	Description
+	+	+ or −	Heat flows into the system, and the system does work so the internal energy could increase or decrease.
+	−	+	Heat flows into the system, and work is done on the system so the internal energy increases.
−	+	−	Heat flows out of the system, and the system does work on its surroundings so the internal energy decreases.
−	−	+ or −	Heat flows out of the system, and work is done on the system so the internal energy could increase or decrease.

The ΔH values for these two reactions are negative, indicating that these reactions are exothermic. In both of these reactions, the products are formed from their basic elements so the ΔHs represent heats of formation. The third reaction is the familiar reaction for photosynthesis. The energy needed for this endothermic reaction is supplied by the sun.

Applications of Heat of Reactions

We have seen that energy changes are an integral part of chemical processes. We close this chapter with a brief look at some of the applications of energy changes in everyday life. All life and our society are dependent on a multitude of exothermic reactions. The metabolic breakdown of food in organisms supplies the energy they require to exist. For humans, the exothermic combustion of fossil fuels powers our modern industrial society. It is interesting to look at the energy content of different fuels. Table 10.5 lists the energy content of several fuels.

The figures in Table 10.5 indicate that the biomass fuel ethanol, proposed as an alternative to fossil fuels, yields significantly less energy per gram than octane. This means that driving a vehicle with ethanol will decrease the fuel efficiency of

Table 10.5

Energy Content of Selected Fuels

Fuel	Energy Content in kJ per gram
Carbohydrate	17.2
Protein	23.8
Ethanol	29.7
Fat	38.9
Octane	47.9
Methane	56.0

a car. It should be noted, though, that ethanol is sometimes used as an additive to decrease the amount of carbon monoxide emissions. The table also shows that the energy content of fat is over twice that of carbohydrates, and also almost twice that of protein. The higher energy content of fats and the fact that they are more readily stored in the body explain why hibernating organisms build up a high fat content during the year.

One common application of thermochemistry in everyday life is the use of hot and cold packs. A variety of these are used to treat injuries and provide warmth. One type of pack used to provide heat utilizes the exothermic oxidation of iron to produce heat. The reaction can be represented as

$$4Fe_{(s)} + 3O_{2(g)} \rightarrow 2Fe_2O_3 \ \Delta H = -1650 \ kJ$$

A hot pack is activated when someone initiates the reaction by some physical action such as shaking or breaking a seal on the pack. The actual hot pack reaction involves several other chemicals, but the primary reaction is the oxidation of iron. The iron in the hot pack is a fine powder to increase the efficiency of the reaction. As expected, cold packs depend on some type of endothermic process. One common cold pack is based on the mixing of ammonia nitrate and water. In this case, the heat of solution is positive, indicating an endothermic reaction:

$$NH_4NO_{3(s)} + H_2O_{(l)} \rightarrow NH_4^+{}_{(aq)} + NO_3^-{}_{(aq)} \ \Delta H = +26kJ$$

When the reaction is initiated by breaking a barrier separating the ammonia nitrate and water contained in the pack, heat is absorbed from the surroundings. The immediate surroundings in this case happen to be the body part to which the cold pack is applied.

In this chapter, we have examined the energy aspects of chemical processes. While our focus on chemistry until this chapter has been on matter, we must realize that a complete study of chemical processes must also include energy. The principles of thermochemistry can help us answer fundamental questions such as how organisms maintain themselves or what is the ultimate fate of the universe. On a more practical level, phase changes and thermochemistry form the basis of many common devices. Car radiators, refrigerators, power plants, air conditioners, and heat pumps all employ phase changes to the benefit of modern society. The combustion of fuels powers modern society. Now that we have laid the groundwork for including energy considerations in our examination of chemical principles, we include these principles in coming chapters.

11

Solutions

Introduction

Much of the world around us is composed of **solutions**. Air, oceans, steel, gasoline, and soda pop are just a few examples of common substances that exist as solutions. A solution is nothing more than a homogeneous mixture in which the particles range in size from 0.2 to 2.0 nm. There are other homogeneous mixtures with smaller particle sizes. These are known as colloids and suspension, but our main focus in this chapter is on solutions. When considering solutions, it helps to think in terms of one substance dissolved in another substance. The simplest solutions consist of two components. The component in the greater amount is referred to as the **solvent**, and the component in the lesser amount is called the **solute**. Therefore, a solution can be thought of as a homogeneous mixture containing a solute dissolved in a solvent:

solute + solvent = solution

By far, the most familiar solutions consist of solid solutes dissolved in water. The oceans are giant solutions containing hundreds of solutes; the major ones exist in ionic form and include sodium, chloride, magnesium, iodide, calcium, and carbonate (Table 11.1). Although the most familiar solution consists of a solid dissolved in a liquid, solutions can consist of numerous combinations of phases. Table 11.2 summarizes different types of solutions based on the state of the solute and solvent.

Table 11.1
Major Ions Present in Seawater

Ion	Percent	ppt
Cl^-	1.9	19
Na^+	1.05	10.5
SO_4^{2-}	0.265	2.65
Mg^{2+}	0.135	1.35
Ca^{2+}	0.040	0.4
K^+	0.038	0.38
HCO_3^-	0.014	0.14
Br-	0.0065	0.065

Table 11.2
Examples of Solutions

Solute	Solvent	Example
solid	liquid	mineral water
solid	solid	alloys such as brass, which is a solution of zinc in copper
liquid	solid	dental amalgam, mercury in silver
liquid	liquid	antifreeze in radiator, ethylene glycol in water
gas	solid	hydrogen gas in palladium metal
gas	liquid	carbonated beverage, CO_2 in water
gas	gas	air in the atmosphere, many gases in nitrogen

Table 11.2 shows that many different types of solutions exist. In this chapter, we are primarily concerned with solutions in which water is the solvent. These are known as aqueous solutions.

The Solution Process

Why is it that some substances readily mix to form solutions while others do not? Whether one substance dissolves in another substance is largely dependent on the intermolecular forces present in the substances. For a solution to form, the solute particles must become dispersed throughout the solvent. This process requires the solute and solvent to initially separate and then mix. Another way of thinking of this is that the solute particles must separate from each other and disperse throughout the solvent. The solvent may separate to make room for the solute particles or the solute particles may occupy the space between the solvent particles. Determining whether one substance dissolves in another requires examining three different intermolecular forces present in the substances—between the

solute particles, between the solvent particles, and between the solute and solvent particles. The formation of a solution can occur when the magnitude of all three of these forces are similar. Conversely, if the intermolecular forces for solute particles are much stronger than their attraction for solvent particles, then mixing is not favored. The tendency of substances with similar intermolecular forces to mix leads to the general rule of thumb: "like dissolves like." Substances with "like" intermolecular forces tend to form solutions. As a simple analogy for the solution process, assume two situations take place in a stadium in a town with two rival teams. In the first situation, consider a football game between the two cross-town rivals. The fans from one side of town will occupy one side of the stadium, while the fans from the opposite side of town will sit together on the other side. There is no attraction of either side for the other, and this corresponds to the situation when two substances do not mix to form a solution. On the other hand, consider a second situation in which a rock concert is held in the stadium. In this case, people from the opposite sides of towns will mix throughout the stadium. Fans from both sides of town are equally attracted to the music, and there is an equal desire to get the best seats. This second situation is analogous to what happens when two substances mix to form a solution.

The "like dissolves like" rule provides a general guideline for determining which type of substances will form solutions. To see how this rule applies, let's look at the solubility of some of the major types of compounds. Ionic compounds tend to dissolve in polar solvents. Because water is the most common polar solvent, ionic compounds tend to form aqueous solutions. For example, NaCl dissolves in water. Consider what happens when a few grains of salt are sprinkled into water. Each salt grain is a crystalline lattice structure containing millions of unit cells of NaCl. Although the ionic bond holding the NaCl unit cells together is strong, the polarity of water causes the positive hydrogen end of water to be attracted to the chloride ion (Cl^-) and the negative oxygen end to be attracted to the

sodium ion (Na^+). Water molecules initially interact with the NaCl crystals on the outer surface of the grains. Water molecules act collectively to pull individual ions out of the crystalline structure (Figure 11.1). The general process by which a solute becomes dispersed in a solution is known as **solvation**. When water is the solvent, the process is known as **hydration**.

While many ionic compounds are soluble in water, many are not. The term "solubility" is somewhat subjective. There are actually degrees of solubility. A substance is considered soluble if 0.1 moles of it can dissolve in 1 liter of water. If less than 0.001 mole of the substance dissolves in water, a substance is considered insoluble. Partially soluble substances fall between these two extremes. Table 11.3 summarizes the solubility of some major groups of ionic compounds in water.

Whether an ionic compound dissolves in water depends on the strength of the ionic bond holding the compound together. Water must have sufficient strength to break the ionic bond. The strength of the

Figure 11.1
Diagram of Hydration of NaCl (Rae Déjur)

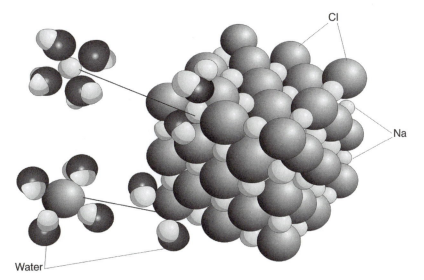

Table 11.3
Solubility of Ionic Compounds in Water at 25°C

1. Compounds containing alkali metal cations or ammonium are soluble.
2. All nitrate (NO_3^-) compounds are soluble.
3. Most compounds containing Cl^-, Br^- and I^- are soluble. Notable exceptions are for compounds containing Ag^+ or Pb^{2+}.
4. Most hydroxides (OH^-) are insoluble. Notable exceptions are for compounds containing an alkali metal or barium.
5. All carbonates (CO_3^{2-}) phosphates (PO_4^{3-}) and sulfides (S^{2-}) are insoluble. Notable exceptions are for compounds that contain an alkali metal.

bond depends on the size and charge of the ions in the compound. Stronger ionic bonds occur when the ions are smaller and the ions carry multiple charges. Because of this, many ionic compounds may not be soluble in water, as is seen by noting the exceptions listed in Table 11.3.

Polar covalent molecules may or may not dissolve in water depending on whether they have the ability to form hydrogen bonds. For example, many alcohols will dissolve in water because the OH group characteristic of alcohols gives them the ability to hydrogen bond with water. The attraction between alcohol and water is demonstrated when equal volumes of the two are mixed to give a total less than the sum of their individual volumes, for example, 100 mL of alcohol + 100 mL of water produces less than 200 mL of solution. Nonpolar substances tend not to dissolve in water, but they do dissolve in nonpolar solvents. Fats, oil, grease, and gasoline, for example, do not dissolve in water, but they form a layer on top of water. When substances do not mix but form distinct layers, they are referred to as **immiscible**. Nonpolar substances can be dissolved in a nonpolar solvent such as benzene or carbon tetrachloride.

Electrolytes

A useful characteristic in classifying solutions is their ability to conduct electricity. Solutions that are good conductors of electricity are said to contain **strong electrolytes**. Strong electrolytes have the ability to produce ions when dissolved in water; the greater the degree of ionization of a substance is, the stronger the electrolyte. Strong electrolytes include soluble salts, strong acids, and strong bases. Many substances, such as weak acids and weak bases, only partially ionize when dissolved in water. These substances are referred to as **weak electrolytes**. Still other substances do not ionize at all, but they dissolve as whole molecules and are referred to as **nonelectrolytes**. Sugar is an example of a nonelectrolyte. When sugar dissolves, the covalent bonds holding the molecule together are sufficiently strong to keep the molecule together. Table sugar, $C_{12}H_{22}O_{11}$, contains a number of oxygen atoms making it a polar molecule. Its high number of oxygen atoms allows it to hydrogen bond to many water molecules, and therefore, sugar is quite soluble.

Soluble ionic compounds tend to be strong electrolytes, while alcohols and organic compounds are nonelectrolytes. Remember that classification as a strong electrolyte, weak electrolyte, or nonelectrolyte is somewhat subjective. Freshwater can be either a weak electrolyte or a nonelectrolyte depending on its purity. The important consideration in classifying a substance is to what extent an aqueous solution of the substance will conduct electricity.

Concentration of Solutions

The amount of solute and solvent in a solution can be quantitatively expressed using numerous concentration units. The choice of a particular concentration unit depends largely on practice and convenience. We have probably all made solutions using recipes or directions that tell us to add so much water to a substance. In the field of chemistry, the most common concentration units are molarity, molality, percent by mass, and "parts per." Each of these is defined here:

Molarity is the moles of solute per liter of solution. It is abbreviated with a large "M." A 1 M solution contains 1 mole of the substance dissolved in 1 liter of solution. It is important to realize molarity is the amount of solute per liter of solution. If a 1 M solution of sodium hydroxide were to be prepared, 40 grams of NaOH would be diluted to 1 liter. Special flasks of various volumes are made for just this purpose. Adding 1 liter of water to 40 grams would not produce a 1 M solution, but a solution of slightly less than 1 M because adding 1 liter of water to 40 grams of NaOH makes the final volume of the solution greater than 1 liter.

Molality is the number of moles of solute dissolved in 1 kilogram of solvent. It is abbreviated by "m." A 1 m solution of NaOH contains 40 grams of NaOH dissolved in 1 kilogram of water. Molality is especially important in working with **colligative** properties such as **boiling point elevation** and **freezing point depression**.

Percent by mass is the mass of the solute divided by the mass of solution multiplied by 100.

Parts per a particular quantity is used extensively in environmental work and is useful when expressing the concentration of dilute aqueous solutions. Common concentration expressions using this unit include parts per thousand, parts per million, and parts per billion. Ocean salinity is expressed in parts per thousand (ppt). Typical ocean salinity is 35 ppt and can be translated to mean that for every 1,000 grams of ocean water there are 35 grams of dissolved material. Parts per million (ppm) is used frequently in environmental work and to express the solubility of gases in water. For example, the solubility of oxygen in a lake may be given as 10 ppm. Because the density of water is approximately 1 mg/L, a ppm is the same as mg/L for dilute aqueous solutions. A part per billion (ppb) is also frequently used for water standards. The recommended drinking water standard for lead is 15 ppb. A part per billion is the same as 1 μg/L. Modern instruments have allowed chemists to measure progressively smaller concentrations of substances in the environment, some at parts per trillion concentrations. Even though it is possible to detect toxic substances at very low concentrations, this should not be interpreted as meaning these substances present a threat to human health. A low concentration may be statistically equivalent to zero, presenting no problem whatsoever.

While concentration units indicate how much solute and solvent make up a solution at any given time, often it is desirable to know the concentration when the solution is **saturated**. A solution is saturated when the maximum amount of solute has dissolved in the solvent at a particular temperature. When the solvent holds less than its maximum amount of solute, the solution is said to be **unsaturated**; when the solution holds more than the maximum amount, it is said to be **supersaturated**. The amount of solute that can dissolve in a solvent can be expressed using different measures of solubility. Molar solubility is the number of moles of solute in 1 liter of saturated solution. Handbooks generally report the solubility of aqueous solutions by giving the amount of solute that will dissolve in 100 grams of water at a specific temperature. The solubility of NaCl is 35.7 grams per 100 grams of water at 0°C and 37.3 grams per 100 grams of water at 60°C. Once a solution becomes saturated, any solute added will settle out or precipitate. The additional solute that settles out is called the precipitate. For example, using the solubility of NaCl at 60°C, the solution would be undersaturated until 37.3 g NaCl have been added to 100 grams of water. At this point, the solution becomes saturated and any additional salt added will precipitate out as solid.

In any discussion of solubility, it is important to remember solubility is temperature dependent. Generally, the solubility of solids in liquids increases with temperature. The variation of solubility in water varies greatly for different solutes. Figure 11.2 demonstrates that the solubility may increase significantly with temperature for some solutes, may be relatively constant (NaCl), and may even decrease with temperature. The variation in solubility with temperature should not be confused with the rate at which a solute dissolves. Increasing the temperature causes the solute to dissolve faster as does stirring and increasing the surface area of the solute, but just because a substance dissolves faster does not mean more of it dissolves.

In some cases a solution may become supersaturated. A good example of this is sodium acetate, CH_3COONa. When a solution containing sufficient sodium acetate is heated until the solid dissolves and allowed to cool slowly with minimal disturbance, the sodium acetate does not crystallize out as the solution cools. For the sodium acetate to settle out, a small seed crystal must be added to the solution to initiate the crystallization process.

Solubility of Gases

The focus in this chapter has primarily been on solid solutes in liquid solvents. The solubility of gases in water and other liquids

Figure 11.2
The solubility of several ionic compounds varies according to temperature.

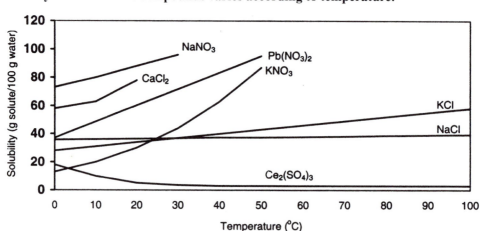

(such as in blood) varies in several respects from that of solids. One of the most obvious differences is the degree of solubility. Gases are not very soluble in water. The saturated value for oxygen in freshwater is 14.6 mg/L at 0°C and 9.2 mg/L at 20°C. The saturation values for oxygen illustrate that the solubility of gases is inversely related to temperature. The fact that gas solubility decreases with increased temperature is the reason for thermal pollution associated with power plants. Power plants use vast quantities of water to condense steam in the generation of electricity. As seen in the last chapter, condensation is an exothermic process. The heat given up by the steam raises the temperature of the cooling water. When the warmed cooling water is released into the environment, for example into a river or lake, the elevated temperatures cause a drop in the oxygen solubility of the water. The lowered oxygen levels may place a significant stress on aquatic organisms living in the vicinity of the power plant. This stress is not only due to the decreased oxygen content, but also due to the impact of elevated temperatures on the metabolism of cold-blooded organisms.

Another factor that differentiates the solubility of gases from solids and liquids is the effect of pressure. The effect of pressure on gas solubility was studied extensively by a contemporary and close associate of John Dalton named William Henry (1775–1836). Henry's Law states that the solubility of a gas is directly proportional to the partial pressure of that gas over the solution. Stated mathematically, Henry's Law is $c = kP$, where c is the concentration of the dissolved gas in moles per liter, k is Henry's law constant for the solution, and P is the partial pressure of the gas above the solution. Henry's Law is demonstrated every time a carbonated beverage is opened. During the carbonation process, carbon dioxide is dissolved in the beverage under increased pressure. When the beverage container is sealed, a small space above the beverage fills with carbon dioxide along with other gases (air, water vapor). The partial pressure of the carbon dioxide above the beverage in the closed container is many times higher than that found in the atmosphere, where it is only about 0.0004 atmosphere. When the container is opened, the partial pressure of the CO_2 will immediately drop. According to Henry's Law, a drop in the partial pressure of CO_2 above the solution results in a decreased solubility of carbon dioxide in the beverage. This is seen whenever a carbonated beverage is opened and the CO_2 is observed escaping as a steady stream of bubbles. Henry's Law also explains why the beverage goes flat if it is left open to the atmosphere for any length of time.

Colligative Properties

Anyone who has ever made ice cream knows that the addition of rock salt to ice causes it to melt and produce a liquid-ice solution below 0°C. This is just one example of how the physical properties of a solution differ from those of a pure solvent. Properties that depend on the amount of solute present in a solution are termed colligative properties. Colligative means collective properties. These properties are termed colligative, because the properties depend on the collective number of particles present in solution rather than the types of particles. The major colligative properties and how they affect solutions compared to their pure solvents are summarized in Table 11.4.

To examine the four properties listed in Table 11.4, we can use the simple case of a nonelectrolyte, nonvolatile solute. The lowering of the vapor pressure is a consequence of nonvolatile solute particles occupying positions at the surface of the

Table 11.4
Colligative Properties

Property	Effect
Vapor Pressure	Vapor pressure of solution with nonvolatile solute is lower than that of solvent
Boiling Point	Boiling point of solution is higher than solvent
Freezing Point	Freezing point of solution is lower than solvent
Osmotic Pressure	Solvent particles migrate through semipermeable membrane toward solution

solution. Figure 11.3 shows two identical containers, one containing an aqueous solution, for example a sugar solution, and the other water. The surface of the solution is occupied by both solute and solvent particles. Because the solute is nonvolatile, solute particles do not vaporize but stay in solution and reduce the surface area occupied by the solvent water molecules. The reduction in the surface area for the solution means that fewer molecules are able to enter the vapor phase as compared to the pure solvent. The solute particles can be thought of as a barrier that impedes the solvent from entering the vapor phase, and therefore, reducing the vapor pressure. The higher the concentration of the solution, the more the vapor pressure is lowered.

Figure 11.3
In a solution, solute particles at the surface reduce the amount of solvent that can enter the vapor phase resulting in a lower vapor pressure.

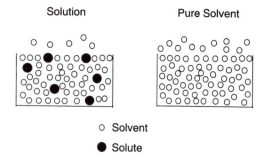

Solution   Pure Solvent

○ Solvent
● Solute

The boiling point is defined as the temperature at which the vapor pressure of a liquid is equal to the external pressure, typically 1 atmosphere. If adding a solute lowers the vapor pressure, then a higher temperature is required to reach the point where the vapor pressure of the liquid is equal to the external pressure. Therefore, the boiling point is higher for a solution as compared to its pure solvent. The change in boiling point due to addition of a solute is known as boiling point elevation. The boiling point elevation can be calculated using the equation $\Delta T_b = K_b m$, where ΔT_b is the increase in boiling point for the solution, K_b is the molal boiling point elevation constant, and m is the molality of the solution. Molal boiling point elevation constants are given in reference books for different solvents. Water's K_b is equal to 0.51 °C/m. Using this value, a 2.0 molal sugar solution boils at a temperature of 101.02°C rather than 100°C.

A solution also exhibits a depression in its freezing point. The freezing point depression is the decrease in the temperature of the freezing point due to the addition of a solute. It is calculated using the equations $\Delta T_f = K_f m$, where ΔT_f is the decrease in freezing point for the solution, K_f is the molal freezing point depression constant, and m is the molality of the solution. Water's K_f value is 1.86°C/m.

A relevant application of freezing point depression and boiling point is the addition of antifreeze to a car's radiator. Ethylene glycol, $CH_2(OH)CH_2(OH)$, is a popular antifreeze. Producers of antifreeze recommend that a 50% antifreeze and a 50% water mixture be used for protection. An average capacity for the cooling system of a car is about 14 quarts. Seven quarts of water has a mass of approximately 6.6 kg. The density of ethylene glycol is about 1.1 g/mL, and its molar mass is 62 g. These values are used to determine that there are about 117 moles contained in 7 quarts of ethylene glycol. The molality of the water-antifreeze mixture would be 117 moles/6.6 kg = 18 m. Using the values for K_b and K_f, gives the freezing point of the water-antifreeze solution as −33°C and the boiling point as 109°C. These temperatures would be the same for any 50% antifreeze solution independent of the cooling system capacity.

The last of the four colligative properties in Table 11.4 is **osmotic pressure**. **Osmosis** occurs when solvent molecules move through a **semipermeable membrane** from the pure solvent or diluted solution to a more concentrated solution. A semipermeable membrane acts like a sieve or filter. It allows movement of solvent particles to pass through, but it blocks solute particles (Figure 11.4). Figure 11.4 shows only the net movement of solvent particles from the solvent to the solution. It is important to realize that solvent particles pass through the membrane in both directions, but the rate of movement from the solvent to solute is greater than in the reverse direction. The pressure required to stop the net movement of solvent particles across the semipermeable membrane is called the osmotic pressure of the solution. The osmotic pressure is given by the formula $\pi = MRT$, where π is the osmotic pressure in atmospheres, M is the molarity of the solution, R is the ideal

Figure 11.4

In osmosis, there is a net movement of solvent from a region of higher concentration to a region of lower concentration.

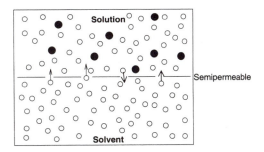

gas law constant, and T is the absolute temperature.

Osmosis is extremely important in physiology. Cell membranes, capillaries, and the lining of digestive tracts act as semipermeable membranes. Osmosis constantly occurs through cell membranes. Cells generally have the same concentration as the fluids surrounding them. In instances where the surrounding fluid has a higher or lower solvent concentration compared to the cell, water flows into or out of the cell, respectively. The former condition is known as **plasmolysis** and the latter as **crenation**.

A process similar to osmosis is **dialysis**. In dialysis, a dialyzing membrane allows both solvent and minute solute particles of a certain size to pass. The passage of these particles, like osmosis, is from a region of high concentration to a region of low concentration. Our kidneys are a dialysis system responsible for the removal of toxic wastes from the blood. In artificial dialysis, blood is circulated through dialysis tubes made of membranes, which are immersed in a "clean" solution that lacks the wastes in the blood. The impurities are filtered out of the blood as they move across the walls of the dialyzing tubes.

Reverse osmosis is a process in which freshwater is obtained from saltwater by forcing water from a region of low freshwater concentration to a region of high freshwater concentration. This is opposite of the natural process, and hence the name reverse osmosis. In a typical reverse osmosis process, a pressure of approximately 30 atmospheres is required to force freshwater to move from seawater across a semipermeable membrane (Figure 11.5).

In the discussion of colligative properties, it was assumed that the solutions were nonelectrolytes. When electrolyte solutions are considered, remember that colligative properties are dependent on the number of particles in solution. While a 1 m concentration of a nonelectrolyte, such as sugar, yields 1 mole of particles per kg of solvent, an electrolyte such as NaCl theoretically dissociates to give 1 mole of sodium ions and 1 mole of chloride ions. In reality, a 1 m NaCl solution produces slightly less than 1 mole of sodium and chloride ions. This is because water is not perfectly efficient in hydrating all ions, and some Na^+ and Cl^- are attracted to each other in solution and act as a single particle. This single particle is known as an **ion pair**. The greater the concentration of the solution is, the greater the chance of ion pairs forming. For dilute concentrations of NaCl, it is assumed there are about 1.9 moles of particles for a 1 m solution of NaCl. The 1.9 should be used as a correction factor in the boiling point elevation, freezing point, and osmotic pressure equations. To see how this works, let's determine the osmotic pressure of seawater, which is 30 atm. The equation is $\pi = MRT$. If it is assumed that the salinity of seawater is 35 ppt, this means there are 35 grams of salt per 1,000 grams of seawater. Assume that all the salt is in the form of NaCl, and the density of seawater is approximately 1 g/mL; this gives the number of moles of salt as 35g/58 g/mole = 0.60 mole. The molarity of seawater is found by taking this value and dividing by the number of liters of solution: molarity = .60 mole/ 1.0 L = 0.60 M. If we use the equation for osmotic pressure, we find π by multiplying (0.60m)(0.082 atm-L/m)(298K). This value is slightly less than 15 atmospheres. The calculated value for seawater does not take into account that we get almost two ions for every NaCl particle. Therefore, our calculated value must be doubled to give 30 atmospheres for the osmotic pressure of seawater. To determine the boiling point elevation or freezing point depression of a salt solution, the factor 1.9 would be used in the appropriate equations. Correction factors, called **van't Hoff factors**, for a number of different electrolytes can be found in chemical reference books.

Reactions of Solutions

Most reactions in chemistry involve solutions. Reactions between solutions continually take place in the atmosphere, ocean, and natural environment. A host of chemical processes such as the refining of petroleum, production of steel, purification

Figure 11.5
In reverse osmosis, a pressure equal to the osmotic pressure of seawater is applied to obtain freshwater from seawater.

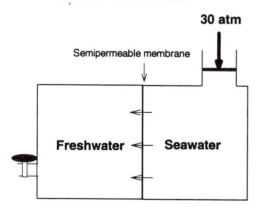

30 atm

Semipermeable membrane

Freshwater ← Seawater

of water supplies, and refining of minerals involve solutions. In fact, solution reactions not only occur all around us, but also within us. Our survival depends on numerous solution reactions taking place within our bodies. The digestion of food, the transport of oxygen to cells, and the removal of waste products are just a few body functions involving solution reactions. We conclude this chapter on solutions by taking a brief look at precipitation reactions. A precipitation reaction takes place when two solutions are mixed, and a solid, called a precipitate, separates from the solution.

To understand how a precipitation reaction works, let's examine the reaction between two aqueous ionic solutions, silver nitrate ($AgNO_3$) and potassium chloride (KCl). According to the solubility rules presented in Table 11.3, both these ionic compounds are soluble in water. When these solutions are mixed, a white precipitate immediately forms. Referring again to Table 11.3, rule 3 indicates chlorides are soluble, but that a notable exception is for chloride compounds containing Ag^+. Based on this information, the white precipitate must be silver chloride, AgCl. The mixing of the two solutions is presented in Figure 11.6. The two original solutions contain the potassium, chloride, silver, and nitrate ions in aqueous solution. When these are mixed, the silver and chloride react to produce a precipitate. The potassium and chloride ions remain in solution and the process can be represented with the **molecular equation**:

$$KCl_{(aq)} + AgNO_{3(aq)} \rightarrow AgCl_{(s)} + KNO_{3(aq)}$$

Rather than write the chemical equation using whole units, a more accurate picture of the reaction is represented by writing the **complete ionic equation**:

$$K^+_{(aq)} + Cl^-_{(aq)} + Ag^+_{(aq)} + NO_3^-_{(aq)} \rightarrow$$
$$AgCl_{(s)} + K^+_{(aq)} + NO_3^-_{(aq)}$$

Figure 11.6
When two solutions are mixed, a solid precipitate may form when the ions form an insoluble compound.

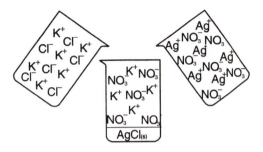

Writing the reaction using ions shows that the potassium and nitrate ions are present on both sides of the equation. These ions do not take part in the overall reaction and are called **spectator ions**. The general ionic equation can be simplified by dropping the spectator ions and writing the **net ionic equation**:

$$Ag^+_{(aq)} + Cl^-_{(aq)} \rightarrow AgCl_{(s)}$$

The net ionic equation shows only the ions that are actually taking part in the reaction.

Precipitation reactions follow the same general pattern as potassium chloride and silver nitrate. The solubility rules in Table 11.3 can be used to predict whether a precipitate will form. In many cases, solutions do not form a precipitate when mixed. In these cases, it is assumed all ions stay in solution. Precipitation reactions can be used to our advantage. In water quality work, one method of removing pollutants is to precipitate the pollutant out of the water in a treatment plant. The pollutant can then be recycled or disposed of rather than being released into the environment. For example, phosphates causing eutrophication (rapid aging) in freshwater lakes can be removed by adding $FeCl_3$ and alum to the water. Another use of precipitation reaction is to identify the presence of specific ions in water. For instance, if a water sample was suspected of containing lead, Pb^{2+}, the sample could be mixed with a solution containing iodide (or chloride, bromide) to see if a precipitate formed. The method of **qualitative analysis** involves conducting a systematic series of reactions using different solutions to determine the presence or absence of particular ions. By noting which solutions produce precipitates and which do not, many unknowns can be identified.

A number of precipitation reactions are detrimental and cause problems. Scaling in pipes due to the precipitation of calcium is a problem in regions of hard water. Hard water is defined as water having a high, typically greater than 100 ppm, concentration of calcium and/or magnesium. Water heated in hot water tanks and hot water heating systems results in the precipitation of calcium carbonate according to the following reaction:

$$Ca^{2+}_{(aq)} + 2HCO_3^-_{(aq)} \rightarrow CaCO_{3(s)} + CO_{2(aq)} + H_2O_{(l)}$$

This reaction shows calcium ion reacting with bicarbonate to yield calcium carbonate, carbon dioxide, and water.

Another example of an undesirable precipitation reaction involves the formation of kidney stones in the human body. The major type of kidney stones consists of calcium in combination with oxalate ($C_2O_4^{2-}$). The reaction can be represented as

$$CaCl_{2(aq)} + H_2C_2O_{4(aq)} \rightarrow CaC_2O_{4(s)} + 2HCl_{(aq)}$$

This reaction shows calcium chloride reacting with oxalic acid to produce calcium oxalate and hydrochloric acid. The calcium oxalate is the kidney stone. Calcium oxalate precipitates out of the urine in the kidneys of all individuals. Normally, the solids formed are grain size and do not cause prob-

lems. In certain individuals, though, the precipitates can grow to appreciable size, causing considerable discomfort and health problems. One recommendation for those with kidney stones is to increase their water intake and reduce the consumption of foods rich in oxalic acid such as chocolate, peanuts, and tea. Other precipitates that can form in the kidney and lead to stone formation are calcium phosphate, $Ca_3(PO_4)_2$, and calcium carbonate, $CaCO_3$.

Colloids and Suspensions

The two other types of mixtures mentioned at the beginning of this chapter were colloids and suspensions. These mixtures can be differentiated from solutions based on particle size. Colloid particles are larger than particles in solution. They range in diameter from 2.0 to 1,000 nm. Colloid particles cannot be filtered out, remain in suspension, and scatter light. Common examples of colloids are fog and milk. Suspensions consist of particles greater than 1,000 **nanometers** in diameter. They are generally opaque to light, separate on standing, and can be filtered. Blood is an example of a suspension.

Colloids were initially differentiated from solutions by Thomas Graham (1805–1869) as a result of his work concerning the diffusion of particles through membranes. Graham observed that certain substances such as starches, glues, and gelatins did not diffuse through membranes like ionic solutions. Graham used the term "colloids," which means glue in Greek, to describe these substances. Graham's work on the diffusion of materials through membranes also led him to the discovery that membranes allowed small molecules and ions to pass, but blocked the movement of colloids. Graham coined the word dialysis to describe this process.

Table 11.5
Types of Colloids

Colloid	Dispersed Phase	Dispersing Medium	Example
Sol	Solid	Liquid	Jelly
Liquid Emulsion	Liquid	Liquid	Milk
Solid Emulsion	Solid	Liquid	Butter
Aerosol	Liquid	Gas	Fog
Aerosol	Solid	Gas	Smoke
Foam	Gas	Liquid	Shave Cream
Solid Foam	Gas	Solid	Marshmallow

A colloid is also called a colloidal dispersion. A colloid dispersion consists of two components similar to a solution. The particles themselves are the **dispersed phase** and are analogous to the solute in a solution. The **dispersing medium** is similar to the solvent. Some examples of different types of colloids are summarized in Table 11.5.

Colloid particles tend to acquire similar charges on their surface. This results in the particles repelling each other, and therefore, they stay dispersed and tend not to settle. Another characteristic of colloids is their ability to scatter light. This phenomenon is known as the **Tyndall effect**, named after the English physicist John Tyndall (1820–1893). This effect is observed when colloidal dust and water particles in the air scatter light from a movie projector or spotlight making the light beam visible.

As the examples in Table 11.5 indicate, many familiar products are colloids. The cleaning ability of soap is due in part to dirt particles emulsifying in the soap. Food and other nutrients in our blood are transported to cells as colloidal particles. The process of jelly making involves producing a sol that sets up as a semisolid. A similar process occurs when making gelatin. Sols that set up like jellies are technically called **gels**.

Just as solutions precipitate, there are processes that cause colloidal particles to group together and settle. The process in which colloidal particles group together is called coagulation. Coagulation can be initiated by heating or adding an acid to a colloid. The coagulation of colloids is observed every time an egg is fried. The colloids in the albumen of the egg coagulate producing a solid white mass when heated.

Kinetics and Equilibrium

Introduction

Two important aspects of chemical reactions are how fast they occur and to what extent the reaction takes place. The area of chemistry that deals with the speed of chemical reactions is known as chemical **kinetics**. In many reactions, reactants are only partially converted into products. In these reactions a state is reached in which the concentrations of reactants and products remain constant. At this point, the reaction is said to have reached **equilibrium**. In this chapter, we explore each of these important areas of chemistry.

Reaction Rates

It is well known that reactions take place at different rates. Some reactions occur in an instant, while others can be measured in seconds, minutes, and hours. The explosive combustion of gasoline in the cylinder of a car occurs in a fraction of a second. Reactions used to develop film take place over seconds to minutes. Chemical reactions responsible for changing the color of fall leaves take days to weeks, and the oxidation of metals takes place over years. Whenever reactions occur, chemical kinetics play a

vital role. Kinetics dictate how fast products can be produced in the chemical industry, how long it takes for medicines and drugs to act in our bodies, how quickly the depletion of the ozone layer occurs, and how long it takes food to spoil. Knowledge of kinetics gives us insights into not only how fast they occur, but also how they occur.

Because kinetics deals with the rates of chemical reactions, it is important to establish a definition for reaction rate. Because most reactions occur in or between solutions, it is logical to define reaction rates in terms of changes in concentration. Reaction rate is typically defined as the change in concentration of a reactant or product over time. For example, hydrogen peroxide decomposes slowly over time. Its decomposition is accelerated by heat and exposure to light, and therefore, it is stored in opaque containers. The reaction is represented by the equation:

$$2H_2O_{2(l)} \rightarrow 2H_2O_{(l)} + O_{2(g)}$$

Hydrogen peroxide sold in stores is typically a 3% solution. This solution is equivalent to approximately 0.90 M. The rate of the decomposition of H_2O_2 can be expressed by the change in hydrogen peroxide concentration over time:

$$\text{Rate} = \frac{\Delta \text{ Concentration}}{\Delta \text{ time}}$$

If after one year the concentration was 1.5% or 0.45 M, then the reaction rate would be equivalent to –0.45 M/year. Rather than looking at the disappearance of hydrogen peroxide, the reaction rate could be expressed in terms of the production of water or oxygen. Expressing the reaction rate in terms of the amount of oxygen formed, the stoichiometry of the reaction could be used to determine the rate of formation of oxygen. The reaction shows 1 mole of molecular oxygen is formed for every 2 moles of hydrogen peroxide that decompose. The amount of oxygen formed after one year would be half of 0.45 M or about +0.22 M/yr. Notice that the plus sign indicates a species is forming and a negative sign indicates a species is disappearing.

The Collision Theory

An understanding of reaction rates can be explained by adopting a collision model for chemical reactions. The collision theory assumes chemical reactions are a result of molecules colliding, and the rate of the reaction is dictated by several characteristics of these collisions. An important factor that affects the reaction rate is the frequency of collisions. The reaction rate is directly dependent on the number of collisions that take place, but several other important factors also dictate the speed of a chemical reaction.

Consider a piece of coal, which for our purposes can be considered pure carbon, sitting on the table. The coal is exposed to billions upon billions of collisions with oxygen molecules, O_2 each second. Yet, the reaction $C + O_2 \rightarrow CO_2$ does not occur. The coal could sit on the desk for years and experience countless collisions with oxygen without reacting. Obviously, merely the presence of collisions does not mean a reaction occurs. It's similar to saying we bruise when we come into contact with an object. While contact with an object may be a necessary condition to bruising, we do not get a bruise every time we touch something.

To understand why most collisions do not cause a reaction, the energetics of a reaction must be considered. As noted several times, reactions involve the breaking and formation of bonds. Energy is required to break bonds, and energy is released when bonds form. According to the collision theory, the energy needed to break bonds comes from the kinetic energy of the colliding molecules. During a collision between molecules, kinetic energy is converted to potential energy. More often than not, there is simply not enough kinetic energy to break bonds and initiate a reaction. The situation can be compared to dropping a rubber ball from a certain height onto the floor. As the ball falls, gravitational potential energy is converted to kinetic energy. The ball's kinetic energy reaches a maximum value just before it contacts the floor. When it strikes the floor, the ball is deformed as kinetic energy is converted to potential energy. This potential energy is referred to as elastic potential energy and is similar to the energy stored when a spring is compressed. When the ball has reached its maximum deformation, it momentarily stops and has no kinetic energy. At this point, the elastic potential energy stored in the rubber ball is at a maximum. This energy is converted back into kinetic energy as the ball bounces back up. If the ball were thrown with sufficient energy, it would break apart when it struck the floor. This is analogous to molecules colliding. Most of the time they just bounce off each other, but if they have enough kinetic energy, they can break apart.

During chemical reactions, the kinetic energy of the molecules are converted into potential energy during collisions. The collisions distort and stretch the bonds; the bonds can be viewed as acting like tiny springs or rubber bands. If the energy of the collision is sufficient, bonds break, similar to the breaking of a rubber band if it is stretched too far. In a chemical reaction, the minimum energy needed to break bonds and initiate a reaction is known as the **activation energy**. No matter how many collisions occur between reactants, a reaction does not occur unless the energy of the collision is equal to the activation energy. Therefore, while the carbon in the lump of coal experiences an endless number of collisions with oxygen, these collisions are not sufficient to provide the activation energy for the combustion of coal. The energy in the collisions is insufficient to break the double bond of the oxygen molecule.

The activation energy is often depicted as an energy hill or barrier that reactants must overcome for a reaction to take place.

Two situations are shown in Figure 12.1 corresponding to exothermic and endothermic reactions. When the reaction is exothermic, the total potential energy of the products is less than the potential energy of the reactants. The difference in potential energies of the reactants and products equals the change in internal energy of the system. Because energy flows from the system to the surroundings in an exothermic reaction, the internal energy decreases and ΔE is negative. In an endothermic reaction, the products have a higher potential energy than the reactants, energy flows from the surroundings into the system, and ΔE is positive.

At the top of the energy hill a transition state exists called the **activated complex**. The activated complex represents the stage in the reaction where reactants may be transformed into products. The idea of an activated complex was first proposed by Svante August Arrhenius (1859–1927) around 1890. Arrhenius received the Nobel Prize in chemistry in 1903. Reactants may reach the activated complex stage and still not be converted into products, but "slide" down the activation energy hill back to reactants. On

Figure 12.1
The activation energy, E_a, acts as an energy barrier. For a reaction to occur, the amount of kinetic energy converted to potential energy during a collision must be greater than the activation energy.

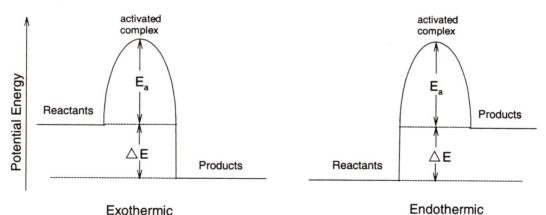

the other hand, when a reaction does occur, then the activated complex stage is where bonds are broken and formed as reactants are changed into products.

Even though molecules may have enough kinetic energy to supply the activation energy, a reaction may still not take place. A collision between molecules must not only possess sufficient energy, but the molecules must also have the proper orientation during the collision. To appreciate the importance of orientation, think of the damage caused when two cars collide. A head-on collision results in appreciable damage compared to a sideswipe or indirect collision. Similarly, the energy transferred when molecules collide can vary significantly depending on how they collide. Experiments demonstrate that the rate of a chemical reaction is often less than what is expected at a particular temperature. One major reason for this disparity is that many of the collisions do not have the correct orientation to break the bonds of reactants.

Factors Affecting Reaction Rates

The collision theory can be used to examine those factors that affect reaction rates. The most obvious factor is temperature. There is a direct relationship between temperature and kinetic energy. Kinetic energy is equal to $\frac{1}{2} mv^2$ and is also a measure of temperature. By increasing the temperature at which a reaction occurs, the kinetic energy of the molecules also increases. Recall from Chapter 9 the distribution of velocities at two different temperatures. Figure 12.2 shows that as the temperature increases, there is a greater proportion of molecules with higher velocities, and therefore, more molecules with sufficient kinetic energy to overcome the activation energy barrier. The higher the

Figure 12.2

As the temperature increases, a greater proportion of particles reach higher velocities and therefore higher kinetic energies. (Rae Déjur)

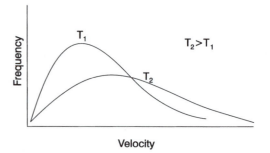

temperature is, the greater the proportion of particles with higher kinetic energies.

It's often taken for granted that increasing the temperature increases the reaction rate. Hot water is used for cleaning, elevated temperatures to cook and bake, and sparks and flames to ignite fuels. One general rule of thumb regarding temperature and reaction rate is that for every increase of 10°C the reaction rate doubles. This rule is sometimes referred to as the van't Hoff rule. Jacobus Hendricus van't Hoff (1852–1911) was one of the foremost chemists of his time. Van't Hoff's did work on thermochemistry, solutions, and organic chemistry. He received the first Nobel Prize awarded in chemistry in 1901. Van't Hoff's rule can also be stated in reverse by saying that the reaction rate is halved for every decrease of 10°C.

Every time an item is placed in the refrigerator we depend on lower temperatures to slow reaction rates to prevent food spoilage. The effect of temperature on reaction rates is also illustrated by its impact on human survival. Normal body temperature is 37°C. An increase of body temperature of just a few degrees to produce a fever condition increases the metabolic rate, while lowering the body temperature slows down metabolic processes. The slowing of human metabolic processes explains why individuals who have fallen through ice and been submerged under cold water for long periods can sometimes be revived. In some cases, people have been under cold water for 20 to 30 minutes and been revived with little or no permanent damage. Under normal conditions the brain requires a steady supply of oxygen. If this supply is cut off for a few minutes, permanent brain damage results. In cold icy waters, the need for oxygen is greatly reduced; therefore, someone may be able to survive a lack of oxygen for an extended period. This phenomenon is also applied in some surgical procedures, such as heart surgery, where the body temperature is lowered during an operation to mitigate reduced oxygen flow to the brain.

Another factor that affects the rate of a chemical reaction is the concentration of reactants. As noted, most reactions take place in solutions. It is expected that as the concentration of reactants increases more collisions occur. Therefore, increasing the concentrations of one or more reactants generally leads to an increase in reaction rate. The dependence of reaction rate on concentration of a reactant is determined experimentally. A series of experiments is usually conducted in which the concentration of one reactant is changed while the other reactant is held constant. By noting how fast the reaction takes place with different concentrations of a reactant, it is often possible to derive an expression relating reaction rate to concentration. This expression is known as the rate law for the reaction.

The decomposition of hydrogen peroxide illustrates the dependence of reaction rate on concentration. One way to determine how fast hydrogen decomposes is to measure the amount of oxygen generated. By performing a series of trials using different concentrations of hydrogen peroxide, it is found that the reaction rate increases in direct proportion to the original concentration of H_2O_2. This is expressed mathematically as

$$\text{rate} = k[H_2O_2]$$

This equation is the **rate law** for the decomposition of hydrogen peroxide. The constant k is referred to as the rate constant and the brackets indicate that the concentration of hydrogen peroxide is expressed in molar units. When the concentration is directly proportional to the concentration of a reactant, the reaction is **first order** with respect to that reactant. Therefore, the decomposition of H_2O_2 is first order with respect to H_2O_2. The reaction order tells us how the reaction rate is related to the concentration. As another example, consider the reaction of hydrogen and iodide to produce hydrogen iodide:

$$H_{2(g)} + I_{2(g)} \rightarrow 2HI_{(g)}$$

The rate law for this reaction has been determined to be

$$\text{rate} = k[H_2][I_2]$$

Therefore, for this reaction, the rate is first order with respect to hydrogen and first order with respect to iodide. The overall order of a reaction is found by summing the individual orders. Therefore, the overall order for the above reaction is 2.

The reaction order is equal to the exponent of the concentration in the rate law. A general expression for the rate law can be written using the standard reaction: A + B → products, where A and B represent reactants. The general rate law for this expression can be written as

$$\text{rate} = k[A]^x[B]^y$$

In this expression, A and B represent molar concentrations of the reactants and x and y are the orders with respect to each reactant. The overall order of the reaction would be equal to x + y. Reaction orders typically have values like 0, 1, and 2, although it is not unusual to have fractional reaction orders. A reaction order of 0 for a reactant means that the rate does not depend on the concentration of that reactant. This makes sense if it's remembered that any value to the 0 power is one. For example, if the reaction did not depend on the concentration of reactant A, then $[A]^0 = 1$ and the rate law becomes rate = $k[B]^y$.

As stated previously, the rate law for a chemical reaction is determined experimentally. Because rates are not constant with time but change during the course of the reaction, rates are measured during the initial stages of the reaction. Once the rate law for a reaction is determined, an expression can be derived using calculus that expresses the change in concentration of reactant with time. One of the most widely used forms of these expresses the change in concentration using **half-life**. The half-life is simply the time it takes for the original concentration of a reactant to drop to half its original value. A shorter half-life means that the reaction is faster. Half-life also is used extensively in areas such as nuclear reactions and toxicology. Half-life is covered more extensively in Chapter 17.

A third important factor in determining how fast a reaction occurs involves the surface area of the reactants. Because collisions take place on the surface of reactants, increasing the surface area increases the rate of chemical reactions. The more a substance is divided, the greater the amount of exposed surface area (Figure 12.3).

Dividing a reactant into small pieces makes a significance difference in the reaction rate. A familiar example is the difference between igniting a log versus igniting wood chips. One situation that illustrates

Figure 12.3
Dividing a substance into smaller pieces increases the amount of exposed surface area. The volume of the large cube and the total volume of the eight small cubes are the same, but the small cubes have twice the total surface area.

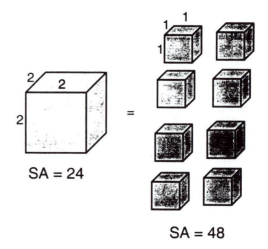

how the reaction rate depends on surface area is the occasional violent explosion of grain dust. Finely powdered grain is as explosive as dynamite due to its tremendous surface area. In grain elevators and storage bins, ventilation and other methods of dust control are used to reduce the chances of a violent explosion.

Reaction Mechanisms and Catalysis

Another method of increasing the reaction rate is by employing a **catalyst**. A catalyst is a substance that speeds up the reaction but is not consumed in the reaction. A substance that slows down or stops a reaction in known as an **inhibitor**. To understand how catalysts work and their role in reaction kinetics requires knowledge of reaction mechanisms. A **reaction mechanism** is the series of reactions or steps involved in the conversion of reactant to products. When a chemical reaction is written, the equation often represents the sum of a series of reactions. For example, the depletion of ozone in the atmosphere can be represented with the reaction

$$O_3 + O \rightarrow 2O_2$$

According to this reaction, ozone, O_3, combines with atomic oxygen to produce molecular oxygen. One mechanism to explain this reaction is

$$
\begin{array}{ll}
Cl + O_3 \rightarrow ClO + O_2 & 1 \\
+ \quad ClO + O \rightarrow Cl + O_2 & 2 \\
\hline
O_3 + O \rightarrow 2O_2 &
\end{array}
$$

The reaction mechanism shows that while the sum of reactions 1 and 2 results in the original reaction given for ozone depletion, the ozone does not react directly with atomic oxygen but with chlorine. Steps 1 and 2 are called the **elementary steps** in the mechanism. Summing the elementary steps in a reaction mechanism gives the overall or net reaction. Chemical reactions are generally presented as the net reaction, and the elementary steps are typically omitted. It is important to remember that while the net reaction gives the reactants and products, the mechanisms show how the reactants became products. We can think of the reactants and products as the start and destination of a trip. We may start a trip in New York City and end in Los Angeles, but there are numerous paths we could use to make our trip. We might go directly from New York to L.A., but we just as easily could have stopped in St. Louis and Denver on the way. Similarly, reactants may go directly to products, but there may also be intermediate reactions along the way.

The reaction mechanism shown for ozone depletion includes chorine. Chlorine in this reaction acts as a catalyst. A principal source of this chlorine is from the ultraviolet breakdown of CFC (chlorofluorocar-

bons) in the upper atmosphere. A catalyst works by altering the reaction mechanism so that the activation energy of the reaction is lowered. If the reaction mechanism is considered as a path, a catalyst can be viewed as providing a shortcut around the activation energy barrier. As an analogy, consider a group of bicycle riders who need to climb a steep mountain. One path might be a road straight to the top, while another might be a tunnel bored through the mountain. Only the heartiest cyclists may be able to make it over the mountain by going up the road, but many more will be able to get to the other side by using the tunnel. To see how a catalyst works in a chemical reaction, reconsider the decomposition of hydrogen peroxide. As pointed out at the start of this chapter, hydrogen peroxide slowly decomposes into water and oxygen. Iodide serves as a catalyst for this reaction according to the following mechanism:

$$
\begin{array}{l}
H_2O_{2(aq)} + I^-_{(aq)} \rightarrow H_2O_{(l)} + OI^-_{(aq)} \\
+ \quad H_2O_{2(aq)} + OI^-_{(aq)} \rightarrow H_2O_1 + O_{2(q)} + I^-_{(aq)} \\
\hline
2H_2O_{2(aq)} \rightarrow 2H_2O_{(l)} + O_{2(g)}
\end{array}
$$

In the decomposition of hydrogen peroxide, the iodide catalyst reacts in the first elementary step, and it is regenerated in the second step.

In a reaction mechanism with two or more steps, the slowest step will control the rate of the net reaction. This step is referred to as the **rate determining step**. The rate determining step in a reaction mechanism can be compared to the slowest step in a series of activities. For example, say we were mailing out letters and set up an assembly line of several people that included the following tasks: 1) take envelope out of box, 2) place stamp on envelope, 3) put letter in envelope, 4) address envelope, and 5) seal envelope. All the steps except step 4 could be done in a matter of seconds. It might take a minute or two to address the envelope, and this step essentially controls how fast the letters can be prepared. In a similar fashion, the slowest reaction in a series of reactions making up the mechanism controls the overall rate of the reaction.

The importance of catalysts in chemical reactions cannot be overestimated. In the destruction of ozone previously mentioned, chlorine serves as a catalyst. Because of its detrimental effect to the environment, CFCs and other chlorine compounds have been banned internationally. Nearly every industrial chemical process is associated with numerous catalysts. These catalysts make the reactions commercially feasible, and chemists are continually searching for new catalysts. Some examples of important catalysts include iron, potassium oxide, and aluminum oxide in the Haber process to manufacture ammonia; platinum and rhodium in the Ostwald synthesis of nitric

acid; and nickel when vegetable oil is hydrogenated to produce saturated fats such as margarine. Catalytic converters in automobiles rely on platinum, rhodium, and palladium as catalysts. The catalytic converters aid in the complete combustion of emissions from engines by converting carbon monoxide to carbon dioxide, nitric oxides into nitrogen and oxygen, and hydrocarbons into carbon dioxide and water. Because catalytic converters must perform both oxidation and reduction (see Chapter 14 on electrochemistry), catalytic converters generally work in two stages. The hot exhaust first passes through a bed of rhodium catalyst to reduce nitric oxides and then through a platinum catalyst to oxidize CO and hydrocarbons (Figure 12.4).

Enzymes are catalysts that accelerate metabolic processes in organisms. The word "enzyme" comes originally from the Greek term "enzymos" for leavening. Enzymes are

Figure 12.4
Catalytic Converter (Rae Déjur)

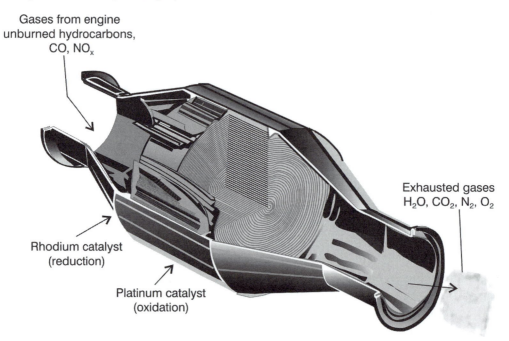

Gases from engine
unburned hydrocarbons,
CO, NO_x

Exhausted gases
H_2O, CO_2, N_2, O_2

Rhodium catalyst
(reduction)

Platinum catalyst
(oxidation)

a critical ingredient in bread, wine, yogurt, and beer. Without enzymes, the biochemical processes would be much too slow. Enzymes have the ability to speed chemical reactions tremendously. It is not unusual for enzymes to increase the reaction rate by a factor of a million. Like any other catalyst, enzymes work by lowering the activation energy. Enzymes, though, are very specific with respect to the compound they interact with and the reactions they catalyze. The compound associated with a particular enzyme is known as the **substrate**. A general model to explain how enzymes work is the **lock-and-key model** (Figure 12.5). In this model, the enzyme, which is typically a large protein molecule, has a "keyhole" or region known as an active site. The active site has the specific shape and chemical characteristics to act on a particular substrate. The substrate enters the active site just as a key would enter a keyhole. The enzyme-substrate forms a complex where the chemical reaction takes place. The enzyme catalyzes the substrate into the product and then the product separates from the enzyme. The human body contains thousands of enzymes involved in all biochemical processes such as digestion, respiration, and reproduction.

Enzymes in humans work best at temperatures of 37°C. When temperatures climb too high, the efficiency of an enzyme can be greatly reduced. This is one example where an increase in temperature can retard the reaction rate. Another problem is that certain substances can disrupt enzymes by blocking active sites and preventing the substrate from bonding with the enzyme. Substances that disrupt enzymes are known as inhibitors. Many poisons and drugs fit in this category.

Chemical Equilibrium

In the discussion of reactions in Chapter 5, all reactions were written as **complete reactions**. Complete reactions are written with a single arrow pointing to the right ($\rightarrow$), indicating reactants are converted into products. For complete reactions, reactants are converted into products until one of the reactants disappears. Many reactions are actually **reversible reactions**. Reversible reactions are written with a double arrow ($\rightleftharpoons$ or $\leftrightarrow$). Reversible reactions actually consist of two reactions called the forward reaction and the reverse reaction. The forward reaction represents the conversion of reactants into products, while the reverse reaction represents the conversion of products back to reactants. The reaction of hydrogen and nitrogen to form ammonia is a reversible reaction:

$$N_{2(g)} + 3H_{2(g)} \leftrightarrow 2NH_{3(g)}$$

If it's assumed that there is no ammonia when the reaction starts, then the reactants initially combine according to the forward reaction to produce ammonia. As the ammonia is

Figure 12.5
Lock-and-Key Model of Enzyme (Rae Déjur)

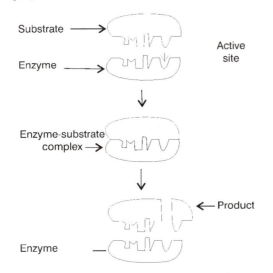

formed, some of this product decomposes back into hydrogen and nitrogen as indicated by the reverse reaction. Eventually, a point is reached where the rate of the formation of ammonia is equal to the rate of decomposition of ammonia. It is also true that both nitrogen and hydrogen are forming and disappearing at a constant rate. At this point, the concentrations of reactants and products do not change and the reaction has reached equilibrium. Figure 12.6 shows how the concentration of nitrogen, hydrogen, and ammonia change with time. According to the reaction, 1 mole of nitrogen reacts with 3 moles of hydrogen to produce 2 moles of ammonia. Therefore, hydrogen disappears three times as fast as nitrogen, while ammonia is formed at twice the rate that nitrogen disappears.

Whenever equilibrium is considered, certain assumptions concerning the reaction are made. One assumption is that the reaction takes place in a closed system, that is, it is assumed reactants are mixed in a closed container. It is also assumed that the reaction takes place under constant temperature and pressure conditions. When conditions under which the reaction takes place change, the equilbium is affected. This topic is addressed shortly, but first the equilbrium state is examined.

It is important to realize that once equilibrium is reached the reaction does not

Figure 12.6
Change in H_2, N_2, NH_3 (Rae Déjur)

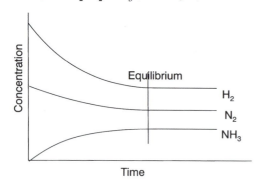

stop. At equilibrium, reactants continue to change into product as products are converted back into reactants. As long as the reaction is at equilibrium, the rates of the forward and reverse reactions are equal and the concentrations remain constant. Equilbrium can be compared to a department store during the day of a big sale. We can consider the store to be empty or contain only a few people before it opens. Similarly, before a reaction starts we can assume there are little or no products. When the store opens, people rush in and start to fill the store; very few, if any, people exit the store at this point. This corresponds to the initial stage of a reaction when the concentration of reactants is at its highest. As the day progresses, people continue to enter, but now at a slower rate than when the store opened. The number of people exiting also increases as the day progresses. Eventually, the rate at which people enter and exit is equal. At this point, the number of people in the store remains constant and we can consider this to be equilibrium. This equilibrium condition exists until closing time approaches. Now the rate of people exiting the store increases and the rate of people entering decreases. Finally, at closing time the store is nearly empty again. Equilibrium was established in the store once the number of people entering equaled the number of people exiting. Equilibrium exists in the store until some condition or event changes the equilibrium. In our example, we can consider the store to be initially at equilibrium with only a few salepeople (product) and lots of people outside the store (reactants). The original condition exists until the store opens, and then a new equilibrium point is established that is maintained throughout most of the day. Equilibrium again changes as the closing hour approaches. Other conditions during the day may cause the equilibrium in the store to shift. For example, if a fire alarm sounded, the store would quickly empty or there may be an increase in shoppers around lunch-time. Chemical reactions are similar to our store with the forward and reverse reactions representing shoppers entering and exiting. Equilibrium must always be considered with respect to the existing conditions. If these conditions change, then the equilibrium shifts until a new equilibrium position is reached.

Le Châtelier's Principle

The equilibrium position of a chemical reaction changes when the conditions under which the reaction takes place change. These conditions include adding or removing reactants or products, changing the temperature, and changing the pressure. A change in any one of these or several of these simultaneously shifts the equilibrium. A change in conditions can favor the forward reaction or reverse reaction. In the former case, more reactants form into products, while the latter causes more products to change to reactants. When the equilibrium favors the forward reaction, the reaction shifts to the "right," and when the reverse reaction is favored, the equilibrium shifts to the "left." The terms "right" and "left" are often used to designate how the equilibrium shifts.

How a chemical system at equilibrium changes when conditions change was first stated by Henri Louis Le Châtelier (1850–1936) in 1884. Le Châtelier was a professor at a mining school in France who worked on both the theoretical and practical aspects of chemistry. His research on the chemistry of cements led him to formulate a principle to predict how changing the pressure affected a chemical system. In the publication *Annals of Mines in 1888*, Le Châtelier stated the principle that bears his name: "Every change of one of the factors

of an equilibrium occasions a rearrangement of the system in such a direction that the factor in question experiences a change in sense opposite to the original change." According to **Le Châtelier's Principle**, changing the conditions of a chemical system at equilibrium places a stress on the system. A stress is just a change in conditions such as temperature or pressure. The equilibrium shifts to the right or left to relieve the stress placed on the system. To see how Le Châtelier's Principle works, consider the reaction used to prepare hydrogen gas:

$$C_{(s)} + H_2O_{(g)} \leftrightarrow CO_{(g)} + H_{2(g)} \quad \Delta H = +131 \text{ kJ}$$

Let's assume this reaction is at equilibrium and use Le Châtelier's Principle to predict what happens when we change the concentration, pressure, and temperature of the system. Le Châtelier's Principle says that when a stress is placed on the system, the equilibrium shifts to relieve the stress. One way to stress the system is by adding or removing reactants or products, that is, changing the concentration of reactants or products. In the hydrogen gas reaction if more steam (H_2O) was added, the reaction would shift to the right producing more hydrogen and carbon monoxide. The reaction shifts to cause more hydrogen to react. This shift counteracts the stress of adding more hydrogen to the system. Likewise, by removing hydrogen and carbon monoxide, the equilibrium shifts to the right, or if more of these products are added, the reaction shifts to the left.

Another way to stress a chemical system is by changing the pressure. What if the pressure of the system increased in the above reaction? How might this stress be relieved? The key to this lies in the stoichiometry of the reaction. The reaction shows 1 mole of solid carbon reacting with 1 mole of steam to yield 1 mole of carbon monoxide and 1 mole of hydrogen. To understand the effect of pressure on equilibrium, remember that pressure is the force per unit area, and this force results from the collisions of gas particles with the container walls. The reaction shows that there is 1 mole of gas on the left side of the equation and two moles of gas on the right side. Therefore, when the pressure is increased, the reaction shifts toward the side with the fewer number of moles of gas. This means the reaction shifts to the left. Each time a carbon monoxide molecule combines with a hydrogen molecule to produce a water molecule and solid carbon, there is a loss of one gas molecule. A decrease in the number of gas molecules means there are fewer collisions, and fewer collisions translates into a pressure decrease. Using the same reasoning, decreasing the pressure on the system when it is at equilibrium shifts the reaction to the right.

The effect of pressure on equilibrium depends on the number of moles of gas particles on the right and left sides of the balanced equation. In reactions where the numbers of moles of gas are equal, a change in pressure has no effect on the equilibrium. Changes in pressure are directly related to changes in volume. This is because a decrease in volume gives the same result as increasing the pressure, while an increase in volume is the same as a decrease in pressure. This situation exists when equilibrium occurs in a cylinder with a movable piston, and the volume is changed by moving the piston.

Pure solids and liquids do not need to be considered when using Le Châtelier's Principle, because the concentrations of pure solids and liquids are constant during the course of the reaction. For example, consider the solid carbon in the hydrogen gas reaction. Figure 12.7 shows two situations with different amounts of carbon. Assume

Figure 12.7
When more solid carbon is added to the system at equilibrium, the equilibrium is not affected. The concentration of carbon remains constant.

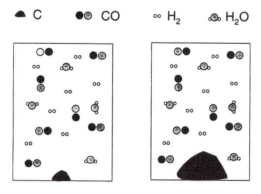

the system is at equilibrium with a small amount, say 12 g, of carbon. The concentration of carbon can be calculated using this information. The molar concentration of a substance is the number of moles of that substance divided by the volume it occupies. Twelve grams of carbon are equivalent to 1 mole of carbon. The volume occupied by the carbon can be found by dividing the mass, 12 g, by its density, 2.6 g/cm^3, and converting to liters. This gives a volume of 0.0046 L. The molar concentration of carbon is found by dividing 1 mole by 0.0046 L, which equals 217 M. If more carbon is added so that the amount of carbon in the system is 120 g, then there are now 10 moles of carbon in the container. The volume of carbon is also ten times greater. Therefore, the molar concentration remains constant at 217 M. Whenever pure solids and liquids are present in the system, they do not need to be considered in the equilibrium. Changing the amount of hydrogen in our reaction shifts the reaction because the concentration of hydrogen changes. Adding or removing carbon has no effect on the equilibrium because the concentration of carbon does not change.

The final stress to be considered is a change in temperature. To apply Le Châtelier's Principle with temperature changes, the sign of ΔH for the reaction needs to be known. The ΔH in our example is $= +131$ **kilojoules**. This indicates that the forward reaction is endothermic and the reverse reaction is exothermic. When the temperature of a system at equilibrium is increased, the equilibrium will favor the endothermic reaction. One way to think of the effect of temperature is to think of energy as a reactant or product. This is seen when the forward and reverse reactions are written as two separate reactions:

Forward Reaction

$$131 \text{ kJ} + C_{(s)} + H_2O_{(g)} \rightarrow CO_{(g)} + H_{2(g)}$$
Endothermic

Reverse Reaction

$$CO_{(g)} + H_{2(g)} \rightarrow C_{(s)} + H_2O_{(g)} + 131 \text{ kJ}$$
Exothermic

In an endothermic reaction, energy can be considered a reactant, and in an exothermic reaction, energy can be considered a product. When the temperature of a system at equilibrium is increased, energy flows into the system. To relieve this stress, energy is used in the endothermic reaction to establish a new equilibrium. The effect of adding energy is no different from the effect of adding another reactant, for example, steam. Decreasing the temperature of the system removes energy from the system; to counteract this stress, the exothermic reaction is favored.

The Equilibrium Constant

Two Norwegians named Cato Maximilian Guldberg (1836–1902) and Peter Waage (1833–1900) carried out a comprehensive study of equilibrium between 1864 and 1879. Their work led to what is known as the **law of mass action**. For a general reversible reaction given by the equation: aA + bB $\leftrightarrow$ cC = dD, the law of mass action states an **equilibrium constant**, K_{eq}, can be defined at a particular temperature:

$$K_{eq} = \frac{[C]^c[D]^d}{[A]^a[B]^b}$$

The small letters are the coefficients in the balanced equation and the brackets indicate the molar concentrations (moles per liter) of the reactants and products. Applying the law of mass action to the hydrogen gas equation gives

$$K_{eq} = \frac{[CO][H_2]}{[H_2O]}$$

Notice that solid carbon does not appear in the reaction. Again, this is because the concentration of solid carbon is constant and is assumed to be part of K_{eq}. The value for K_{eq} is approximately 2 at 1,000°C.

Values for equilibrium constants have been tabulated for numerous reactions occurring at various temperatures. Because many of these constants are associated with industrial processes, the temperature at which a K_{eq} is reported varies considerably. That is, it might seem odd that a K_{eq} value for a reaction is given at what seems like an arbitrary temperature, but it should be remembered that these constants have probably been determined with regard to a specific industrial process, for example, 1,000°C for the production of hydrogen gas from steam and coke (carbon).

Values for equilibrium constants vary widely. For example, consider the following two reactions:

$$N_{2(g)} + O_{2(g)} \leftrightarrow 2NO_{(g)}$$
$$K_{eq} = 1.0 \times 10^{-31} @ 25°C$$

$$H_{2(g)} + Cl_{2(g)} \leftrightarrow 2HCl_{(g)}$$
$$K_{eq} = 1.8 \times 10^{33} @ 25°C$$

The value of an equilibrium constant indicates whether reactants or products dominate at equilibrium. The small K_{eq} value for the first reaction above shows that, for all practical purposes, nitrogen and oxygen do not react to form nitric oxide, NO, at 25°C. Conversely, hydrogen and chloride are quite reactive and react almost completely at 25°C. Remember that the equilibrium constant is reported at a specific temperature. It is known that nitrogen and oxygen do not react at room temperature from the fact that the reaction is not occurring while you sit here and read these words. Nitric oxide, though, is a significant air pollutant that is a by-product of automobile exhaust and power plant combustion. The higher temperatures associated with fuel combustion shifts the equilibrium to the right and increases the K_{eq} value for the formation of nitric oxide. The value of K_{eq} for this reaction is approximately 1.7×10^{-3} at 2,000°C.

Many different types of reversible reactions exist in chemistry, and for each of these an equilibrium constant can be defined. The basic principles of this chapter apply to all equilibrium constants. The different types of equilibrium are generally denoted using an appropriate subscript. The equilibrium constant for general solution reactions is signified as K_{eq} or K_c, where the c indicates equilibrium concentrations are used in the law of mass action. When reactions involve gases, partial pressures are often used instead of concentrations, and the equilibrium constant is reported as K_p (p indicates that the constant is based on partial pressures). K_a and K_b are used for equilibria associated with acids and bases, respectively. The equilibrium of water with the hydrogen and hydroxide ions is expressed as K_w. The equilibrium constant used with the solubility of ionic compounds is K_{sp}. Several of these different K expressions will be used in chapters to come, especially in the next chapter on acids and bases.

Ammonia and the Haber Process

Similar to chemical kinetics, chemical equilibrium plays a critical role in all areas of chemistry. The chemical industry constantly uses equilibrium principles to increase product yield, design efficient chemical processes, and ultimately increase profits. One example that illustrates how equilibrium and Le Châtelier's Principle apply to the chemical industry is in the production of ammonia. Ammonia is widely used in fertilizers, refrigerants, explosives and cleaning agents and to produce other chemicals, such as nitric acid. Approximately 18 million tons of ammonia are manufactured annually in the United States, and each year it ranks as one of the top ten chemicals produced. During ancient times, ammonia was produced by the distillation of animal dung, hooves, and horn. In fact, the name is derived from the ancient Egyptian diety Amun who was known to the Greeks as Ammon. Greeks observed the preparation of ammonia from dung in Egypt from sal ammoniac (ammonium chloride) near the temple of Amun. Sal ammoniac or "salt of Ammon" is an ammonium-bearing mineral (ammonium chloride) of volcanic origin.

The importance of nitrogen fertilizers in agriculture was established during the mid-1800s, and this coupled with the growth of the chemical industry, provided incentive to find a method for fixing nitrogen. Nitrogen fixation is a general term to describe the conversion of atmospheric nitrogen, N_2, into a form that can be used by plants. One method to fix nitrogen mimicked the natural fixation of nitrogen by lightning. The process involved subjecting air to a high voltage electric arc to produce nitric oxide:

$$N_{2(g)} + O_{2(g)} \rightarrow 2NO_{(g)}$$

The nitric oxide was then converted into nitric acid and combined with limestone to produce calcium nitrate. The problem with this process was that the procedure was energy intensive, and it was only economical where there was a steady and cheap supply of electricity.

Another way to fix nitrogen is by the synthesis of ammonia. The conversion of nitrogen and hydrogen into ammonia had been studied since the mid-1800s, but serious work on the subject did not occur until the turn of the century. In 1901, Le Châtelier attempted to produce ammonia by subjecting a mixture of nitrogen and hydrogen to a pressure of 200 atmospheres and a temperature of 600°C in a heavy steel bomb. Contamination with oxygen led to a violent explosion, and Le Châtelier abandoned his attempt to synthesize ammonia. Toward the end of his life, Le Châtelier was quoted as saying "I let the discovery of the ammonia synthesis slip through my hands. It was the greatest blunder of my scientific career."

By 1900, the increasing use of nitrate fertilizers to boost crop production provided ample motivation for finding a solution to the nitrogen fixation problem. During the last half of the nineteenth century and the early twentieth century, the world's nitrate supply came almost exclusively from the Atacama Desert region of northern Chile. This area contained rich deposits of saltpeter, $NaNO_3$, and other minerals. The War of the Pacific (1879–1883) between Chile and Bolivia and its ally Peru was fought largely over control of the nitrate-producing region. Chile's victory in the war (sometimes referred to as the "nitrate war") forced Bolivia to cede the nitrate region to Chile, and as a result, Bolivia not only lost the nitrate-rich lands but also became landlocked.

After Le Châtelier's failed attempt to synthesize ammonia, the problem was tackled by the German Fritz Haber (1868–1934). As early as 1905, Haber's progress on the problem indicated it might be possible to make the process commercially feasible. Haber continued to work throughout the next decade and by 1913, on the eve of World War I, had solved the problem. The implementation of Haber's work into a commercial process was carried out by Karl Bosch (1874–1940) and became known as the **Haber** or **Haber-Bosch process**. Haber's work was timely because Germany acquired a source of fixed nitrogen to produce nitrate for fertilizer and explosives to sustain its war effort. Without this source, a blockade of Chilean nitrate exports would have hampered German's war effort severely. Haber received the Nobel Prize in chemistry in 1918 for his discovery of the process that bears his name. Karl Bosch received the Nobel Prize in 1931 primarily in conjunction with his work on the production of gasoline.

The Haber process for the synthesis illustrates several concepts presented in this chapter. The reaction is represented as follows:

$$N_{2(g)} + 3H_{2(g)} \leftrightarrow 2NH_{3(g)} \quad \Delta H = -92.4\,kJ$$

According to Le Châtelier's Principle, the production of ammonia is favored by a high pressure and a low temperature. The Haber process is typically carried out at pressures between 200 and 400 atmospheres and temperatures of 500°C. While Le Châtelier's Principle makes it clear why a high pressure would be favorable in the Haber process, it is unclear why a high temperature would be desirable because the reaction is exothermic. An increase in temperature shifts an exothermic reaction to the left. Even though the equilibrium shifts to the left at high temperatures, the reaction is too slow at lower temperatures. This demonstrates the trade-off between equilibrium and kinetics in this reaction. It is fruitless to keep the temperature low to produce a higher yield if it takes too long. For instance, it would make more sense to accept a 10% yield if you could get this in 30 minutes as opposed to a 60% yield that took 12 hours. In the commercial production of ammonia, NH_3 is continually removed as it is produced. Removing the product causes more

nitrogen and hydrogen to combine according to Le Chatelier's Principle. Unreacted, nitrogen and hydrogen are separated from the product and reused.

An important aspect of the Haber process is the use of several catalysts in the reaction. A major problem in Haber's attempt to synthesis ammonia was identification of suitable catalysts. Haber discovered that iron oxides worked as the primary catalysts in combination with smaller amounts of oxides of potassium and aluminum.

Acids and Bases

Introduction and History

Two of the most important classes of chemical compounds are acids and bases. A small sampling of acids and bases found around the home demonstrates their importance in daily life. A few of these include fruit juice, aspirin, milk, ammonia, baking soda, vinegar, and soap. Beyond their presence in numerous household items, acids and bases are key ingredients in the chemical process industry. More sulfuric acid is produced than any other chemical in the United States with an annual production of 40 million tons. While the commercial applications of acids and bases illustrate their importance in everyday life, on a more fundamental level each one of us inherited our characteristics and genetic make-up through the acid DNA, deoxyribonucleic acid.

Human use of acids and bases dates back thousands of years. Probably the first acid to be produced in large quantities was acetic acid, $HC_2H_3O_2$. Vinegar is a diluted aqueous solution of acetic acid. This acid is an organic acid that forms when naturally occurring bacteria called *acetobacter aceti* convert alcohol to acetic acid. Ancient Sumerians used wine to produce vinegar for use in medicines and as a preservative. A significant advance in chemistry occurred around the year 1200 when alchemists discovered how to prepare strong mineral acids. These acids include sulfuric, nitric, and hydrochloric acid. Mineral acids were prepared by distilling various vitriols (sulfate compounds) or virtriol mixtures. Distilling green vitriol ($FeSO_4$) and dissolving the vapor in water produced sulfuric acid, formerly called oil of vitriol. The distillation of vitriol and saltpeter (KNO_3) resulted in nitric acid. *Aqua regia* or royal water consists of a mixture of one part nitric acid and three parts hydrochloric acid. The name *aqua regia* denotes the ability of this mixture to dissolve precious metals such as gold. The word "acid" comes from the Latin word "acere," which means sour.

The use of bases or alkalines also dates back thousands of years. Bases were no doubt created as prehistoric humans carried out their daily activities. Bases are a key ingredient of soap; some of the first soap recipes date back to 2800 B.C. from the Babylonian period. The Egyptians combined lime (calcium oxide) and soda ash (sodium carbonate) and evaporated the product to produce caustic soda (NaOH). They used this base to produce cleansers and dyes for materials and in the preparation of papyrus. Bases were used in ancient China to prepare paper. The use of lye (NaOH) to produce soap dates back to at least Roman times, and it is still used today to prepare soap. Alkaline is derived from the Arabic "al qualy," which means to roast. Evidently the term referred to roasting or calcinating plant material and leaching the ash residue to prepare a basic carbonate solution.

Acids were of critical concern to Lavoisier as he gathered evidence disputing the phlogiston theory. In fact, Lavoisier mistakenly thought that oxygen was a common chemical in all acids and adopted the term "oxygen," which means "acid former," for this element. Lavoisier's oxygen theory of acids persisted for two decades. Chemical analyses conducted by Humphrey Davy and several other prominent chemists around 1810 demonstrated that oxygen was not present in many acids, for example, hydrochloric acid (HCl). This led Davy to abandon the oxygen theory of acids and suggest it was hydrogen that characterized acids. Most chemists had abandoned Lavoisier's oxygen acid theory and accepted Davy's hydrogen theory by 1820. In 1838, Justus Liebig (1803–1873) advanced Davy's concept of an acid and defined an acid as a compound that contains hydrogen that can be replaced by a metal. While chemists throughout the 1800s continued to improve their description of an acid, they were unable to furnish an adequate definition of a base. For the most part, a base was considered a substance that neutralized an acid.

It was not until the last decade of the nineteenth century that chemists had an adequate theoretical description of acid and bases. Until then, most acids and bases were classified according to their general properties. Chemists knew acids and bases displayed contrary properties, and these were adequate for identifying many acids and bases. Some of these properties are listed in Table 13.1.

Chemical Definitions of Acids and Bases

A general chemical definition for acids and bases was proposed by Svante Arrhe-

Table 13.1
General Properties of Acids and Bases

Acids	Bases
Turn blue litmus paper red	Turn red litmus paper blue
Dissolve certain metals to form hydrogen gas	Have a slippery or soapy feel
Taste sour	Taste bitter
Strong acids react with strong bases to form a salt and water	Strong bases react with strong acids to form a salt and water
Are electrolytes	Are electrolytes

nius in 1887. Arrhenius defined an acid as a substance that dissociates in water to give hydrogen ions (H^+) and defined a base as a substance that dissociates in water to give hydroxide ions (OH^-). A hydrogen ion is simply a hydrogen atom minus its electron. Because a hydrogen atom consists of a single proton and a single electron, removing an electron leaves just the proton. Therefore, a hydrogen ion is equivalent to a proton, and both can be symbolized as H^+. If the general formula of an acid is represented as HA and a base as BOH, Arrhenius' definitions for an acid and base can be represented by the following general reactions:

$$\text{Acid HA} + H_2O \rightarrow H^+_{(aq)} + A^-_{(aq)}$$

$$\text{Base BOH} + H_2O \rightarrow OH^-_{(aq)} + B^+_{(aq)}$$

When an acid dissociates to produce hydrogen ions in water, the hydrogen ions do not remain as individual ions but are attracted to the polar water molecules represented by the following reaction:

$$H^+ + H_2O \rightarrow H_3O^+$$

The H_3O^+ ion is called the **hydronium** ion. To be technically correct, Arrhenius' definition for an acid should state an acid is a substance that produces hydronium ions in solution. Several water molecules may actually be associated with a single hydrogen ion to produce ions such as $H_5O_2^+$ or $H_7O_3^+$ in acids. The important thing to remember is that hydrogen ions do not exist as individual ions in water, but become hydrated. While you will often see the symbol H^+ in reactions involving acids, the H^+ should be interpreted as a hydronium ion rather than a hydrogen ion. Throughout this chapter hydrogen ions and hydronium ions should be considered synonymous.

The Arrhenius definition implies that acids contain hydrogen ions and bases con-

tain the hydroxide ion. This is illustrated in the chemical formulas for some common acids and bases displayed in Table 13.2.

Table 13.2 illustrates the presence of hydrogen in acids. It is also apparent that bases contain hydroxide ions, but the weak base ammonia seems to be an exception. Ammonia illustrates one of the shortcomings of the Arrhenius definition of acids and bases; specifically, bases do not have to contain the hydroxide ion to produce hydroxide in aqueous solution. When ammonia dissolves in water, the reaction is represented by:

$$NH_{3(g)} + H_2O_{(l)} \leftrightarrow NH_4^+_{(aq)} + OH^-_{(aq)}$$

Table 13.2
Some Common Acids and Bases

Acid or Base	Formula
Sulfuric	H_2SO_4
Hydrochloric	HCl
Nitric	HNO_3
Acetic	$HC_2H_3O_2$
Citric	$HC_6H_7O_7$
Ascorbic	$HC_6H_7O_6$
Acetylsalicylic	$HC_9H_7O_4$
Carbonic	H_2CO_3
Sodium Hydroxide	NaOH
Potassium Hydroxide	KOH
Calcium Hydroxide	$Ca(OH)_2$
Ammonia	NH_3

This reaction shows that the hydroxide ions come from ammonia pulling a hydrogen away from water resulting in the formation of OH^-. Therefore, a compound does not have to contain hydroxide to be a base. In addition to this limitation, the Arrhenius definition limits acids and bases to aqueous solutions.

Recognizing these limitations, the Danish chemist Johannes Brønsted (1879–1947) and the English chemist Thomas Lowry (1874–1936) independently proposed a new definition for acids and bases in 1923. Brønsted and Lowry noted that in certain reactions substances acted like acids even though hydrogen ions were not present. Brønsted and Lowry viewed acids and bases not as isolated substances but as substances that interacted together as a pair. According to the definition, an acid is a proton donor and a base is a proton acceptor. An analogy helps to explain the idea behind the Brønsted-Lowry theory. Acids and bases can be compared to throwing a football. A quarterback throws a football to a receiver. In this example, the quarterback represents the acid, the football the proton, and the receiver the base. Applying the Brønsted-Lowry definition to nitric acid, it is seen that HNO_3 donates a proton to water, Water in this case acts as a base.

$$\underset{Acid}{HNO_{3(aq)}} + \underset{Base}{H_2O_{(l)}} \rightarrow H_3O^+_{(aq)} + NO_3^-$$
$$\underset{\hspace{1.2cm}H^+}{\rule{2.5cm}{0.4pt}}$$

When ammonia dissolves in water, water acts as an acid when it donates a proton to the base ammonia:

$$\underset{Base}{NH_{3(g)}} + \underset{Acid}{H_2O_{(l)}} \rightarrow NH_4^+{}_{(aq)} + OH^-_{(aq)}$$
$$\underset{\hspace{1.2cm}H^+}{\rule{2.5cm}{0.4pt}}$$

Just as a quarterback and receiver act as a pair, Brønsted-Lowry acid and bases act as pairs (Figure 13.1). Acids and bases are

Figure 13.1
Football Analogy (Rae Déjur)

defined in terms of reactions and the interaction of two substances, rather than as individual substances. In the two examples it is seen that water, which is not commonly considered an acid or base, is an acid when it reacts with ammonia and a base when it reacts with nitric acid.

To complete our discussion of the Brønsted-Lowry theory, we see that when a proton is transferred, two new species are formed. In the nitric acid reaction, water is transformed into the hydronium ion and nitric acid into the nitrate ion, NO_3^-. The hydronium and nitrate ions formed are themselves an acid and base, respectfully. Again, using our quarterback and receiver example, once the receiver has the football, he or she can act as an acid and pass it on to someone else. Likewise, the quarterback can act as a base by receiving the football. The $HNO_3–NO_3^-$ and $H_2O–H_3O^+$ are referred to as **conjugate acid-base pairs.** Notice that the only difference between the substances making up a conjugate acid-base pair is the presence of a proton. In the ammonia reaction, the conjugate acid base pairs are NH_3(base)–NH_4^+ (acid) and H_2O (acid)–OH^- (base).

In the same year that Brønsted and Lowry proposed their definition for acids and bases, the American G. N. Lewis proposed an alternative definition based on the valence electron structure of substances. Brønsted-Lowry bases must contain at least one unshared electron pair to accept a proton. We can see this if we look at the Lewis dot structure of several bases:

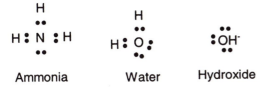

| Ammonia | Water | Hydroxide |

Lewis defined a base as an electron pair donor and an acid as an electron pair acceptor. Lewis' electron pair donor was the same as Brønsted-Lowry's proton acceptor, and therefore, was an equivalent way of defining a base. Lewis' acids were defined as a substance with an empty valence shell that could accommodate a pair of electrons. This definition broadened the Brønsted-Lowry definition of an acid. The three definitions of acids and bases are summarized in Table 13.3.

Each of the three definitions expands our concept of acids and bases. Arrhenius' basic definition is adequate for understanding many of the properties of acids and bases. It is important to recognize, though, that acids and bases are not fixed labels that can be applied to a substance. Brønsted-Lowry and Lewis showed that acid-base characteristics are dependent on the reactions that take place between substances. A

Table 13.3
Definitions of Acids and Bases

Arrhenius Acid	gives H$^+$ in aqueous solution
Arrhenius Base	gives OH$^-$ in aqueous solution
Brønsted-Lowry Acid	proton donor
Brønsted-Lowry Base	proton acceptor
Lewis Acid	electron pair acceptor
Lewis Base	electron pair donor

substance can act as an acid or base depending on the conditions. When a substance has the ability to act as an acid or a base, it is called **amphoteric.** We have already seen that water is an amphoteric substance. Furthermore, substances not typically classified as acids or bases, such as water, act as acids and bases in many instances. Throughout the remainder of this chapter, the Arrhenius' definition will be used. This simple definition, though it has limitations, is useful for understanding most of the basic concepts of acids and bases.

Strengths of Acids and Bases

Many of us start our day by consuming acid in the form of fruit juice and washing ourselves with a base-like soap. We are also aware that contact with some acids and bases causes severe burns. Our personal experiences with different acids and bases point out the wide differences in the strengths of these compounds. The strength of acids and bases is a function of their ability to ionize. Strong acids and bases can be considered 100% ionized or completely dissociated. Weak acids and bases only partly ionize in water. We can apply our concepts of equilibrium from Chapter 12 to differentiate between strong and weak acids or bases. For example, nitric acid is a strong acid so HNO_3 dissociates completely into H^+ and NO_3^-. When we represent the dissociation of HNO_3 in aqueous solution with the equation $HNO_{3(aq)} + H_2O_{(l)} \rightarrow H_3O^+_{(aq)} + NO_3^-_{(aq)}$, the equation is written with a single arrow pointing to the right to indicate 100% ionization (complete dissociation). On the other hand, a weak acid such as acetic acid dissociates very little. Only about 0.4% of a 1 M acetic acid solution dissociates into ions, that is, most of the acetic acid remains in the combined form of $HC_2H_3O_2$

in solution. Similarly, in an aqueous solution of the weak base NH_3 the concentrations of NH_4^+ and OH^- ions are less than 1% of that of the concentration of NH_3. Equations representing the dissociation of weak acids or bases in solution are written with a double arrow to indicate that equilibrium exists between the acid or base and its ions. The equilibrium condition for weak acids or bases is such that the equilibrium favors the combined states of acids or bases rather than their dissociated ion state:

weak acid or weak base $\leftrightarrow$ partially dissociated ions

strong acid or base $\rightarrow$ total dissociation

A quantitative measure of the degree of dissociation is given by the equilibrium constant for the acid or base. The higher the equilibrium constant is, the greater the percent dissociation of the acid or base. Therefore, a higher equilibrium constant means a stronger acid or base. Equilibrium constants, K_a and K_b, are listed for several common weak acids and bases in Table 13.4.

Table 13.4
K_a and K_b for Weak Acids and Bases at 25°C

Acid or Base	Constant
Ascorbic acid	$K_a = 8.0 \times 10^{-5}$
Acetic	$K_a = 1.8 \times 10^{-5}$
Acetylsalicylic	$K_a = 3.0 \times 10^{-4}$
Carbonic	$K_a = 4.4 \times 10^{-7}$
Caffeine	$K_b = 4.1 \times 10^{-4}$
Ammonia	$K_b = 1.8 \times 10^{-5}$

A final point concerning our discussion of acid/base strengths deals with the strength of conjugate acid-base pairs. When an acid donates a proton, what remains is a base, and when a base accepts a proton, the product is an acid. Whenever we consider Brønsted-Lowry acid-base pairs, a general rule is that a strong acid or base produces a corresponding weak base or acid. This means that strong acids produce weak bases, weak bases produce strong acids, and so on. To illustrate this point, let's consider a strong acid (HCl), a weak acid ($HC_2H_3O_2$), and a weak base (NH_3). Equations for their solution in water are

$$HCl_{(aq)} + H_2O_{(l)} \rightarrow H_3O^-_{(aq)} + Cl^-$$
$$HC_2H_3O_{2(aq)} + H_2O_{(l)} \leftrightarrow H_3O^+_{(aq)} + C_2H_3O_2^-$$
$$NH_{3(aq)} + H_2O_{(l)} \leftrightarrow NH_4^+_{(aq)} + OH^-_{(aq)}$$

In the first reaction, HCl donates a proton to water and the **conjugate base** of HCl is Cl^-. Because HCl is a strong acid (indicated by the one-way arrow), Cl^- is a weak base. In the next reaction, acetic acid donates a proton, creating the acetate ion, $C_2H_3O_2^-$. The acetate ion is the conjugate base of acetic acid and is a strong base, because it is formed from a weak acid. Finally, in the ammonia reaction, the base NH_3 accepts a proton to become the ammonium ion, NH_4^+. Ammonium is the **conjugate acid** of ammonia. The ammonium ion is a strong acid, because it is produced from the weak base ammonia. There is an inverse relationship between the strength of an acid or base and the conjugate formed when a proton is transferred.

Acid Concentration

It is important not to confuse acid/base strength, which depends on the degree of ionization, with concentration. Concentration measures how much acid or base is in solution, and it is typically measured in moles per liter or molarity. For instance, acetic may have a high concentration, but because it is a weak acid most of it will exist in solution as $HC_2H_3O_2$. Thus, a 10 M acetic acid solution is highly concentrated, but only a small fraction of the acid ionizes to produce hydrogen ions; the concentration of hydrogen ions is perhaps only 0.1 moles per liter. Conversely, a 10 M solution of hydrochloric acid (HCl) ionizes 100% to produce 10 moles each of hydrogen and chloride ions per liter.

Another concentration unit often associated with acids and bases is **normality.** Normality is similar to molarity, but rather than a measure of the number of moles per liter it measures equivalents per liter (N). For acids and bases, an equivalent can be considered the mass that produces 1 mole of hydrogen or hydroxide ions, respectively. To understand normality, we have to consider acids and bases that yield more than 1 mole of H^+ or OH^- when they dissociate. Sulfuric acid (H_2SO_4), for example, will yield 2 moles of hydrogen ions for every mole of sulfuric acid that dissociates, and each mole of calcium hydroxide, $Ca(OH)_2$, gives 2 moles of hydroxides when it dissociates. An equivalent is found by dividing the molar mass by the number of moles of hydrogen or hydroxide ions that the molecule produces. For H_2SO_4, an equivalent would be equal to 49 grams because the molar mass of sulfuric acid is 98 grams and each mole of sulfuric acid produces 2 moles of hydrogen ions in solution. Thus, a 1 normal solution of H_2SO_4 contains 49 grams of H_2SO_4 in 1 liter of solution. It can easily be shown that a 1 M H_2SO_4 solution is the same as a 2 N H_2SO_4 solution:

1 M H_2SO_4 contains 1 mole or 98 g H_2SO_4 per liter

98 grams is the same as 2 equivalents of H_2SO_4

Table 13.5
Stock Solutions of Acids and Bases

Acid or Base	Concentration	
	Molarity	Normality
HCl	12	12
HNO_3	16	16
H_2SO_4	18	36
$HC_2H_3O_2$	17	17
NH_4OH	15	15

so there are 2 equivalents of H_2SO_4 per liter which = 2 N

Many acids such as HCl and NaOH yield a mole of ions per every mole of acid or base. In these cases, the normality and molarity are equal.

Commercial acids and bases are sold in standard high concentrations and diluted to the desired concentration. The standard concentrated forms are referred to as stock solutions. Concentrations of stock solution are given in Table 13.5.

pH and pOH

The basic criterion for distinguishing acids and bases in aqueous solutions is the concentration of hydrogen and hydroxide ions. When an acid dissolves in water, the water acts as a base and accepts protons to form the hydronium ion. Water acts as an acid to produce hydroxide ions when it donates protons to a base dissolved in water. While we can considered water to be amphoteric when it interacts with other acids and bases, we can also consider pure water to be both an acid and base. The reason for this is that water itself actual dissociates to a very small extent to produce hydrogen and hydroxide ions:

$$H_2O_{(l)} \leftrightarrow H^+_{(aq)} + OH^-_{(aq)}$$

The above reaction depicts water as an Arrhenius acid and base. Treating water in terms of the Brønsted-Lowry theory, a more appropriate reaction would be

$$H_2O_{(l)} + H_2O_{(l)} \leftrightarrow H_3O^+_{(aq)} + OH^-_{(aq)}$$

At 25°C the concentrations of hydronium and hydroxide ions in water are both equal to 1.0×10^{-7} M. To put this concentration into perspective, consider that about two out of every billion water molecules dissociates into ions. The equilibrium constant for the above reaction is known as the **ion product constant** of water, symbolized by K_w, and is equal to the product of the H^+ and OH^- molar concentrations:

$$K_w = [H^+][OH^-] = (1.0 \times 10^{-7})(1.0 \times 10^{-7}) = 1.0 \times 10^{-14}$$

The equation for K_w applies to both pure water and aqueous solutions.

Because the ion concentrations are small and the negative exponents make them tedious to work with, Soren Peer Lauritz Sorenson (1868–1939), a Danish biochemist, devised the pH concept in 1909 to express the hydrogen ion concentration. The abbreviation pH comes from the French *pouvoir hydrogène* meaning power of hydrogen. The pH of a solution is given by the equation:

$$pH = -\log[H^+]$$

Remember that brackets are used to signify the molar concentration of a substance. This equation states that pH is equal to the negative log of the hydrogen ion molar concentration, but it should be remembered it is really the hydronium ion concentration in solution that is being measured. In a similar fashion, the pOH of a solution can be defined as:

$$pOH = -\log[OH^-]$$

Using these definitions, the pH and pOH of a neutral solution at 25°C are both equal to 7. We can see from the expression for K_w that $[H^+]$ and $[OH^-]$ are inversely related, and consequentially pH and pOH are inversely related. We can picture pH and pOH as sitting on opposite sides of a seesaw, as one goes up, the other always goes down. The product of the hydrogen and hydroxide concentrations will be equal to 1.0×10^{-14}, while the sum of the pH and pOH will be equal to 14. In an acidic solution, the hydrogen ion concentration increases above 1.0×10^{-7}, the hydroxide concentration decreases, and the pH value gets smaller. The relationship between the type of solution, pH, pOH, and ion concentrations is shown in Table 13.6. The pHs of a number of common substances are presented in Table 13.7.

Because the pH and pOH scales are based on logarithms, a change in 1 pH or pOH unit represents a change in ion concentration of a factor of ten. Coffee with a pH of 5 has approximately 100 times the hydronium ion concentration as tap water with a pH of 7.

The pH of a solution can be measured using several methods. One popular method

Table 13.6
Solution, pH, pOH, and Ion Concentrations

Solution	pH	pOH	[H⁺]	[OH⁻]
Neutral	7	7	1.0×10^{-7}	1.0×10^{-7}
Acid	<7	>7	$>1.0 \times 10^{-7}$	$<1.0 \times 10^{-7}$
Base	>7	<7	$<1.0 \times 10^{-7}$	$>1.0 \times 10^{-7}$

Table 13.7
Approximate pH Values of Common Substances

battery acid	0.5	coffee	5.0	blood	7.4
stomach acid	1.5	rain	5.6	seawater	8.3
lemon juice	2.3	urine	6.0	baking soda	8.3
vinegar	2.9	milk	6.6	toothpaste	10
cola	3.0	saliva	6.8	milk of magnesia	10.5
apple	3.5	tap water	7.0	ammonia cleaner	12

is to use an **indicator.** An indicator is a substance that "indicates" the condition of another substance to which the indicator has been added. Indicators used to determine the pH of a substance are known as acid-base indicators. Many common acid-base indicators come from plant pigments. Litmus consists of blue and red dyes extracted from lichens originally found in the Netherlands. While litmus paper only gives a general measure of pH by identifying a substance as an acid or a base, there are hundreds of other indicators that can be used to more accurately measure pH. Simple indicators can be obtained by boiling plants such as beets, cabbage, and spinach. Several indicators are often mixed to produce a universal indicator. Universal indicators show various colors and shades across a wide pH range and can measure pH in the range from 2–12. Numerous synthetic indicators have also been produced. One of the most common is phenolphthalein, which is colorless in acidic solutions and turns pink in basic conditions.

The way indicators work can be understood if we consider indicators to be weak acids. If we let HIn represent the general formula of an indicator, then we can write the following equilibrium expression:

$$HIn_{(aq)} \leftrightarrow H^+_{(aq)} + In^-_{(aq)}$$

color   different color

Indicators have different colors in the combined and dissociated forms. The equilibrium of the indicator in solution shifts according to Le Châtelier's principle as an acid or base is added to the solution. The shift in equilibrium to the right or left causes the color to change accordingly.

Accurate and precise measures of pH can be made with a pH meter. Typical pH meters usually contain a glass electrode and reference electrode arranged similar to an electrochemical cell. We discuss electrochemical cells in Chapter 14. For now, though, consider a pH meter as essentially a modified voltmeter in which the voltage measured is directly proportional to the hydrogen ion concentration of the solution. Simple pH meters are capable of measuring pH within $\pm$ 0.1 pH units, while more sophisticated instruments are precise to within 0.001 pH units.

Neutralization Reactions

When acidic and basic solutions containing hydroxide are mixed, a **neutralization** reaction occurs in which the acid and base lose their characteristic properties. Acid and base solutions combine during neutralization to form a salt and water. As an example, consider the reaction between hydrochloric acid and sodium hydroxide solutions:

$$HCl_{(aq)} + NaOH_{(aq)} \rightarrow H_2O_{(l)} + NaCl_{(aq)}$$

For complete neutralization to take place, the proper amounts of acid and base must be present. The salt formed in the above reaction is NaCl. If the water were evaporated after completing the reaction, we would be left with common table salt. Sodium chloride is just one of hundreds of salts that form during neutralization reactions. While we commonly think of salt, NaCl, as a seasoning for food, in chemistry a salt is any ionic compound containing a metal cation and a nonmetal anion (excluding hydroxide and oxygen). Some examples of salts that result from neutralization reactions include potassium chloride (KCl), calcium fluoride (CaF_2), ammonium nitrate (NH_4NO_3), and sodium acetate ($NaC_2H_3O_2$).

Neutralization reactions are very important in both nature and industry. Because pH has a profound effect on many chemical processes, neutralization can be used to control the pH. A common application of neutralization reactions is in the use of antacids. As noted in Table 13.7, the pH of stomach acid is approximately 1.5. The hydrochloric acid present in gastric juices helps us digest food and activates specific enzymes in the digestive process. Our stomach and digestive tract are protected from the hydrochloric acid secreted during digestion by a protective mucous lining. Cells lining our digestive system constantly regenerate this protective layer. Stomach ulcers develop when the protective mucous lining is weakened or excess stomach acid is generated due to health problems. Antacids are a common method used to neutralize excess stom-

ach acid. An antacid is nothing more than a base or salt that produces a basic solution when ingested. Some popular antacids and their associated bases are shown in Table 13.8.

The reactions in Table 13.8 show that carbon dioxide is a common product in many neutralization reactions. This is clearly displayed when a drop of vinegar (acetic acid) is added to baking soda (sodium bicarbonate). Some aspirin includes an antacid in their formulation to neutralize some of the acidity imparted by the aspirin (acetylsalicylic acid). These are commonly referred to as buffered aspirins.

A detrimental form of neutralization occurs when **acid precipitation** reacts with building materials to accelerate weathering. Natural precipitation (rainfall not affected by pollution) is acidic with a pH of about 6. The acidic pH is due to the presence of CO_2 in the atmosphere and the production of carbonic acid when carbon dioxide dissolves in water. Acid precipitation or acid rain is generally classified as rain in which the pH is lower than its natural level due to human effects. In extreme cases, acid rain may have a pH as low as 2 or 3. The primary cause of acid rain is the burning of fossil fuels. The combustion of sulfur-bearing coals in power plants results in the production of sulfur dioxide, SO_2. This sulfur dioxide is then converted into sulfurous acid (H_2SO_3) and sulfuric acid (H_2SO_4) in the atmosphere. Natural gas combustion and automobiles produce nitrogen oxides. These can be converted into nitrous and nitric acids in the atmosphere. Reactions summarizing the formation of acids in rain are displayed in Figure 13.2. Acid rain causes millions of dollars of damage each year to buildings and structures made of marble and limestone. In areas impacted by acid precipitation, such as the northeast United States and northern Europe. Limestone and marble structures suffer significant damage due to

Table 13.8
Several Popular Antacids

Product	Active Ingredient	Active Ingredient Formula	Neutralizing Reaction
Alka-Seltzer	Sodium bicarbonate	$NaHCO_3$	$H_3O+ HCO_3^- + \rightarrow$ $CO_2 + 2H_2O$
Tums	Calcium carbonate	$CaCO_3$	$2H_3O+ CO_3^{2-} \rightarrow$ $CO_2 + 3H_2O$
Milk of magnesia	Magnesium hydroxide	$Mg(OH)_2$	$2H_3O^+ + 2OH^- \rightarrow 4H_2O$
Rolaids	Calcium carbonate Magnesium hydroxide	$CaCO_3 + Mg(OH)_2$	$4H_3O^+ + 2OH^- + CO_3^{2-} \rightarrow 7H_2O$ $+ CO_2$

Figure 13.2
Acid rain occurs when reactions in the atmosphere lead to the formation of various acids lowering the pH below the natural pH of rain, which is approximately 5.5. The acidic pH of natural rain water is due to the formation of carbonic acid by the reaction:
$$CO_{2(g)} + H_2O \rightarrow H_2CO_3.$$

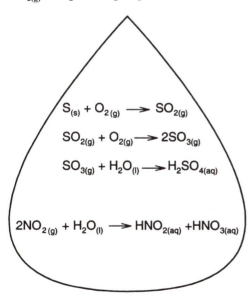

its effects. Limestone and marble are calcium carbonate; a typical reaction representing the destruction of limestone by acid rain would be

$$CaCO_{3(s)} + H_2SO_{4(aq)} \rightarrow Ca^{2+}_{(aq)} + SO_4^{2-}_{(aq)} + H_2O_{(l)} + CO_{2(g)}$$

In this reaction, calcium carbonate reacts with sulfuric acid resulting in the dissolution of calcium carbonate.

In addition to the damage acid rain causes to structures, acid rain also affects natural environments. Significant loss of spruce forests due to the burning of spruce needles by acid rain has occurred in Scandinavia. Acid rain also extracts a heavy toll on aquatic systems and associated organisms. Most adult fish cannot tolerate pHs much lower than 5.0, and even the most tolerant species will not survive below a pH of 4.0. Fish larvae are even more susceptible to low pH levels. Insects and their larvae also perish when pH approaches 4.0 in aquatic systems. Numerous lakes in upstate New

York and New England are devoid of fish due to the effects of acid rain. Indirect effects of the low pH values associated with acid rain also affect organisms. As noted in Table 13.1, one of the properties of an acid is the ability to dissolve certain metals. This has a profound effect on soil subjected to acid rain. Acid rain can mobilize metal ions such as aluminum, iron, and manganese in the basin surrounding a lake. This not only depletes the soil of these cations disrupting nutrient uptake in plants, but also introduces toxic metals into the aquatic system.

The detrimental effects of acid rain are a major reason why legislation such as the Clean Air Act places strict limitations on sulfur and nitrogen emissions. It is also a reason why low sulfur coal is preferred over high sulfur coal. To reduce sulfur dioxide emissions, industry also uses a technique call **scrubbing.** Industrial scrubbers employ a variety of physical and chemical processes to remove sulfur dioxide from emissions. Another technique used to combat acidification of lakes is to treat these systems with lime. The lime acts to neutralize the acid, but such techniques are usually costly and are only a temporary remedy for combating the problem.

Liming an acidic lake is similar to the process many people use to maintain a pH balance in their soil for lawn maintenance. Plants have an optimum pH range in which they strive. Acidic conditions often develop in soils for several reasons. Rain tends to leach away basic ions, weak organic acids develop from the carbon dioxide produced by decaying organic matter, and strong acids, such as nitric acid, can form when ammonium fertilizers oxidize. To neutralize these acids, different forms of lime such as quicklime, CaO, and slaked lime, $Ca(OH)_2$, are used to neutralize the acid and increase the pH of the soil. Table 13.9 shows how much fertilizer is wasted when applied to

Table 13.9
Percentage of Fertilizer Wasted

pH	Nitrogen	Phosphate	Potash
4.5	70	77	67
5.0	47	66	48
5.5	23	52	23
6.0	11	48	0
7.0	0	0	0

lawns with different soil pH values. Much of a fertilizer is wasted unless the soil's pH is maintained at close to a neutral pH 7 value.

Buffers

A buffer is a solution that resists changes in pH when an acid or base is added to the solution. Buffers are important for limiting the change in pH within specific limits. Many commercial and natural products contain buffers for pH control, for example, shampoo, medicines, and blood. The most common buffers consist of a weak acid and its conjugate base. The acid part of the buffer neutralizes excess base added to the solution while the conjugate base neutralizes acid added to the solution. Buffers may also consist of a weak base and its conjugate acid. Buffers are prepared by combining a weak acid (or base) and a salt containing its conjugate base.

To illustrate how a buffer works, consider the simple buffer consisting of acetic acid and the acetate ion: $HC_2H_3O_2/C_2H_3O_2^-$. This buffer is prepared by combining acetic acid and sodium acetate. In solution the two components give the following reactions:

$$HC_2H_3O_{2(aq)} + H_2O_{(l)} \leftrightarrow H_3O^+_{(aq)} + C_2H_3O_2^-_{(aq)}$$

$$NaC_2H_3O_{2(s)} \rightarrow Na^+_{(aq)} + C_2H_3O_2^-_{(aq)}$$

Acetic acid partially dissociates to give hydronium and acetate ions, and sodium acetate dissociates completely into sodium and acetate ion. Acid added to this solution is neutralized by the acetate ion, while any base added is neutralized by the hydronium ion. When these neutralization reactions take place, the equilibrium in the first reaction shifts to replenish the reacted hydronium or acetate ions according to Le Châtelier's Principle. Therefore, the pH remains constant. The ability of buffers to resist changes in pH is limited. Buffers do actually experience very slight changes in pH when small amounts of acid or base are added to the buffer. Eventually, when a certain amount of acid or base is added, a buffer is no longer able to control the pH within certain limits. Buffers are generally prepared to control pH over a limited range. This range is dictated by the concentrations of acid and salt used in preparing the buffer. The amount of acid or base that a buffer can absorb before its pH moves out of its buffer range is known as the buffering capacity.

One of the most important buffer systems in the human body is that which keeps the pH of blood around 7.4. If the pH of blood fall below 6.8 or above 7.8, critical problems and even death can occur. There are three primary buffer systems at work in controlling the pH of blood: carbonate, phosphate, and proteins. The primary buffer system in the blood involves carbonic acid, H_2CO_3 and its conjugate base bicarbonate, HCO_3^-. Carbonic acid is a weak acid that dissociates according to the following reaction:

$$H_2CO_3 + H_2O \leftrightarrow H_3O^+ + HCO_3^-$$

The equilibrium in this reaction lies to the right and the amount of bicarbonate is approximately ten times the amount of carbonic acid. The bicarbonate ion neutralizes excess acid and the carbonic acid neutralizes excess base in the blood:

$$H_3O^+ + HCO_3^- \leftrightarrow H_2O + H_2CO_3$$

$$OH^- + H_2CO_3 \leftrightarrow H_2O + HCO_3^-$$

The excessive amount of bicarbonate in the blood means that blood has a much greater capacity to neutralize acids. Many acids accumulate in the blood during strenuous activity, for example lactic acid. Excretion of bicarbonate through the kidneys and the removal of carbon dioxide through respiration also regulate the carbonic acid/bicarbonate blood buffer.

Many aspirins are advertised as buffered aspirin, and this has led to the misunderstanding that these aspirin are buffers. In reality, as mentioned previously, buffered aspirins actually contain an antacid to reduce the problems brought on by the effects of aspirin, which is the acid called acetylsalicylic acid. Because of its widespread use and importance to humans, let's take an extended look at the history and chemistry of aspirin.

Aspirin is the most widely used medication. Over 10,000 tons of aspirin are used in this country annually, and worldwide the annual consumption is 35,000 tons. The history of acetylsalicylic acid actually goes back thousands of years. Hippocrates (460–377 B.C.) and the ancient Greeks used powdered willow bark and leaves to reduce fever (**antipyretic**) and as a pain reliever (**analgesic**). Native American populations also used willow and oil of wintergreen for medication. The chemicals responsible for the medicinal properties in willow and oil of wintergreen are forms of salicylates. Willows (genus *Salix*) contain salicin and oil of wintergreen contains methyl salicylate.

While the use of willow bark and oil of wintergreen as an accepted antipyretic and analgesic has been around for at least 2,000 years, by the nineteenth century medicines were starting to be synthesized in chemical laboratories. In 1837, Charles Frederic Gerhardt (1819–1856) prepared salicylic acid from salinin, but abandoned work in the area because of difficulties he encountered in trying to synthesize this compound. The German Adolph Wilhelm Hermann Kolbe (1818–1884) synthesized salicylic acid in 1853. Salicylic acid was used for pain relief, fever, and to treat rheumatism, but it caused gastrointestinal problems and had an unpleasant taste so its use was limited. In 1860, salicylic acid was identified as the agent in plants that resulted in pain relief. Chemists sought to improve medicines by synthesizing different compounds containing salicylic acid. Sodium salicylate was first used around 1875, and phenyl salicylate, known as salol, appeared in 1886, but both these produced the undesirable gastrointestinal side effects.

The discovery of aspirin or acetylsalicylic acid as a pain reliever is credited to Felix Hoffman (1868–1946) who was looking for a substitute for sodium salicylate to treat his father's arthritis. Hoffman uncovered and continued the work of Gerhardt from forty years before. Hoffman reacted salicylic acid with acetic acid to produce acetylsalicylic acid in 1897 Figure 13.3.

Hoffman's acetylsalicylic acid competed with the other salicylate compounds after its initial synthesis. In 1899 Hoffman's employer, the Bayer Company, founded in 1861 by Friedrich Bayer (1825–1880), began to market acetylsalicylic acid under the name Aspirin. The term "aspirin" was derived by combining the letter "a" from acetyl and "spirin" from spiric acid. Spiric acid is another name for salicylic acid found in plants of the genus *Spirea*. Bayer pro-

Figure 13.3
Acetylsalicylic Acid

$$C_6H_4(OH)COOH + HC_2H_3O_2 \rightarrow C_9H_8O_4 + H_2O$$

Salicylic acid + Acetic acid
$\rightarrow$ Acetylsalicylic acid + Water

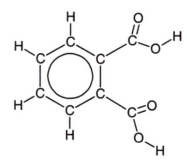

moted aspirin widely at the turn of the century and rapidly established a large market for its use. Bayer also acquired a trademark to the name Aspirin, but relinquished these rights to England, the United States, and France as part of the Treaty of Versailles ending World War I. After Bayer's patent on aspirin expired, other companies started to produce aspirin.

Aspirin's original use as an analgesic, antipyretic, and to reduce inflammation continues to this day, and more recently some evidence has been found that it may lessen the chance of heart attacks due to its effect as a blood "thinner." Just as aspirin continues to provide the same benefits as a century ago, it also produces some of the same problems. The major problem is that it can upset the stomach. In the acidic environment of the stomach, aspirin can diffuse through the protective mucous lining of the stomach and rupture cells and produce bleeding. Under normal doses, the amount of blood loss in most individuals is only a milliliter or two, but in some individuals who take heavy doses bleeding can be severe. To counterattack this side effect, manufacturers include an antacid such as aluminum hydroxide and call the aspirin a buffered aspirin. As noted previously, this term is misleading because the antacid does not buffer the solution, but neutralizes some of the acidic effects of the aspirin. Another type of aspirin called enteric aspirin dissolves in the intestines where the environment is more basic.

Summary

We have seen in this chapter that acids and bases describe two broad classes of compounds that have significant impacts on our lives. Our bodies contain scores of these compounds. We are exposed to hundreds of acids and bases daily. It was only at the start of the twentieth century that an adequate definition for acids and bases was proposed by several individuals. The most common acids and bases are those that occur in aqueous solutions. A solution's pH defines whether a solution is an acid or base and is the one of the most common measurements made in chemistry. As we refer to acids and bases in the pages ahead, we will have a much better understanding of these important compounds.

Electrochemistry

Introduction

Whenever you start a car, use a battery-powered device, apply a rust inhibitor to a piece of metal, or use bleach to whiten your clothes, you deal with some aspect of electrochemistry. Electrochemistry is that branch of science that involves the interaction of electrical energy and chemistry. Many of our daily activities use some form of electrochemistry. Just imagine how your life would be in a world without batteries. What immediately comes to mind is the loss of power for our portable electronic devices. While this would certainly be an inconvenience, consider the more critical needs of those with battery-powered wheelchairs, hearing aids, or heart pacemakers. In this chapter, we examine the basic principles of electrochemistry and some of their applications in our lives.

History

The phenomenon called electricity has been observed since the earliest days of humans. Lightning struck fear in our prehistoric ancestors and led to supernatural explanations for the displays accepted today as a natural atmospheric occurrence. The ancient Greeks observed that when a tree resin they called *elektron* hardened into amber, it possessed a special attractive property. When amber was rubbed with fur or hair, it attracted certain small objects. The Greek Thales observed this attractive property and noted that this property also existed in certain rocks. Thales' attractive rock was lodestone (magnetite), which is an iron-bearing mineral with natural magnetic properties. The modern study of electricity commenced with the publication of William Gilbert's (1544–1603) *De Magnete* in 1600. Gilbert coined the word "electricity" from the Greek "elektron" and proposed that the Earth acted as a giant magnet.

During the seventeenth and eighteenth centuries, several individuals made significant contributions to the science of electricity. Otto von Guericke (1602–1686) created one of the first **electrostatic** generators by building a sphere of sulfur that could be rotated to accumulate and hold an electric charge. Numerous devices were built to generate static electricity through friction, and other instruments in the form of plates, jars, and probes were constructed to hold and transfer charge (Figure 14.1). One of the

Figure 14.1
Numerous devices were used to generate and hold charge: a) Guericke's generator, b) Leyden jar, and c) Wimhurst generator. (Rae Déjur)

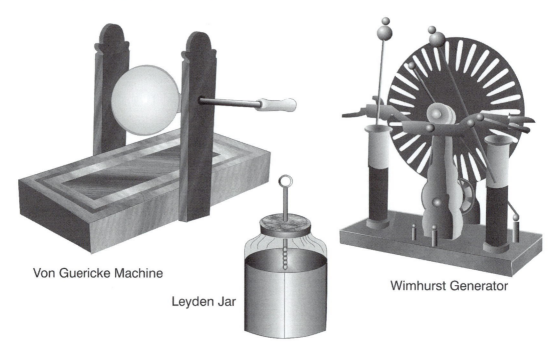

Von Guericke Machine

Leyden Jar

Wimhurst Generator

most popular of these devices was the Leyden jar, which was perfected at the University of Leyden by the Dutch physicist Pieter van Musschenbroek (1692–1761). The original Leyden jar was a globular-shaped glass container filled with water fitted with an insulated stopper. A nail or wire extended through the stopper into the water. The Leyden jar was charged by contacting the end of the protruding nail with an electrostatic generator. Leyden jars are still used in science laboratories.

Charles Du Fay (1698–1739) observed that charged objects both attracted and repelled objects and explained this by positing that electricity consisted of two different kinds of fluids. Du Fay called the two fluids vitreous and resinous electricity and said each of these two different fluids attracted each other, but each repelled itself. Benjamin Franklin (1706–1790) also con-

sidered electricity a fluid, but he considered electricity a single fluid. Franklin considered the fluid either present, positively charged, or absent, negatively charged, and said that the electrical fluid flowed from positive to negative. The practice of defining electricity as flowing between positive and negative poles can be attributed to Franklin.

The birth of electrochemistry paralleled the birth of modern chemistry. Both occurred at the end of the eighteenth century, and both grew out of conflicting theories supported by eminent scientists. The battle over phlogiston between Priestley and Lavoisier contributed largely to the development of modern chemistry. During the same period that Lavosier and Priestley were advancing their ideas, the two prominent Italian scientists Luigi Galvani (1737–1798) and Alessandro Volta (1745–1827) held opposite views on the subject of animal electricity. Galvani, physician and professor of anatomy at the University of Bologna, devoted most of his life to the study of the physiology of muscle and nerve response to electrical stimuli. Galvani's work was a natural consequence of his profession and the period in which he lived. Electrical shock was a popular therapeutic treatment in the late eighteenth century. Patients were subjected to static electrical shocks as a means to treat all sorts of ailments including gout, rheumatism, infertility, toothaches, and mental disorders (Figure 14.2). The area for which this "shock therapy" seemed to hold the most promise was in treating paralysis. The hope of reversing paralysis grew out of experiments in which dissected body parts of animals responded to electrical discharge through them. Within this framework, Galvani and other researchers sought a theoretical explanation for the contraction of muscles when nerves were electrically stimulated.

Galvani conducted hundreds of experiments starting in the late 1770s. These experiments consisted of using dissected body parts and subjecting them to charge held on a device such as a Leyden jar

Figure 14.2
Galvani's Study of Animal Electricity (Courtesy of the Bakken Library, Minneapolis)

(Figure 14.1). Researchers would use various conductors and configurations to transfer the electricity to the tissue and note the response of the limb. A favorite subject was using frog legs and the attached spinal cord. Over the years, Galvani developed his theory of animal electricity. He explained the response of limbs to static electricity by proposing that animals contained a special electrical fluid within their muscles or nerves. According to Galvani, limbs moved when the animal electricity flowed between areas of positive and negative accumulation.

Galvani published his theory on animal electricity in 1792. Alessandro Volta, a physics professor at the University of Pavia, was a respected colleague of Galvani who initially supported Galvani's theory, but began to raise questions on the work soon after its publication. In addition to questioning Galvani's animal electricity theory, Volta commenced his own series of experiments on the subject. Volta believed muscle contraction was not due to a flow of nervoelectrical fluid, but was a result of an electrical flow between dissimilar metals used in the experiment. Volta's experiments demonstrated that the stimulation of nerves caused the contractions. A key part of Volta's experiments involved placing dissimilar metals along small segments of nerves, providing a connection between the metals, and observing that the entire limb responded when the connection was made. Volta's theory became known as the contact theory for animal electricity.

Galvani's death in 1798 left his theory to be defended by his proponents, but his theory was ultimately proven incorrect. Modern science, though, has shown that Galvani was correct in believing that animal electricity could be generated from within the organism. Today it's known that electrical impulses are generated by chemical processes associated with neurons and

Figure 14.3
The Arrangement of Disks in Volta's Pile.

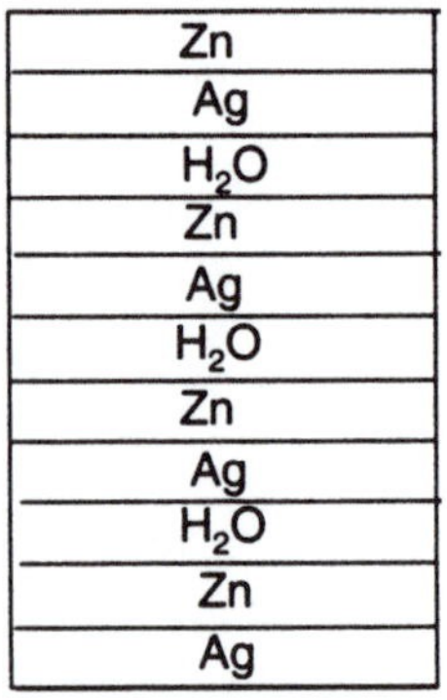

| Zn |
| Ag |
| H₂O |
| Zn |
| Ag |
| H₂O |
| Zn |
| Ag |
| H₂O |
| Zn |
| Ag |

chemicals known as neurotransmitters. Nerves can also be stimulated by external stimuli. In summary, although both Galvani's and Volta's ideas had flaws, they provided fundamental knowledge on how muscles and nerves work.

The most important result of the Galvani-Volta controversy was a device that Volta created to study how dissimilar metals in contact with each other generated electricity. Volta's pile, which was actually the first battery created, consisted of stacks of metal disks about the size of a quarter (Figure 14.3). Two different metals were used to make the disks such as silver and zinc, and the disks were arranged so that one metal rested on another forming a pair. Between each pair of metals Volta inserted a paper or felt disk that had been soaked in water or brine solution (Figure 14.3). Volta's original pile contained about 20 Ag-Zn pairs. Besides the pile configuration, Volta also used an arrangement called the crown of cups. The crown of cups consisted of nonconducting cups filled with a water or brine solution. The cups were connected by bimetallic conductors (Figure 14.4).

If a date were to be placed on the birth of electrochemistry, that date would be March 20, 1800. This is the date of a letter Volta sent to Sir Joseph Banks, president of

Figure 14.4
Volta's Crown of Cups (top) and Pile

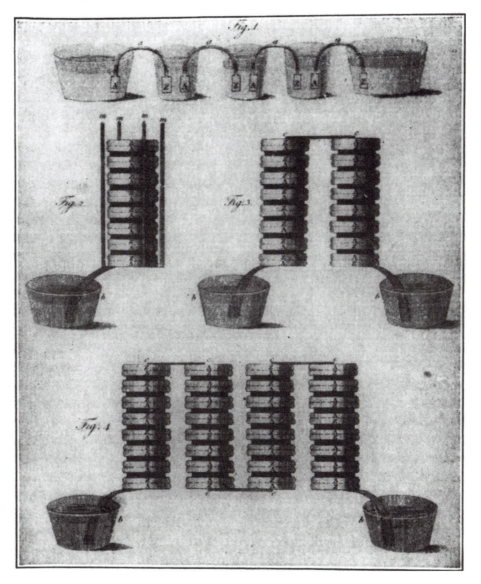

the Royal Society of London. Included in the letter was a description of Volta's pile and his findings on how an electrical current could be generated by connecting metals. Volta observed that he received a noticeable shock when he placed two wet fingers on opposite ends of the pile. The same was true for his crown of cups, where he noted the shock progressively increased as the number of cups between his fingers increased.

Even before its official publication, Volta's findings received considerable attention. Here was a device that could provide a steady, although small, current of electricity. Until this time, scientists used the process of generating static electricity and transferring this charge in pulses. Volta's results were published in English (the original letter was written in French) in *The Philosophical Magazine*, September 1800. Other scientists immediately constructed their own versions of Volta's pile to study this new type of electricity. One of the first experiments using a pile was done by William Nicholson (1753–1815) and Anthony Carlisle (1768–1840). They created a pile and used it to **electrolyze** water into hydrogen and oxygen.

The most important figure that emerged in the new science of electrochemistry was Humphrey Davy. Volta had attributed the current generated in his pile to the direct contact between metals. His work focused on the physical relationship between the metals, and for the most part neglected the chemical aspects of the electrical current. Davy believed a chemical reaction was the basis of the current produced in the piles or electrochemical cells. Davy discovered that no current was generated in the cells if pure water was used, and better results were obtained when an acid solution was used to separate metals. Davy constructed a variety of cells. He used cups of agate and gold and experimented with various metals and solutions to construct cells that were more powerful. Using a huge battery constructed by connecting over two hundred cells, Davy applied current to various salts and his work led to the discovery of a number of new elements. Potassium was discovered when Davy ran current through potash in 1807 and several days later he isolated sodium from soda in a similar fashion. The next year he discovered magnesium, strontium, barium, and calcium. Davy determined oxygen was not a product in the **electrolysis** of hydrochloric acid leading him to refute Lavoisier's oxygen theory of acids. Because of his work on HCl, Davy is sometimes credited with the discovery of chlorine that had been isolated a generation earlier by Scheele.

The work of Davy was continued and expanded upon by the great English scientist Michael Faraday (1791–1867). Faraday's primary studies in electrochemistry took place between 1833 and 1836. Faraday is responsible for giving us much of our modern electrochemical terminology. The terms "electrode," "anode," "cathode," "electrolyte," "anion," "cation," and "electrolysis" are all attributed to Faraday. Even more important than his qualitative description of electrochemistry, Faraday did quantitative studies that led to his formulation of electrochemical laws. These laws provided a means to determine the relationship between current and the amount of materials reacting in an electrochemical reaction. Because Faraday's major contributions are still used today, they are covered in the principles of electrochemistry later in the chapter rather than in this historical section.

Oxidation and Reduction

Early pioneers in the field of electrochemistry had no knowledge of electrons or the structure of the atom. In fact, early studies involving electrochemistry were contemporaneous with Dalton's proposed atomic theory. It was not until Thomson's discovery of the electron at the end of the nineteenth century that a true understanding of electrochemical processes could begin. Chemical reactions involve the breaking and formation of bonds. These bonds consist of electron pairs that rearrange as atoms separate and recombine during a chemical reaction. Many reactions involve the transfer of electrons from one substance to another. The transfer of electrons is responsible for ionic bonding. As a brief review, consider the simple reaction of magnesium and oxygen to form magnesium oxide:

$$2Mg + O_2 \rightarrow 2MgO$$

In this reaction, each of the two magnesium atoms donates two electrons to the two oxygen atoms making up the oxygen molecule. In the process, each magnesium atom becomes Mg^{2+} and each oxygen atom becomes O^{2-}

$$Mg \rightarrow Mg^{2+} + 2e^-   O + 2e^- \rightarrow O^{2-}$$

Because the balanced equation involves two magnesium and oxygen atoms, the previous equations are more appropriately written as

$$2Mg \rightarrow 2Mg^{2+} + 4e^-   O_2 + 4e^- \rightarrow 2O^{2-}$$

The two magnesium ions with a positive charge are attracted to the two negatively charged oxygen atoms to form the ionic compound magnesium oxide, MgO.

The reaction of magnesium and oxygen is an example of an **oxidation reaction**. The combination of an element with oxygen was the traditional way to define an oxidation reaction. This definition of oxidation has been broadened by chemists to include reactions that do not involve oxygen. Our modern definition for **oxidation** is that oxidation takes place when a substance loses electrons. Anytime oxidation takes place and a substance loses one or more electrons, another substance must gain the electron(s). When a substance gains one or more electrons, the process is known as **reduction**. Reactions that involve the transfer of one or more electrons always involve both oxidation and reduction. These reactions are known as oxidation-reduction or **redox reactions**.

Redox reactions always consist of one oxidation reaction and one reduction reaction. The separate oxidation and reduction reactions are known as **half reactions**. The sum of the two half reactions gives the overall reaction. When the half reactions are summed, there is an equal number of electrons on each side of the equation. In our example, the sum of the two half reactions is

$$2Mg \rightarrow 2Mg^{2+} + 4e^-   \text{oxidation half reaction}$$
$$\underline{+ O_2 + 4e^- \rightarrow 2O^{2-}   \text{reduction half reaction}}$$
$$2Mg + O_2 + 4e^- \rightarrow 2Mg^{2+} + 2O_2^- + 4e^-$$

When writing the overall reaction, it is customary to cancel the number of electrons on both side of the chemical equation.

In the formation of magnesium oxide, magnesium undergoes oxidation and oxygen undergoes reduction. Another way of saying this is that magnesium was oxidized and oxygen was reduced. Because oxygen accepted electrons causing magnesium to be oxidized, oxygen was the **oxidizing agent** in the reaction. In a similar manner, magnesium donated electrons and caused the oxygen to be reduced so it was the **reducing agent**. The terms associated with redox reactions are summarized in Table 14.1.

Oxidation Numbers

In the oxidation of magnesium, each magnesium atom loses two electrons and acquires a charge of +2, and each oxygen atom accepts these two electrons and acquires a charge of –2. The charge or apparent charge an atom has or acquires is called its **oxidation number**. The oxidation number of magnesium in MgO is +2, and the oxidation number of O is –2. The concept of oxidation number is used in chemistry as a form of electron accounting. In the magnesium oxide reaction, electrons are transferred from magnesium to oxygen. The magnesium and oxygen start in their uncombined elemental forms so their oxidation numbers are originally both zero. As noted already after Mg and O combine, their oxidation numbers become +2 and –2, respectively. Likewise, when sodium combines with chlorine to give NaCl, the oxidation number of Na changes from 0 to +1

Table 14.1
Terms Associated with Redox Reactions

oxidation	loss of electrons
reduction	gain of electrons
oxidize	to lose electrons
reduce	to gain electrons
reducing agent	substance that causes reduction = oxidized substance
oxidizing agent	substance that causes oxidation = reduced substance

and chlorine changes from 0 to –1. Assigning oxidation numbers to atoms taking part in a redox reaction gives us another means to identify the oxidizing and reducing agents. The reducing agent's (substance being oxidized) oxidation number will increase, and the oxidizing agent's oxidation number will decrease.

The definition for oxidation numbers states that it is the charge or apparent charge of an atom. Apparent charge means the charge of an atom if it is assumed that the electrons are not shared but are assigned to the more electronegative element in a molecule. By applying the concept of apparent charge, oxidation numbers can be assigned to atoms in covalently bonded molecules even though the atoms' electrons are shared. In essence, the oxidation number concept assumes that the bonds are ionic rather than covalent in substances. For example, in the water molecule each hydrogen atom shares its single electron with oxygen to form H_2O. To assign oxidation numbers to hydrogen and oxygen in water, it is assumed that oxygen, because it is the more electronegative element, possesses both electrons. It's as though H_2O acted like an ionic compound and each hydrogen atom donates its electron to oxygen. When this happens, oxygen acquires a –2 charge and each hydro-

gen atom obtains a charge of +1. It must be stressed that when oxidation numbers are assigned it is assumed that electrons are not shared. This false assumption is made for the sole purpose of being able to assign oxidation numbers to elements in compounds.

The determination of oxidation numbers can be a bit tricky, but fortunately there are some general guidelines to help us. These general guidelines are:

1. The oxidation number of an atom in its elementary uncombined state is 0.

2. The oxidation of a monatomic ion is equal to the charge of the ion.

3. The oxidation number of hydrogen is usually +1.

4. The oxidation number of oxygen is usually –2.

5. The oxidation number of alkali metals is always +1.

6. The oxidation number of alkali earth metals is always +2.

7. The sum of the oxidation numbers of all atoms in a neutral compound is equal to zero. In a polyatomic ion, the sum of the oxidation numbers of all atoms is equal to the charge of the ion.

It can now be seen how these rules apply to the previous examples. For MgO, Mg has an oxidation number of $+2$ (rule 6) and O has an oxidation of -2 (rule 3). In water, hydrogen has an oxidation of $+1$ (rule 1) and oxygen has an oxidation of -2 (rule 3). These rules can be used for a few more substances to illustrate how to assign oxidation numbers:

For ammonia, NH_3, assign $+1$ to hydrogen (rule 3). According to rule 7, the sum of the oxidation numbers of all atoms must be equal to 0; therefore, the oxidation number of N must be -3. Remember oxidation numbers are charges. To maintain neutral charge, the 3 hydrogen atoms each with a $+1$ charge must be balanced by a -3 charge for nitrogen.

In sulfuric acid, H_2SO_4, H is assigned $+1$ (rule 3), O is assigned -2 (rule 4). Because the molecule is neutral, the two hydrogens and four oxygens means that S must have a charge of $+6$ because $(2 \times +1) + (1 \times +6) + (4 \times -2) = 0$.

For bicarbonate, HCO_3^-, H is $+1$ (rule 3), O is -2 (rule 4), and C is $+4$. Notice in this example the sum is equal to -1, because the bicarbonate ion has a charge of -1.

Oxidation numbers are used when writing redox equations to account for the transfer of electrons. For example, consider the simple reaction representing the rusting of iron:

$$\overset{0}{4Fe} + \overset{0}{3O_2} \rightarrow \overset{+3\ -2}{2Fe_2O_3}$$

The oxidation numbers are written above the elements. It is seen that iron's oxidation number changes from 0 to $+3$ indicating that each iron has lost three electrons. Oxygen's oxidation goes from 0 to -2 indicating each oxygen atom gained two electrons. Because there are a total of 4 iron atoms in the reaction, 12 electrons have been transferred from iron to the 6 oxygen atoms, each oxygen atom acquiring 2 electrons resulting in a neutral compound of iron oxide.

Electrochemical Cells

The concept of oxidation has been expanded from a simple combination with oxygen to a process in which electrons are transferred. Oxidation cannot take place without reduction, and oxidation numbers can be used to summarize the transfer of electrons in redox reactions. These basic concepts can be applied to the principles of electrochemical cells, electrolysis, and applications of electrochemistry.

Placing a strip of zinc in a beaker of copper sulfate solution, $CuSO_{4(aq)}$ results in several changes taking place. The zinc immediately darkens, and as time passed, the blue color of the $CuSO_4$ solution lightens. After a period, clumps of a reddish brown precipitate are seen clinging to the zinc strip and falling and accumulating on the bottom of the beaker. At this point, the blue color may have disappeared altogether, producing a clear solution. The changes observed result from a spontaneous chemical reaction involving the oxidation of zinc and the reduction of copper. The basic concepts of oxidation and reduction can be used to explain our observations. The copper in solution exists as Cu^{2+} ions. These ions are reduced to solid copper as the zinc metal is oxidized to Zn^{2+} ions. The solid copper appears as a reddish-brown precipitate. The characteristic blue color of the copper solution fades as the copper ions leave solution. This simple experiment demonstrates a simple redox reaction involving zinc and copper:

$$Zn_{(s)} + CuSO_{4(aq)} \rightarrow Cu_{(s)} + ZnSO_{4(aq)}$$

In the above reaction, the exchange of electrons occurs directly between the copper and zinc on the surface of the metal. The transfer of electrons in redox reactions

Figure 14.5
Daniell Cell (Rae Déjur)

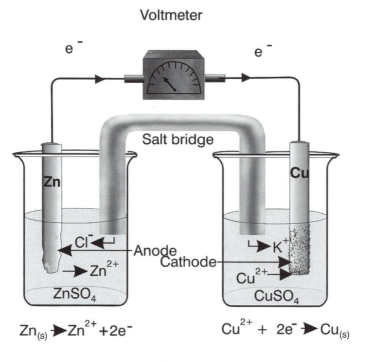

can often be harnessed to do useful work. An **electrochemical cell** is an arrangement in which a redox reaction is used to generate electricity. Electrochemical cells are also known as voltaic or galvanic cells in honor of Volta and Galvani. All batteries are forms of electrochemical cells. The main purpose of an electrochemical cell is to convert chemical energy from a spontaneous chemical reaction into electrical energy.

Figure 14.5 shows the basic arrangement of a electrochemical cell called the Daniell cell. This cell is named for John Frederick Daniell (1790–1845) who constructed this type of cell in 1836. The Daniell cell components include zinc and copper solutions in separate containers. Between the solutions is a **salt bridge**

containing a solution of a strong electrolyte such as KCl. In each respective solution, there is a metal strip corresponding to the cation of the solution. The metal strips are referred to as electrodes. The two metal strips are connected by a wire. A voltmeter can be placed in the circuit to measure the electrical potential of the cell.

In each compartment of the electrochemical cell a half reaction occurs. The two half reactions result in an overall reaction that generates a flow of electrons or current. In one cell compartment, zinc is oxidized according to the reaction: $Zn_{(s)} \rightarrow Zn^{2+} + 2e^-$. The reduction of copper takes place in the other cell's compartment: $Cu^{2+}_{(aq)} + 2e^- \rightarrow Cu_{(s)}$. Notice that these reactions are the same ones that take place when a zinc strip is placed in copper sulfate solution. In the cell arrangement, the electrons lost by zinc when it oxidizes flow through the wire and react with the copper ions on the surface of the copper electrode. If the masses of the copper and zinc electrodes were measured over time, the mass of the zinc strip would decrease because zinc is being oxidized. The mass of the copper electrode would increase as solid copper plates on to it. The salt bridge completes the circuit and maintains a balance of charge in each compartment. As zinc is oxidized creating positive zinc ions in one compartment, the negative ion (Cl^-) in the salt bridge flows into this compartment. As copper is reduced in the other compartment, the positive ion (K^+) diffuses into the solution to replace this loss of positive charge. Therefore, as the reaction continues both solutions remain neutral as ions from the salt bridge diffuse into both compartments.

The Daniell cell illustrates the basic features of an electrochemical cell. Electrochemical cells always involve a redox reaction. Oxidation occurs at the cathode of the cell and reduction takes place at the anode. Electrons always flow from the anode to the cathode. Electrochemical cells come in many arrangements. To gain an appreciation for the variety of electrochemical cells, consider all the types of batteries available.

In the previous example, zinc was more easily oxidized than copper. The ability of one element to donate electrons to another element is based on the electron structure of the element and energy considerations. A spontaneous redox reaction, like any spontaneous reaction, results in a more stable configuration of the chemical system. The following list of elements shows how easily an element is oxidized.

lithium (Li)	▲ more easily oxidized
potassium (K)	
barium (Ba)	
calcium (Ca)	
sodium (Na)	
magnesium (Mg)	
aluminum (Al)	
zinc (Zn)	
chromium (Cr)	
iron (Fe)	
cadmium (Cd)	
nickel (Ni)	
tin (Sn)	
copper (Cu)	
mercury (Hg)	
silver (Ag)	
plantium (Pt)	
gold (Au)	▼ less easily oxidized

A term used to describe how easily a metal is oxidized is **active**. A more active metal is one that is more easily oxidized. A listing of metals in order of activity is known as an **activity series**. The activity series is used to determine which substances will be oxidized and reduced in an electrochemical cell: the element higher on the list will be oxidized. For example, in a cell with aluminum and silver electrodes in their appropriate solutions, aluminum is oxidized and silver is reduced. Therefore, aluminum is the anode and silver is the cathode. If you have ever bitten a piece of aluminum foil and experienced discomfort, you had this electrochemical process occur in your mouth. Silver (or mercury) fillings and the aluminum serve as electrodes and your saliva serves as an electrolyte between the two. The resulting current stimulates the nerves in your mouth resulting in the discomfort.

The Daniell cell generates an electrical current, and hence a voltage is created when the cell operates. Electrical current can be thought of as the flow of electrons or charge in a circuit, but the concept of voltage is a little less obvious. Voltage measures the **electrical potential** or **electromotive force** of a charge. What does this actually mean? Potential energy is the ability of a system to do work by virtue of its position or configuration. When referring to gravitational potential energy, sea level is used as a reference, hills and mountains have positive gravitational potential energy. The higher the elevation of a unit mass (1 kg), the more gravitational potential energy it has. In a similar fashion, electrical potential can be considered as the potential energy of a unit charge in electrical space. This unit charge can be at different levels. When the charge is on an electrical hill or mountain, its electrical potential is higher than when in a valley. Just as the ability of a unit mass to do work is governed by its difference in elevation, the difference in electrical potential dictates the ability of a unit charge to do work. Electrical potential is measured using the units volts.

A voltmeter connected to the circuit of the cell shown in Figure 14.5 would read 1.10 V. This measurement assumes that the concentrations of the solutions are both 1 M and the temperature is 25°C. Under these conditions (and for gases at 1 atmosphere), the measured voltage is referred to as the standard potential of the cell, symbolized E^0. Different electrochemical cells obviously give different E^0 measurements.

To determine the E^0 of different cell arrangements, chemists use what are called **standard reduction potentials** for half-cells. A standard reduction potential is the electrical potential under standard conditions of a cell compared to the **standard hydrogen electrode**. The standard hydrogen electrode is a special half-cell that has been chosen as a reference to measure electrical potential. Just as sea level is a logical elevation for measuring gravitational potential,

Figure 14.6
Standard Hydrogen Electrode (SHE)

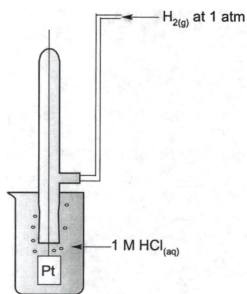

an electrical "sea level" must be chosen for measuring cell potentials. The standard hydrogen electrode is shown in Figure 14.6. Hydrogen gas at 1 atmosphere pressure and 25°C is bubbled through a 1 M solution of hydrochloric acid. At the platinum electrode, the following reduction reaction is defined to have a standard reduction potential of 0:

$$2H^+_{(aq)} + 2e^- \rightarrow H_{2(g)} \quad E^0 = 0 \text{ V}$$

Remember that the standard hydrogen electrode is arbitrarily defined as having a voltage of 0; it cannot be measured. The gravitational potential energy does not have to be defined as 0 at sea level, but using this measurement as a reference allows the gravitational potential energy at any other elevation to be determined. Similarly, the reduction potential of half-cell can be measured with respect to the standard hydrogen electrode.

Chemists have chosen to use standard reduction potentials as the basis of calculat-

Figure 14.7
Two half-cells in the Daniell cell are each connected to the standard hydrogen half-cell (Rae Déjur)

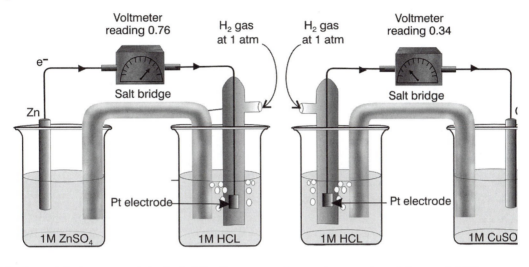

ing E^0. The oxidation potential of a half-cell is obtained simply by changing the sign of the reduction half-cell. This might sound a bit confusing so let's demonstrate how the procedure would work to obtain the 1.10 V of the Daniell cell. Each of the two half-cells are connected to the standard hydrogen electrode as shown in Figure 14.7.

The two reduction reactions and the measured voltages are

$$Cu^{2+}_{(aq)} + 2e^- \rightarrow Cu_{(s)} \quad E^0 = 0.34 \text{ V}$$
$$Zn^{2+}_{(aq)} + 2e^- \rightarrow Zn_{(s)} \quad E^0 = -0.76 \text{ V}$$

There is nothing wrong with having a negative potential. Again, referring to the comparison to gravity, there are places on earth that are below sea level and have a negative gravitational potential. In the Daniell cell, zinc is oxidized. To obtain the oxidation potential of zinc, the reduction reaction must be reversed and the sign of E^0 changed:

$$Zn_{(s)} \rightarrow Zn^{2+} + 2e^- \quad E^0 = +0.76$$

Adding the two half-cell reactions to obtain the overall reaction and the cell E^0 gives

$$Zn_{(s)} \rightarrow Zn^{2+} + 2e \quad E^0 = +0.76V$$
$$+ \ Cu^{2+} + 2e^- \rightarrow Cu_{(s)} \quad E^0 = 0.34 \text{ V}$$
$$\overline{Zn_{(s)} + Cu^{2+}_{(aq)} \rightarrow Cu_{(s)} + Zn^{2+}_{(aq)} \ E^0 = + \ 1.10V}$$

Chemistry handbooks and many texts list the standard reduction potentials for numerous half reactions. A small sampling of these is provided in Table 14.2.

The reduction potential increases moving up the table. This means that substances near the top of the table are more likely to be reduced (are better oxidizing agents) and substances near the bottom are more likely to be oxidized (are better reducing agents). The substances in Table 14.2 correspond to the previous listing of how easily substances are oxidized. When two substances in Table 14.2 take part in a redox reaction, the one higher in the table is the substance that is reduced.

Thinking about the position of elements in the periodic table and their valence electron structure should help in understanding the relative order of the reduction potentials. Fluorine gas is very electronegative and readily accepts electrons to obtain a stable configuration. Conversely, alkalines and

Table 14.2
Standard Reduction Potentials

Half Reaction	E°	
$F_{2(g)} + 2e^- \rightarrow 2F^-_{(aq)}$	2.87 V	more easily reduced
$Cl_{2(g)} + 2e^- \rightarrow 2Cl^-_{(aq)}$	1.36 V	less easily oxidized
$Ag^+_{(aq)} + e^- \rightarrow Ag_{(s)}$	0.80 V	
$Fe^{3+}_{(aq)} + e^- \rightarrow Fe^{2+}_{(aq)}$	0.77 V	
$Cu^{2+}_{(aq)} + 2e^- \rightarrow Cu_{(s)}$	0.34 V	
$Sn^{4+}_{(aq)} + 2e^- \rightarrow Sn^{2+}_{(aq)}$	0.15 V	
$2H^+_{(aq)} + 2e^- \rightarrow H_{2(g)}$	0 V	
$Pb^{2+}_{(aq)} + 2e^- \rightarrow Pb_{(s)}$	-0.13 V	
$Ni^{2+}_{(aq)} + 2e^- \rightarrow Ni_{(s)}$	-0.26 V	
$Fe^{2+}_{(aq)} + 2e^- \rightarrow Fe_{(s)}$	-0.44 V	
$Zn^{2+}_{(aq)} + 2e^- \rightarrow Zn_{(s)}$	-0.76 V	
$Al^{3+}_{(aq)} + 3e^- \rightarrow Al_{(s)}$	-1.66 V	
$Mg^{2+}_{(aq)} + 2e^- \rightarrow Mg_{(s)}$	-2.37 V	
$Na^+_{(aq)} + e^- \rightarrow Na_{(s)}$	-2.71 V	more easily oxidized
$Li^+_{(aq)} + e^- \rightarrow Li_{(s)}$	-3.04 V	less easily reduced

alkali earth metals such as sodium and magnesium readily donate electrons to obtain a stable octet electron configuration; therefore, their reduction potentials are low or equivalently their oxidation potentials are high.

One more example demonstrates how to use standard reduction potentials to determine the standard potential of a cell. Let's say you wanted to construct a cell using silver and zinc. This cell resembles the Daniell cell of the previous example except that a silver electrode is substituted for the copper electrode and a silver nitrate solution is used in place of copper sulfate. From Table 14.2, it is determined that when silver and copper interact silver is reduced and copper oxidized. The two relevant reactions are

$$Ag^+_{(aq)} + e^- \rightarrow Ag_{(s)} \quad E^0 = 0.80V$$
$$Zn^{2+} + 2e^- \rightarrow Zn_{(s)} \quad E^0 = -0.76 \text{ V}$$

Because zinc is oxidized, the zinc reaction must be reversed and the sign of its E^0 changed to give

$$Ag^+_{(aq)} + e^- \rightarrow Ag_{(s)} \quad E^0 = 0.80V$$
$$Zn_{(s)} \rightarrow Zn^{2+} + 2e^- \quad E^0 = +0.76 \text{ V}$$

The number of electrons lost by zinc and gained by silver is not equal. To balance the transfer of charge, the silver half reaction must be multiplied by 2:

$$2Ag^+_{(aq)} + 2e^- \rightarrow 2Ag_{(s)} \quad E^0 = 0.80V$$
$$Zn_{(s)} \rightarrow Zn^{2+} + 2e^- \quad E^0 = +0.76 \text{ V}$$

Notice that when the silver reaction is multiplied by 2 its E^0 value stays the same. The potential of a half-cell does not depend on the coefficients of the equation because the potential of the cell is independent of the quantities of reactants. Adding the two half reactions gives the overall reaction and the cell voltage:

$$Zn_{(s)} + 2Ag^+ \rightarrow Zn^{2+} + 2Ag_{(s)} \quad E^0 = 1.56 \text{ V}$$

Electrolysis

Electrochemical cells produce electrical energy from a spontaneous chemical reaction. In electrolysis, the process is reversed so that electrical energy is used to carry out a nonspontaneous chemical change. A cell arranged to do this is called an electrolytic cell. An electrolytic cell is similar to an electrochemical cell except that an electrolytic cell's circuit includes a power source, for example, a battery. The same electrochemical cell terminology applies to electrolytic cells. Reduction occurs at the cathode and oxidation at the anode.

In an electrolytic process, redox reactions that occur spontaneously in electrochemical cells can be reversed. One of the most common electrolytic procedures demonstrating this is when a battery is charged. During its normal operation, a rechargeable battery functions as an electro-chemical cell. When charging it, electrical energy reverses the process. This concept can be applied to the previous example using silver and zinc. If instead of reducing silver and oxidizing zinc, we want to oxidize silver and reduce zinc, we attach a power supply to the circuit and use a voltage greater than 1.56 V to reverse the current. Electrons now flow from the silver electrode to the zinc electrode where Zn^{2+} is reduced. Oxidation occurs at the solid silver electrode.

On several occasions the electrolysis of water has been mentioned. This was one of the first investigations conducted with Volta's pile. The electrolysis of water can be observed by placing a small 9 V battery in a glass of water and sprinkling in a little salt (to create an electrolyte to conduct current). Very soon a steady stream of bubbles will appear emerging from the positive and neg-ative terminals. The standard half reactions representing the electrolysis of water are

anode:

$$2H_2O_{(l)} \rightarrow O_{2(g)} + 4H^+_{(aq)} + 4e^- \qquad E^0 = -1.23\,V$$

cathode:

$$+\,4H_2O_{(l)} + 4e^- \rightarrow 2H_{2(g)} + 4OH^-_{(aq)} \quad E^0 = -0.83\,V$$

$$6H_2O_{(l)} \rightarrow 2H_{2(g)} + O_{2(g)} + 4H^+_{(aq)} + 4OH^-_{(aq)}\;E^0 = -2.06$$

The reactions show that oxygen gas is pro-duced and the solution becomes acidic at the anode. Hydrogen gas and a basic environ-ment occur at the cathode. The overall reac-tion for the electrolysis of water can be simplified if it's assumed the hydrogen and hydroxide ions combine to form water. Doing this and canceling water from both sides of the equation gives

$$2H_2O_{(l)} \rightarrow 2H_{2(g)} + O_{2(g)}$$

Note that the actual potential to electrolyze pure water is about 1.25 volts. Remember,

the value for E^0, which is –2.06 V, assumes standard conditions where the concentration of hydrogen and hydroxide ions are both 1 M. In reality, the concentration of H^+ and OH^- is only 10^{-7} M, and this changes the actual cell voltage.

Batteries

Electrochemical processes occur all around us. We close this chapter by examin-ing a few of these processes and relating them to the electrochemical principles pre-viously introduced. Batteries are probably the most common example of electrochem-ical applications associated with every-day life. While batteries come in all sizes and shapes, all batteries contain the basic elements common to all electrochemical cells. What differentiates one battery from another are the materials used for cathode, anode, and electrolyte, and how these mate-rials are arranged to make a battery. The standard dry cell battery or alkaline cell is shown in Figure 14.8. Batteries consist of

Figure 14.8
Dry Cell Battery (Rae Déjur)

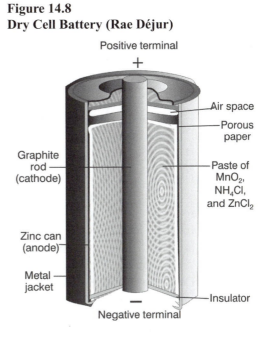

two or more cells connected to deliver a specific charge, although the terms are often used synonymously.

Dry cells are classified as **primary cells** because the chemicals in them cannot be regenerated, that is, primary cells cannot be recharged. The dry cell battery was patented by the French engineer George Leclanché (1839–1882) in 1866. The standard dry cell consists of a zinc container that acts as the anode of the cell. The container contains an electrolyte (this corresponds to the salt bridge) containing a mixture of ammonium chloride (NH_4Cl), zinc chloride ($ZnCl_2$), and starch. A carbon (graphite) rod runs down the axis of the battery. Surrounding the rod is a mixture of manganese dioxide (MnO_2) and carbon black. (Carbon black is basically carbon in the form of soot produced from burning kerosene or another hydrocarbon with insufficient oxygen.) The MnO_2-carbon black mixture acts as the cathode. The MnO_2-carbon black and electrolyte exist as a moist paste, so a dry cell is not actually dry. While several reactions occur in a dry cell, the primary reactions governing its operation can be represented as:

Anode: $Zn_{(s)} \rightarrow Zn^{2+}_{(aq)} + 2e^-$
Cathode: $2MnO_{2(s)} + 2NH_4^+{}_{(aq)} + 2e^-$
$\rightarrow Mn_2O_{3(s)} + 2NH_{3(aq)} + H_2O_{(l)}$
Overall: $Zn_{(s)} + 2NH_4^+{}_{(aq)} + 2MnO_{2(s)}$
$\rightarrow Zn^{2+}_{(aq)} + 2NH_{3(aq)} + H_2O_{(l)}$
$+ Mn_2O_{3(s)}$

When the circuit containing the battery is closed, electrons from the oxidation of zinc flow from the negative terminal through the object being powered to the negative carbon rod that serves as an electron collector. Several reduction reactions occur at the collector; one main reaction is represented earlier. The dry cell with the NH_4Cl electrolyte creates acidic conditions in the battery. The acid attacks the zinc, which accelerates its

deterioration. To extend the life of a dry cell, a basic substance, typically KOH or NaOH, can be used for the electrolyte. Alkaline cells derive their name from the alkaline electrolyte they contain.

Mercury cells, like dry cells, have a zinc anode and a use a mercuric oxide (HgO) cathode. The electrolyte is potassium hydroxide, KOH. These small, flat, metallic cells are widely used in watches, calculators, cameras, hearing aids, and other applications where small size is a premium. The reactions in the mercury cell are:

Anode: $Zn_{(s)} + 2OH^-{}_{(aq)} \rightarrow ZnO_{(s)}$
$+ H_2O_{(l)} + 2e^-$
Cathode: $HgO_{(s)} + H_2O_{(l)} + 2e^- \rightarrow Hg_{(l)}$
$+ 2OH^-{}_{(aq)}$
Overall: $Zn_{(s)} + HgO_{(s)} \rightarrow ZnO_{(s)} + Hg_{(l)}$

The voltage obtained from a mercury cell is about 1.35 V.

A common of battery is the lead storage battery. This battery, composed of several cells, is used mainly in cars. The lead storage battery, the first **secondary** or rechargeable cell, was invented in 1859 by the French physicist Gaston Planté (1834–1889). A lead storage battery consists of a series of plates called grids immersed in a solution of sulfuric acid (approximately 6 M) that serves as the electrolyte (Figure 14.9). Half the plates contain lead and serve as the anode of the cell, and the remaining plates are filled with lead dioxide (PbO_2) and serve as the cathode.

The following reactions represent the two half-cell reactions:

Anode: $Pb_{(s)} + HSO_4^-{}_{(aq)} \rightarrow PbSO_{4(s)}$
$+ H^+ + 2e^-$
Cathode: $PbO_{2(s)} + HSO_4^-{}_{(aq)} + 3H^+{}_{(aq)}$
$+ 2e^- \rightarrow PbSO_{4(s)} + 2H_2O_{(l)}$
Overall: $Pb_{(s)} + PbO_{2(s)} + 2H^+{}_{(aq)} +$
$2HSO_4^-{}_{(aq)} \rightarrow 2PbSO_{4(s)} + 2H_2O_{(l)}$

Figure 14.9
Lead Storage Battery (Rae Déjur)

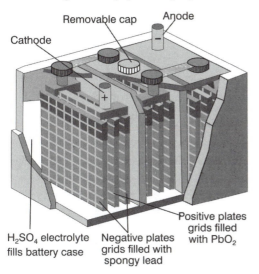

The cell reaction generates a potential of about 2 V. A car battery is created by connecting six cells in series. The reactions show that electrons are created when lead oxidizes at the anode. The lead ions formed in oxidation combine with the sulfate ions in the electrolyte to form lead sulfate. The electrons flow through the car's electrical system back to the battery's cathode plates. Here the electrons combine with PbO_2 and hydrogen ions present in the acid and also form lead sulfate. While driving, the car's alternator continually recharges a battery. The alternator carries out an electrolytic process and reverses the reactions just shown. In this manner, the lead sulfate generated on the plates is converted back into lead and lead dioxide.

Advances in the past few decades have improved car battery technology immensely. Many lead batteries currently manufactured are labeled as maintenance free. This refers to the fact that the acid level in them does not have to be checked. The addition of water to a lead battery is necessary because the charging process causes water to undergo electrolysis. This process creates hydrogen and oxygen gas that escapes from the battery vents. Modern battery technology has greatly reduced and even eliminated this problem by perfecting new types of electrodes, for example, lead-calcium and lead-selenium electrodes. Maintenance-free batteries are sealed and water does not have to be added to them. Prior to this development, it was generally recommended that the fluid level in batteries be checked on a regular basis. Additionally, a hydrometer could be used to check the density of the fluid in the battery to indicate its condition. Many manufacturers now claim a 10–15 year lifetime of batteries as opposed to the typical 3–5 year lifetime of twenty years ago. While improved technology has produced batteries that are more reliable, you still may experience the problem of starting a car on a cold day. The inability to start a vehicle on a cold day results from charge being impeded as the electrolyte turns viscous in the cold conditions.

Another familiar secondary cell is ni-cad or nickel cadmium cell. In these cells, the anode is cadmium and the cathode consists of a nickel compound $(NiO(OH)_{(s)})$. One of the problems with ni-cad cells is the high toxicity of cadmium. An alternative to ni-cad cells currently gaining a foothold in the market is lithium ion cells. Lithium is much less toxic than cadmium. Additionally, lithium ion cells do not suffer from the memory effect of ni-cad cells. This effect results in the loss of efficiency when ni-cad cells are recharged before they completely lose their charge. Besides these factors, lithium has several properties that make it a highly desirable material for use in batteries. Table 14.2 shows that lithium has the lowest reduction potential. This is equivalent to saying it has the highest oxidation potential. Lithium's high oxidation potential means that a greater voltage can be obtained using lithium cells. Another property of lithium is that it is light, which is especially important for portable applications such as cell phones and laptop computers. A final advantage of lithium ions is that they are small and move more readily through materials than large ions.

Original lithium cells employed lithium metal as the anode, and these cells have been used in the military and space program for years. Lithium, though, is highly reactive presenting fire and explosive hazards. Lithium ion cells retain the advantages of original lithium batteries, but they eliminate the hazards associated with them. In lithium ion cells, the lithium metal anode is replaced by a substance with the ability to absorb lithium ions such as graphite. The cathode is made of a lithium compound such as $LiCoO_2$, $LiSO_2$, or $LiMnO_2$. The electrolyte separating the electrodes allows the passage of lithium ions but prevents the flow of electrons. The electrons move from the anode to cathode through the external circuit.

Numerous other types of cells exist such as zinc-air, aluminum-air, sodium sulfur, and nickel-metal hydride (NiMH). Companies are on a continual quest to develop cells for better batteries for a wide range of applications. Each battery must be evaluated with respect to its intended use and such factors as size, cost, safety, shelf-life, charging characteristics, and voltage. As the twenty-first century unfolds, cells seem to be playing an ever-increasing role in society. Much of this is due to advances in the consumer electronics and the computer industry, but there have also been demands in numerous other areas. These include battery-powered tools, remote data collection, transportation (electric vehicles), and medicine.

One area of research that will advance in years to come is the study of **fuel cells**. The first fuel cell was actually produced in 1839 by Sir William Grove (1811–1896). A

Figure 14.10
Schematic of Hydrogen-Oxygen Fuel Cell

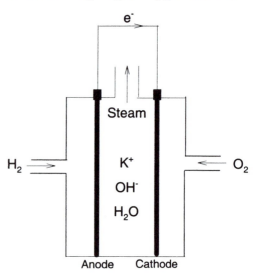

fuel cell is essentially a battery in which the chemicals are continuously supplied from an external source. One simple type of fuel cell is the hydrogen-oxygen fuel cell (Figure 14.10). These cells were used in the Gemini and Apollo space programs. In the hydrogen-oxygen fuel cells, the reactions at the anode and cathode are

Anode: $2H_{2(g)} + 4OH^-_{(aq)} \rightarrow 4H_2O_{(l)} + 4e^-$
Cathode: $O_{2(g)} + 2H_2O_{(l)} + 4e^- \rightarrow 4OH^-_{(aq)}$
Overall: $2H_{2(g)} + O_{2(g)} \rightarrow H_2O_{(l)}$

The electrodes in the hydrogen-oxygen cell are porous carbon rods that contain a platinum catalyst. The electrolyte is a hot (several hundred degrees) potassium hydroxide solution. Hydrogen is oxidized at the anode where the hydrogen and hydroxide ions combine to form water. Electrons flow through the external circuit.

Many different types of fuel cells are currently under development. Many of these are named after the electrolyte or fuel used in the cell. The polymer electrolyte membrane or proton exchange membrane cell (pem) also uses hydrogen and oxygen. The electrolyte in this cell allows hydrogen ions (protons) to pass to the cathode where they combine with oxygen to produce water. This fuel cell has several advantages over the alkaline fuel cell. Its lower operating temperature and relatively lightweight make it suitable for use as a power supply for vehicles. Several prototype vehicles have been built using pem cells.

Corrosion

Corrosion refers to the gradual natural deterioration of a substance due to a chemical or electrochemical reaction. Many metals and nonmetals, such as glass, are subject to corrosion, but the most common form of corrosion is rusting of iron. In rusting, iron does not combine directly with oxygen but involves the oxidation of iron in an electrochemical process. There are two requirements for rust: oxygen and water. The necessity of both oxygen and water is illustrated by observing automobiles operated in dry climates and ships or other iron objects recovered from anoxic water. Autos and ships subjected to these conditions show remarkably little rust, the former due to lack of water and the latter due to lack of oxygen.

Figure 14.11 illustrates the electrochemical rust formation process.

Figure 14.11
The formation of rust involves an electrochemical reaction taking place between two different regions in the metal.

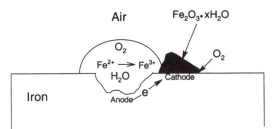

Rust is a product of electrochemical reactions occurring on a piece of iron or steel. Iron metals are never uniform. There are always minor irregularities in both the composition and physical structure of the metal. These minute differences give rise to the anodic and cathodic regions associated with rust formation. During the rusting process, iron is oxidized at the anode according to the reaction:

$$Fe_{(s)} \rightarrow Fe^{2+} + 2e^-.$$

The electrons from this reaction flow through the metal to an area of the metal where oxygen is reduced according to the reaction:

$$O_{2(g)} + 4H^+_{(aq)} + 4e^- \rightarrow 2H_2O_{(l)}$$

The hydrogen ions in this reaction are provided mainly by carbonic acid from the dissolution of atmospheric carbon dioxide in the water. The water acts as an electrolyte connecting the anode and cathode regions of the metal. Iron ions, Fe^{2+}, migrate through the electrolyte and in the process are further oxidized by dissolved oxygen in the water to Fe^{3+}. Iron (III) oxide (Fe_2O_3), the technical term for rust, is formed at the cathode according to the equation:

$$4Fe^{2+}_{(aq)} + O_{2(g)} + (4 + 2x)H_2O_{(l)}$$
$$\rightarrow 4Fe_2O_3 \cdot + xH_2O_{(s)} + 8H^+_{(aq)}$$

The "x" in the above equation indicates a positive whole number corresponding to the variable amount of water molecules that takes part in the reaction.

The area of rust formation is different from the point where oxidation of iron takes place. As noted, water serves as an electrolyte through which iron ions migrate. This explains why vehicles rust much more rapidly in regions where road salts are used to melt winter ice. The salts improve the conductivity of the electrolyte, thereby accelerating the corrosive process.

Most metals undergo corrosion of some form except the so-called noble metals of gold, platinum, and palladium. Why then don't we experience rusting problems with many other metals, for example, aluminum? The key is the layer of the oxide or other product formed in the corrosion process. Iron (III) oxide is highly porous and as a result rust does not protect the underlying metal from further corrosion. Metals such as aluminum and zinc have an even greater tendency to oxidize compared to iron, but the oxide layer on these metals forms an impervious protective layer that protects the metal below from further oxidation. Copper oxidizes in moist air and its surface changes from the characteristic reddish-brown color to green patina. This patina consists of sulfate compounds, $CuSO_4 \cdot xCu(OH)_2$ (where x is a positive whole number).

A number of methods are used to reduce and prevent corrosion. The most common method is to paint iron materials so that the metals are protected from water and oxygen. Alloying iron with other metals is also a common means to reduce corrosion. Stainless steel is an alloy of iron, chromium, nickel, and several other metals. Iron may also be protected by coating it with another metal. **Galvanizing** refers to applying a coating of zinc to protect the underlying metal. Additionally, because it is a more active metal, zinc oxidizes rather than iron.

Connecting iron objects to a more active metal is called **cathodic protection**. Cathodic protection is widely used to protect underground storage tanks, ship hulls, bridges, and buried pipes. One of the most common forms of cathodic protection is to connect the object to magnesium. When magnesium is connected to an iron object, magnesium rather than iron becomes the anode in the oxidation process. In cathodic

Figure 14.12
In cathodic protection, a metal more active than iron such as magnesium or zinc is connected to the iron object. The more active metal is oxidized protecting the iron.

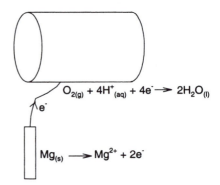

protection, bars of magnesium are connected either directly or by wire to the iron structure (Figure 14.12). Because the metal connected to the iron corrodes over time, it is called the **sacrificial anode**. Sacrificial anodes must eventually be replaced.

Electroplating

Our discussion of batteries and corrosion involves electrochemical cells. One of the most common applications of electrolysis is in the area of **electroplating**. During electroplating, the object to be electroplated serves as the cathode in an electrolytic cell. The anode of the cell is often made of the plating metal. The object is immersed in a bath or solution containing ions of the plating metal and is connected to the negative terminal of a power source. Electrons from the power source flow through the object where they cause the plating metal ions to be reduced on to the surface of the object (Figure 14.13).

Electroplating is another technique used to protect iron from corrosion. The most common plating metal used for this purpose is chromium.

Figure 14.13
Electrolysis is used to electroplate an object by making the object the cathode in an electrolytic cell.

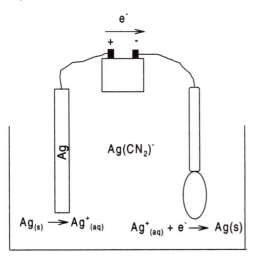

Commercial Applications and Electrorefining

Most metals found in nature exist as ores in combination with other elements such as oxygen and sulfur. Electrolysis can be used to separate metals from their ores and remove impurities from the metals. The general process is known as **electrorefining**. Other electrolytic processes are used to obtain a number of important chemicals.

Chlorine (Cl_2) and sodium hydroxide (NaOH) are two important chemicals produced by electrolysis. Chlorine and sodium hydroxide are generally among the top ten chemicals produced annually in the United States. The electrolysis of brine or aqueous NaCl solution is used to produce chlorine, sodium hydroxide, and hydrogen (Figure 14.14). The chloride ion in the brine solution is oxidized at the anode, while water is converted at the cathode according to the following reactions:

Anode: $2Cl^- \rightarrow Cl_{2(g)} + 2e^-$

Cathode: $2H_2O_{(l)} + 2e^- \rightarrow H_{2(g)} + 2OH^-_{(aq)}$

The anode and cathode sides of the cell are separated by a membrane that allows the sodium ions to migrate from the anode side of the cell to the cathode side.

Electrorefining can be used to purify a metal by using alternate electrodes of a pure and impure metal. Impurities oxidized at the anode, which is made of the impure metal, travel into solution. By arranging the cell appropriately, the ion of the metal to be purified is reduced on the pure metal cathode. For example, copper metal that contains lead and iron may be used as one electrode and pure copper as the other electrode in a cell. When the proper voltage is applied, copper, lead, and zinc will be oxidized and move into the electrolyte. Because copper is more easily reduced compared to zinc and lead, it will be plated out at the pure copper cathode. Therefore, this process effectively removes the zinc and lead impurities from the copper.

An important application of electrolysis is in the production of aluminum. Aluminum is the most abundant metal in the Earth's crust. Until the late 1800s, it was considered a precious metal due to the difficulty of obtaining pure aluminum. The process involved the reduction of aluminum chloride using sodium. In 1886, Charles Martin Hall (1863–1914) in the United States and Paul L. T. Héroult (1863–1914) in France independently discovered a means to produce aluminum electrochemically. Hall actually started his work on aluminum in high school, continued his research through his undergraduate years at Oberlin College, and discovered a procedure for its production shortly after graduating. The problem that confronted Hall, and other chemists trying to produce aluminum using electrochemical means, was that aqueous solutions of aluminum could not be used. This was because water rather than the aluminum in solution was reduced at the cathode. To solve this problem, Hall embarked

Figure 14.14
The electrolysis of aqueous sodium chloride is used to prepare chlorine, sodium, and hydrogen.

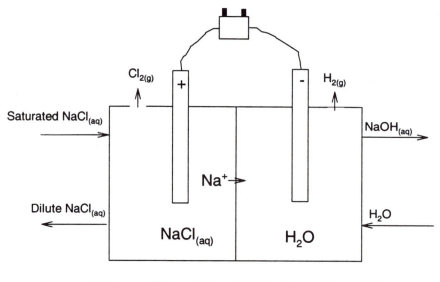

$$2Cl^-_{(aq)} \rightarrow Cl_{2(g)} + 2e^-$$ $$2H_2O + 2e^- \rightarrow H_{2(g)} + 2OH^-_{(aq)}$$

on a search for a substance that could be used to dissolve alumina (Al_2O_3) to use in his electrolytic cells. Hall and Héroult simultaneously found that the substance cryolite could be used. Cryolite is an ionic compound with the formula Na_3AlF_6.

The production of aluminum actually involves several steps. Bauxite is the ore that contains aluminum oxide (Al_2O_3) used to produce aluminum. Impurities of iron, sulfur, silicon and other elements are removed from bauxite using the Bayer process to produce purified alumina. The Bayer process, patented in 1887 by Austrian Karl Josef Bayer (1847–1904), involves pulverizing bauxite and treating it with a hot sodium hydroxide solution to produce sodium aluminate ($NaAlO_2$). Sodium aluminate is then placed in a reactor in which temperature and pressure can be varied to precipitate out impurities. The sodium aluminate solution is then hydrolyzed to produce purified alumina:

$$2NaAlO_2 + 4H_2O \rightarrow Al_2O_3 \cdot 3H_2O + 2NaOH$$

The Hall-Héroult process involves taking the purified alumina and dissolving it in cryolite. This step requires producing a molten mixture. One of the problems Hall had to overcome in developing his process was producing the molten alumina mixture. Pure alumina's melting point is 2,050°C, but Hall was able to produce a cryolite-alumina mixture that melted at approximately 1,000°C. The purified alumina-cryolite melt is electrolyzed in a large iron electrolytic cell that is lined with carbon. The iron container serves as the cathode in the cell. The cell contains a number of carbon electrodes that serve as anodes (Figure 14.15). The exact nature of the reactions at the anode and cathode is not clearly known, but can be considered to involve the reduction of aluminum and oxidation of oxygen:

anode: $6O^{2-} \rightarrow 3O_{2(g)} + 12e^-$
cathode: $4Al^{3+} + 12e^- \rightarrow 4Al_{(s)}$

The oxygen produced reacts with the carbon electrodes to produce carbon dioxide; an

Figure 14.15
In the Hall-Héroult process, a molten alumina is dissolved in cryolite and the mixture is electrolyzed to produce molten aluminum.

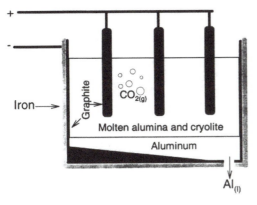

overall equation involving conversion of alumina into aluminum can be written:

$$2Al_2O_{3(l)} + 3C_{(s)} \rightarrow 4Al_{(l)} + 3CO_{2(g)}.$$

Soon after Hall and Héroult developed their process, aluminum plants were constructed to process bauxite into aluminum. Hall became a partner in the Pittsburgh Reduction Company, which eventually became ALCOA. Because the electrolytic production of aluminum was, and still is, energy intensive, aluminum plants were located in areas where electricity was abundant and relatively cheap. These plants were situated near hydroelectric plants to take advantage of their electrical output. The high electrical demand to produce aluminum from bauxite (the aluminum industry is the largest consumer of electrical energy in the United States) has created a large market for recycled aluminum. Aluminum made from raw bauxite is referred to as **primary production** aluminum, while aluminum from recycled aluminum is **secondary production**. The amount of energy for secondary production of aluminum is only about 5% of the primary production requirement. One way of thinking about the energy saving involved with recycling of aluminum is to consider that every time an aluminum can is thrown away that it is equivalent to dumping half that can filled with gasoline. The United States continues to produce the most aluminum in the world, but over the last few decades its position has fallen from 40% of the world market to around 15%. Current world production is approximately 20 million tons.

Summary

The history of electrochemistry has paralleled the history of modern chemistry. The discoveries of Galvani and Volta occurred as Lavoisier was proposing his new chemical system. As the theory of electrochemistry developed, so too did its applications. Society's dependence on electricity is brought to our attention whenever there is a power outage. During these times, we turn to electrochemical energy to supply our electrical needs. There is no reason to think that as the twenty-first century unfolds that our reliance on cells and batteries and other forms of electrical storage devices will lessen. In fact, there is every reason to think just the opposite. The development of vehicles that pollute less, sources to power a mobile society's use of communication devices and computers, and a host of cordless gadgets provide ample motivation to continue the search for more efficient cells. These new technologies will augment traditional chemical industrial processes that are based on electrochemical principles.

15

Organic Chemistry

Introduction and History

Organic chemistry is the study of carbon compounds. While the formal development of this branch of chemistry began in the mid-eighteenth century, organic chemicals have been used throughout human history. All living material, the food we eat, and the **fossil fuels** we burn all contain organic compounds. Numerous organic substances were prepared by ancient civilizations including alcohols, dyes, soaps, waxes, and organic acids. From its modest beginning in the early 1800s, an increasing number of organic compounds have been isolated, synthesized, and identified. By far, organic compounds dominate the number of known chemicals. Currently, there are over twenty million known organic compounds and these make up over 80% of all known chemicals.

The term "organic" was not used until 1807 when Berzelius suggested it to distinguish between compounds devised from living (organic) and nonliving (inorganic) matter. Before Berzelius made this distinction, interest in organic compounds was based primarily on the medicinal use of plants and animals. Plant chemistry involved the study of chemicals found in fruits, leaves, saps, roots, and barks. Animal chemistry focused on compounds found in the blood, saliva, horn, skin, hair, and urine. In contrast to plant and animal chemistry, scientists distinguished those compounds isolated from nonliving mineral material. For example, mineral acids, referred to in Chapter 13, were made from vitriolic sulfur compounds.

Until the mid-eighteenth century, scientists believed organic compounds came only from live plants and animals. They reasoned that organisms possessed a vital force that enabled them to produce organic compounds. The first serious blow to this theory of **vitalism**, which marked the beginning of modern organic chemistry, occurred when Friedrich Wöhler (1800–1882) synthesized urea from the two inorganic substances lead cyanate and ammonium hydroxide:

$$Pb(OCN)_2 + 2NH_4OH \rightarrow 2(NH_2)_2CO + Pb(OH)_2$$
lead cyanate + ammonium hydroxide → urea
+ lead hydroxide

Wöhler was actually attempting to synthesize ammonium cyanate when he discovered crystals of urea in his samples. He first prepared urea in 1824, but he did not identify this product and report his findings until 1828. In a note written to Berzelius, he proclaimed: "I must tell you that I can make urea without the use of kidneys, either man or dog. Ammonium cyanate is urea." While Wöhler's synthesis of urea signaled the birth of organic chemistry, it was not a fatal blow to vitalism. Some argued that because Wöhler used bone material in his preparations that a vital force could still have been responsible for Wöhler's urea reaction. Over the next twenty years, the vitalism theory eroded as the foundation of organic chemistry started to take shape. Hermann Kolbe (1818–1884) synthesized acetic acid from several inorganic substances in 1844. Kolbe's work and that of Marcellin Berthelot (1827–1907), who produced numerous organic compounds including methane and acetylene in 1850, marked the death of the vitalism theory.

Much of the development of modern chemistry in the first half of the eighteenth century involved advances made in organic chemistry. At the time of Wöhler's synthesis of urea, there was much confusion about the proper values of atomic masses, the role of oxygen in acids, and how atoms combined. Some chemists felt organic and inorganic compounds followed different chemical rules and sought a theory to replace vitalism. One of the early explanations of how organic compounds formed involved the radical theory. Lavoisier considered a radical an atom or groups of atoms that did not contain oxygen. Radicals combined to give different substances. Organic substances were formed from the combination of a carbon radical and a hydrogen radical. Berzelius also believed radicals combined to form compound radicals that were equivalent to organic compounds. Berzelius' radicals were held together by electrical forces. According to Berzelius,

organic compounds contained a negative carbon radical and a positive hydrogen radical. Compound radicals could incorporate other atoms such as oxygen and nitrogen into their structures. Chemists employing the radical theory sought to identify radicals. In 1815 Gay-Lussac identified cyanogens gas, $(CN)_2$, as a radical and explained how this radical was present in many other organic compounds. In 1828, Jean-Baptiste André Dumas (1800–1884) and Pierre François Guillaume Boullay (1806–1835) proposed ethylene (C_2H_4) as the radical that formed the base for many different organic compounds. For example, ethyl alcohol (C_2H_5OH) could be formed by adding water to the ethylene radical. Berzelius named the ethylene radical etherin and Dumas and Boullay's idea came to be known as the etherin theory. In the same year, Wöhler and his close friend Liebig proposed the benzoyl radical ($C_{14}H_{10}O_2$). Like the ethylene radical, a number of different organic compounds could be formed by addition to this radical.

Throughout the 1830s, chemists identified other radicals. Dumas, Liebig, and Berzelius believed that organic compounds could be classified based on their component radicals. While the radical was a means to organize the different types of organic compounds, it also led to much confusion. Chemists would often propose alternate radicals as the base for different compounds. Adding to the problem was the disagreement over atomic masses; therefore, the true formulas of many compounds were still unknown. Berzelius continued to believe radicals were held together by electrostatic forces. Some chemists believed that radicals were true substances, while others believed that they were merely hypothetical constructs to explain their observations.

One serious problem with Berzelius' radical theory occurred when Dumas and his student Auguste Laurent (1808–1853) discovered that chlorine replaced hydrogen in a number of reactions. Dumas' work in this area originated when he was presented with a problem by the French royalty. Evidently, the emissions from candles burned during a royal ball irritated the guests. Dumas determined the reason for this was that the gas hydrogen chloride (HCl) was emitted when the candles burned. The chloride was introduced when the candles were bleached with chorine solution. Dumas proceeded to conduct numerous experiments on the chlorination of wax. As Dumas, Laurent, and others reported that chlorine could substitute for hydrogen in organic compounds, Berzelius voiced strong objections. Berzelius and his supporters Liebig and Wöhler believed a negative chlorine atom could not replace a positive hydrogen in an organic compound, because this would mean there was an attraction between two negative atoms: carbon and chlorine.

During the 1840s as more scientists demonstrated that substitution occurred in organic chemical reactions, Berzelius' electrochemical theory of organic radicals was severely questioned. This period also saw several alliances and bitter rivalries develop among the chemists attempting to develop a coherent theory of organic compounds. A main source of controversy was a theory developed by Laurent called the nuclear theory. Laurent believed organic compounds were based on three-dimensional nuclei instead of radicals. According to Laurent, atoms occupied positions on the corners and edges of geometrical shapes. Substitution could take place at these positions or addition of other atoms took place on the face of the figure (Figure 15.1).

Dumas initially supported his student Laurent, but then under pressure from Berzelius and Liebig turned against Laurent. Wöhler and others ridiculed Laurent in

Figure 15.1
August Laurent believed organic structure was based on geometric nuclei. Substitution and addition take place at different positions in the figure.

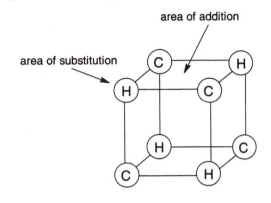

scientific journals. At the same time, Dumas developed his own theory of organic substitution called type theory. Dumas' theory had many similarities to Laurent's nuclear theory. Laurent accused Dumas of stealing his ideas. Laurent, ostracized by the established chemical community and unable to obtain an academic position, joined forces with another chemical outcast and brilliant chemist named Charles Gerhardt (1816–1856). Gerhardt's idea on organic compound formation was called the theory of residues. According to Gerhardt, organic radicals decomposed into residues. Some of these residues recombined to form organic compounds, while other residues formed stable inorganic compounds such as water and ammonia.

The collaboration of Gerhardt and Laurent resulted in the type theory for organic compounds. A type according to Gerhardt and Laurent was an inorganic compound that produced an organic compound when substitution of a hydrogen atom in the inorganic compound occurred. Gerhardt and Laurent defined four types of organic compounds: water type, ammonia type, hydrogen type, and hydrogen chloride type. These four types could be used to explain the formation of all organic compounds. For example, alcohols were a water type that formed when one of the hydrogen atoms in water was replaced by a radical, for example, methyl alcohol formed when the methyl radical replaced one of the hydrogen atoms in water: $H_2O + CH_3 \rightarrow CH_3OH$. A vast array of organic compounds such as **ethers, aldehydes, ketones, amines**, and so on were defined using the four types. Other chemists proposed additional types to explain the ever-increasing array of organic chemicals being discovered. Laurent and Gerhardt's method of classification for organic compounds was based on a specific type that unified a group of organic compounds. This method formed the basis of the modern classification of organic chemicals based on **functional groups**. These functional groups are examined later in this chapter. The death of Berzelius in 1848 and the advancement of the work in organic chemistry contributed to the acceptance of the work of Laurent and Gerhardt. Unfortunately, both of these chemists died just as their work was being accepted and they were being vindicated by the established chemical community.

In the late 1850s, Frankland introduced the concept of valence (see Chapter 7). This set the stage for Archibald Scott Couper (1831–1892) and Kekulé to explain the structure of organic compounds. Couper wrote a paper in 1858 entitled *On a New Chemical Theory*. Couper's paper showed that carbon acted as a tetravalent (capable of combining with four hydrogen atoms) or divalent atom (capable of combining with two hydrogen atoms). Couper also explained how tetravalent carbon could form chains to produce organic molecules. He also presented diagrams in his paper depicting the structure of organic compounds and even proposed ring structures for some organic compounds. Unfortunately for Couper, the

presentation of his paper was delayed, and in the meantime Kekulé's similar, but less developed, ideas appeared in print. Kekulé, who originally studied architecture but switched to chemistry, had close contacts with the great chemists of his day including Liebig, Gerhardt, and Dumas. His work was readily accepted as a solid theory to explain organic structure. Kekulé showed that organic molecules could be constructed by carbon bonding to itself and other atoms. In 1865, Kekulé explained how the properties of benzene (C_6H_6) could be explained using a ring structure. Benzene's structure came to Kekulé during a dream in which he saw atoms as twisting snakes and was inspired when one snake bites its own tail.

While Kekulé received the credit for the modern theory of tetravalent carbon during his day, Couper suffered from a mental breakdown and spent the last thirty years of his life in ill-health. It was not until the 1900s after his death that chemists appreciated the depth of his work and acknowledged he deserved to share credit with Kekulé for the theory of modern organic structure. The work of Couper and Kekulé marked the start of modern organic theory. Radicals, types, and residues were no longer needed to explain organic structure. The revival of Avogadro's hypothesis by Cannizzaro enabled chemists to agree on atomic masses, and the acceptance of atoms as the building blocks of molecules advanced modern organic theory. With this brief history as a background, we can now examine the basic groups of organic compounds, important organic reactions, and the role organic chemicals play in modern society.

Organic Compounds

Of all the existing chemical compounds, the number that contains carbon is overwhelming. Due to the preponderance of organic chemicals produced naturally, the number of organic compounds that exist cannot be determined. Approximately six million organic compounds have been described. The reason for the tremendous number of organic compounds is carbon's ability to covalently bond with itself and many other atoms. This ability is due to carbon's electron structure. The electron configuration of carbon, atomic number 6, is $1s^2 2s^2 2p^2$. Carbon needs four more electrons to complete its outer shell and acquire a stable octet electron configuration. Of all the atoms, only carbon has the ability to form long chains of hundreds and even thousands of carbon atoms linked together. The carbon-carbon bonds in these molecules exist as stable single, double, and triple bonds. Carbon also has the ability to form ring structures and other atoms and groups of atoms can be incorporated into the molecular structures formed from carbon atoms.

Organic compounds contain covalent bonds. In general, compared to inorganic compounds organic compounds have low melting and boiling points, tend to be flammable, have relatively low densities, do not dissolve readily in water, and are primarily nonelectrolytes.

Because of the number and variety of organic chemicals, an elaborate set of rules exists for classifying and naming them. These rules have been established by IUPAC (International Union of Pure and Applied Chemistry). The IUPAC rules governing the naming of organic compounds forms an entire language comparable to a foreign language. All you have to do is look at a label of many commercial products to gain an appreciation for the variety and complexity of organic nomenclature. Like any language, the language of chemistry, and especially organic chemistry, requires daily use to become proficient in its vocabulary and syntax. Some of the major groups of organic compounds will be examined with emphasis on the most common compounds. Chemical nomenclature will be incorporated into our discussion as needed. Common names for organic compounds will be used throughout our discussion, but some general IUPAC rules will be given. These general rules are helpful when you know the name of a compound and want to know to which family it belongs.

Aliphatic Hydrocarbons

Organic compounds containing only carbon and hydrogen are known as **hydrocarbons**. There are several major groups of hydrocarbons. **Aliphatic hydrocarbons** are primarily straight and branched chains of hydrocarbons consisting of alkanes, alkenes, and alkynes. Alkanes are hydrocarbons that contain only carbon-carbon (C—C) single bonds. When only single C—C bonds exist in a hydrocarbon, it is called a **saturated hydrocarbon. Unsaturated hydrocarbons** contain at least one double or triple bond. Alkanes have the general formula C_nH_{2n+2}, where n = 1,2,3 .... The simplest alkane, defined when n = 1 in this equation, is CH_4, methane. Table 15.1 lists the first eight alkanes and gives their molecular and structural formulas. The first four alkanes have common names and the alkanes with more than four carbon atoms take the numerical Greek prefix for their root names.

The structural formulas shown in Table 15.1 are written in their expanded form. The expanded form shows all atoms and all bonds in the formula. Structural formulas can also be written using a condensed form. In the condensed form, groups of atoms are used to convey how the molecule is arranged. For hydrocarbons, essential information is shown using the carbon-carbon bonds and grouping the hydrogen atoms

Table 15.1
The First Eight Alkanes

Name	Alkane Prefix	Molecular Formula	Structural Formula	Isomers
Methane	meth	CH_4		1
Ethane	eth	C_2H_6		1
Propane	prop	C_3H_8		1
Butane	but	C_4H_{10}		2
Pentane	pent	C_5H_{12}		3
Hexane	hex	C_6H_{14}		5
Heptane	hept	C_7H_{16}		9
Octane	oct	C_8H_{18}		18

with carbon atoms. For instance, the condensed formulas of ethane and propane is:

$$H_3C—CH_3 \qquad H_3C—CH_2—CH_3$$

Ethane   Propane

Throughout this chapter, both expanded and condensed structural formulas are used. Molecules may even be shown using both the condensed and expanded forms for different parts of the molecule. Rather than adhere to using one structural formula, our goal is to represent the basic structure of the molecule.

Alkyl groups are a type of hydrocarbon formed by removing a hydrogen atom from an alkane. Their names are formed by using the alkane prefix and attaching "yl." The corresponding alkyl groups for the first eight alkanes are shown in Table 15.2.

The alkyl groups attach themselves to other groups, and their names often arise in organic compounds, for example, methyl alcohol and ethyl alcohol.

The structural formula for butane shows all the carbons linked in a **straight chain**. In reality, the carbons actually line up in a zigzag fashion, so the term "straight chain" simply refers to a continuous arrangement of carbon atoms. Butane and alkanes that

exist as straight chains are referred to as **normal alkanes**. The butane in Table 15.1 is called normal butane or n-butane. It is also possible for butane to exist in a branched form:

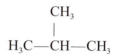

This second branched molecule is called isobutane. Compounds sharing the same molecular formula but having different structures are called **structural isomers**. Normal butane and isobutane have different physical properties. The number of structural isomers for the alkanes is included in Table 15.1. It can be seen in this table that as the number of carbon atoms increases that the number of possible isomers also increases. The fact that numerous isomers exist for most organic compounds is another reason why there are so many organic compounds.

Alkanes are relatively unreactive organic compounds. They do undergo combustion, and several of the alkanes listed in Table 15.1 are familiar fuel sources. Natural gas is primarily methane, propane is used for cooking and heating, butane is used in lighters, and octane is a principal component of gasoline. The octane number of gasoline is an indication of the fuel's ability to resist premature ignition and cause knock in the engine. A fuel's octane number is determined by comparing the fuel's performance in a standard engine to that when an isooctane-heptane mixture is used in the engine. The isooctane used to determine octane rating is 2,2,4 trimethylpentane:

Table 15.2
Alkyl Groups

Structural Formula	Alkyl Group
—CH₃	methyl
—CH₂-CH₃	ethyl
—CH₂-CH₂-CH₃	propyl
—CH₂-CH₂-CH₂-CH₃	butyl
—CH₂-CH₂-CH₂-CH₂-CH₃	pentyl
—CH₂-CH₂-CH₂-CH₂-CH₂-CH₃	hexyl
—CH₂-CH₂-CH₂-CH₂-CH₂-CH₂-CH₃	heptyl
—CH₂-CH₂-CH₂-CH₂-CH₂-CH₂-CH₂-CH₃	octyl

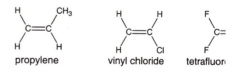

propylene   vinyl chloride   tetrafluor

This isooctane is one of the eighteen isomers of octane, and its octane number is set at 100. Normal heptane's octane number is 0. In general, straight chain alkanes have low octane numbers and branch chain alkanes have high numbers. In fact, normal octane has an octane rating of −20. A gas with a 92 octane rating would have an anti-knock performance equal to a blend that is 92% isooctane and 8% n-heptane. Often on gas pumps the formula R + M/2 is posted. The R and M in this formula stand for research and motor, respectively. Research octane and motor octane refer to different methods to determine octane numbers. Motor octane is found with the engine operating under a heavier load and running at higher rpm, more reflective of driving on a freeway. Research octane is found at lower rpm and better approximates acceleration from rest. The octane rating of the gasoline is an average of these two as indicated by the formula.

Alkanes also undergo substitution where hydrogen atoms in the alkane are replaced by other atoms. Substitution reactions are the most common type of reaction in organic chemistry. Substitution occurs when an atom or group of atoms replaces an atom or group of atoms in an organic compound. **Halogenation** is a common substitution reaction in which hydrogen atoms are replaced by halogen atoms. One group of organic compounds that result from substitution reactions is **chlorofluorocarbons**. Chlorofluorocarbons or CFCs have received widespread attention in recent years because of their role in stratospheric ozone destruction. CFCs are commonly used as refrigerants. The two most common CFCs are trichlorofluoromethane and dichlorodifluoromethane:

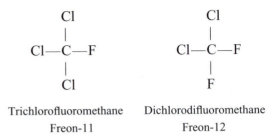

Trichlorofluoromethane   Dichlorodifluoromethane
Freon-11   Freon-12

The trade name Freon is used for the common CFCs.

Alkenes and alkynes are unsaturated hydrocarbons. Alkenes contain at least one carbon-carbon double bond and alkynes have at least one carbon-carbon triple bond. The names of alkenes and alkynes use the alkane prefixes, but add "ene" and "yne" endings, respectively (Table 15.3).

Alkenes have the general formula C_nH_{2n}, where n = 2,3,4,. . . . Alkenes were called olefins, a term primaryly associated with the petrochemical industry. Alkynes have the general formula C_nH_{2n-2}, where n = 2,3,4,. . . .

Alkenes are very reactive organic compounds. They readily undergo addition reactions in which atoms or groups of atoms are added to each carbon in a double bond. During this process, the double bond is converted into a single bond. In addition reactions, there is no loss of atoms from the original compound. A common addition reaction is **hydrogenation** in which hydrogen is added to each carbon in a double bond. Hydrogenation occurs under increased temperature and pressure generally using a nickel or platinum catalyst. Hydrogenation is used to convert unsaturated liquid vegetable oils (cottonseed, corn, and soybean) into saturated solid margarine, shortening, and vegetable spreads. Hydrogenation allows more healthy vegetable oils to be substituted for animal fats. If you examine the ingredient label of many of these products, you will

Table 15.3
Basic Alkenes and Alkynes

Alkene	Molecular Formula	Alkyne	Molecular Formula
Ethene	C_2H_4	Ethyne	C_2H_2
Propene	C_3H_6	Propyne	C_3H_4
1-Butene	C_4H_8	1-Butyne	C_4H_6
1-Pentene	C_5H_{10}	1-Pentyne	C_5H_8
1-Hexene	C_6H_{12}	1-Hexyne	C_6H_{10}
1-Heptene	C_7H_{14}	1-Heptyne	C_7H_{12}
1-Octene	C_8H_{16}	1-Octyne	C_8H_{14}

see the term "partially hydrogenated" referring to this process.

Ethene or ethylene is the most important organic chemical used in commercial applications. Annual production of ethylene in the United States was over twenty-five million tons in the year 2000. Propylene is also used in large quantities with an annual production of over thirteen million tons. Alkenes such as ethylene and propylene have the ability to undergo **addition polymerization**. In this process, multiple addition reactions take place and many molecules link together to form a **polymer**. A polymer is a long chain of repeating units called **monomers**. For example, the addition of two ethylene molecules can be represented as

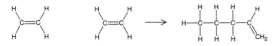

In the production of polyethylene, this process is repeated thousands of times. The basic repeating structure can be pictured as

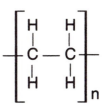

Other examples of addition polymerization of alkenes are the production of polypropylene from propylene, polyvinyl chloride (PVC) from vinyl chloride, and Teflon from tetrafluoroethylene. The structure of the three monomers is depicted in Figure 15.2.

The characteristics of different polymers may be modified by the inclusion of different

Figure 15.2
The polymers polyethylene, polyvinyl chloride, and Teflon are made by repeated addition of their respective monomers.

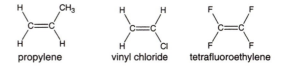

propylene   vinyl chloride   tetrafluoroethylene

atoms or different reaction conditions, for example, varying temperature and pressure. Additives affect properties such as flammability, flexibility, thermal stability, and resistance to chemicals and bacteria. One example of the difference in polymer characteristics is demonstrated by two main types of polyethylene. High density polyethylene (HDPE) consists mostly of straight chain molecules, while low density polyethylene (LDPE) consist of molecules containing many branches. For this reason, HDPE molecules can pack together more tightly producing a hard rigid material. HDPE plastics are use to produce materials where flexibility is not critical such as T.V. cabinets or rigid bottles. Conversely, the branching in LDPE plastics results in a significantly more flexible material. Items such as plastic grocery bags, squeeze bottles, and various food wraps, commonly called plastic wrap, are made of LDPE.

A variety of polymers are classified under the general heading of plastics. While this classification may suffice for everyday usage, recycling of plastics requires them to be separated according to their chemical composition. Failure to separate plastics by chemical composition results in contaminating one type of plastic with another. Because of this, international agreement requires plastic containers greater than eight ounces to contain a symbol designating their chemical composition. The symbol consists of a triangle made of three arrows containing a number. The symbol is generally molded right into the bottom of the container during manufacturing. Table 15.4 summarizes the symbols, their composition, and some uses of several common polymers.

Table 15.4
Plastic Recycling Symbols

Symbol	Composition	Abbreviation	Uses of Plastic
1	polyethylene terephthalate	PET	soda and juice, peanut butter, seasonings
2	high density polyethylene	HDPE	milk, juice, motor oil, bleach
3	polyvinyl chloride	V	residential siding, pipe, flooring
4	low density polyethylene	LDPE	cellophane, squeeze bottles, trash/grocery bags, diaper lining
5	polypropylene	PP	bottle caps, margarine tubs, outdoor and athletic clothing, car battery cases
6	polystyrene	PS	Styrofoam, packing material, coolers, coffee cups, egg cartons, insulation
7	other		

Approximately 90% of the plastic recycled by consumers falls into categories 1 and 2.

Alkynes contain a triple bond. The simplest alkyne is ethyne (C_2H_2) and goes by the common name acetylene:

$$HC\equiv CH$$

Acetylene is the most widely used alkyne. Alkynes undergo the same reactions as alkenes. Because a triple bond connects the two carbon atoms, addition of an atom initially forms a double bond. The addition of a second atom converts the double bond into a single bond. For example, hydrogenation of propyne to propane is represented as:

$$HC\equiv C-CH_3 \xrightarrow{H_2,\ Ni} H_2C=CH-CH_3 \xrightarrow{H_2,\ Ni} H_3C-CH_2-CH_3$$

The alkanes, alkenes, and alkynes discussed thus far are **acyclic**. This means that the hydrocarbon chains are linear and do not form rings that close on themselves. Alkanes, alkenes, and alkynes have the ability to form closed rings. At least three carbon atoms are needed to form a ring so the simplest **cyclic** alkane is a form of propane called cyclopropane. Several simple cyclic alkanes are shown in Figure 15.3.

Cyclic compounds have different general formulas compared to their acyclic counterparts. The general formula for cyclic alkanes for instance is C_nH_{2n}.

Aromatic Hydrocarbons— Benzene

Aromatic hydrocarbons are unsaturated cyclic compounds that are resistant to addition reactions. The aromatic hydrocarbons derive their name from the distinctive odors they exhibited when discovered. Benzene is the most important aromatic compound. Because many other aromatic compounds are derived from benzene, it can be considered the parent of other aromatic compounds. Benzene molecular formula is C_6H_6 and its structural formula is

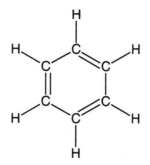

This structure is usually represented without the carbon and hydrogen atoms in either of two ways as shown in Figure 15.4.

Benzene was discovered in 1825 by Michael Faraday who identified it from a liquid residue of heated whale oil. Faraday called the compound bicarburet of hydrogen and its name was later changed to benzin by Eilhardt Mitscherlich (1794–1863) who isolated the compound from benzoin. Benzene's formula indicates it is highly unsaturated. This would suggest benzene

Figure 15.3
Alkane, alkenes, and alkynes can fold themselves to form cyclic compounds. Those represented above are cycloalkanes.

cyclopropane   cyclobutane   cyclopentane   cyclohexane

Figure 15.4
Two Alternate Ways of Representing Benzene

and other aromatic compounds should readily undergo addition reactions like the aliphatic compounds. The fact that benzene did not undergo addition puzzled chemists for a number of years. In 1865 Kekulé, as discussed earlier in this chapter, determined the correct structure of benzene to explain its behavior. Kekulé proposed that benzene oscillated back and forth between two structures so that all carbon-carbon bonds were essentially equivalent. His model for the structure of benzene is represented in Figure 15.5. In Figure 15.5, benzene is shown as changing back and forth between two structures in which the position of the double bonds shifts between adjacent carbon atoms. The two structures are called **resonance structures**.

A true picture of benzene's structure was not determined until the 1930s when Linus Pauling produced his work on the chemical bond. Benzene does not exist as either of its resonance structures, and its structure should not be considered as either one or the other. A more appropriate model is to consider the

structure of benzene as a hybrid of the two resonance structures. Each carbon atom in the benzene ring is bonded to two other carbon atoms and a hydrogen atom in the same plane. This leaves six delocalized valence electrons. These six delocalized electrons are shared by all six carbon atoms. This is demonstrated by the fact that the lengths of all carbon-carbon bonds in benzene are intermediate between what one expects for a single bond and a double bond. Rather than consider a structure that exists as one resonance form or the other, benzene should be thought of as existing as both resonance structures simultaneously. In essence, 1.5 bonds are associated with each carbon in benzene rather than a single and double bond. Using the hexagon symbol with a circle inside is more representative of the delocalized sharing of electrons rather than using resonance structures, although benzene is often represented using one of its resonance structures.

Because of its structure, benzene is a very stable organic compound. It does not readily undergo addition reactions. Addition reactions involving benzene require high temperature, pressure, and special catalysts. The most common reactions involving benzene involve substitution reactions. Numerous atoms and groups of atoms may replace a hydrogen atom or several hydrogen atoms in benzene. Three important types of substitution reactions involving benzene are alkylation, halogenation, and nitration. In alkylation, an alkyl group or groups substitute for hydrogen(s). Alkylation is the primary process involving benzene in the chemical industry. An example of alkylation is the production of ethyl benzene by reacting benzene with ethylchloride and an aluminum chloride catalyst:

Figure 15.5
Kekulé proposed that the structure for benzene resonated between two alternate structures in which the position of the double and single bonds switched positions.

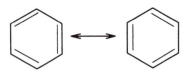

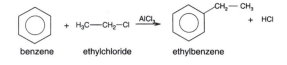

benzene   ethylchloride   ethylbenzene

In halogenation, halogen atoms are substituted for hydrogen atoms. The process of nitration produces a number of nitro compounds, for example, nitroglycerin. Nitration involves the reaction between benzene and nitric acid in the presence of sulfuric acid. During the reaction, the nitronium ion, NO^{2+}, splits off from nitric acid and substitutes on to the benzene ring producing nitrobenzene:

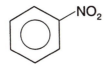

Just as an alkyl group is formed by removing a hydrogen atom from an alkane, removing a hydrogen atom from an aromatic ring produces an **aryl group**. Removing a single hydrogen atom from benzene produces **phenyl**, C_6H_6. The phenyl group bonds with carbon in many organic compounds. Benzene rings may fuse together to form fused ring aromatic compounds. A common fused ring is naphthalene. Naphthalene is used in mothballs and in the manufacture of many chemical products. Naphthalene has the distinctive odor of an aromatic compound. Its structural formula is

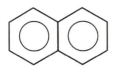

Some fused benzene rings have been implicated as **carcinogens** (cancer-causing agents). Several of these are found in tobacco smoke.

Millions of organic chemicals are derived from benzene. Many of these are familiar chemicals such as aspirin (Chapter 13) and several already mentioned in this section. Figure 15.6 shows some other common benzene derivatives. Besides the benzene

Figure 15.6
Several Common Benzene Derivatives

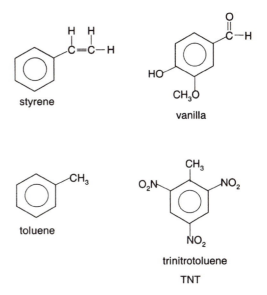

styrene

vanilla

toluene

trinitrotoluene
TNT

derivatives mentioned in this section, countless other products are based on the benzene ring. Cosmetics, drugs, pesticides, and petroleum products are just a few categories containing benzene-based compounds. Thousands of new benzene compounds are added annually to the existing list of millions.

The field of organic synthesis is like working with an atomic Lego set. Using just a few atoms, millions of compounds can be built. This process has occurred naturally throughout Earth's history, and in the last century chemists have been able to mimic it. This section dealt with the two main classes of hydrocarbons called the aliphatic and aromatic hydrocarbons. In the next section the concept of the functional group is examined. Functional groups are used to define organic families.

Functional Groups and Organic Families

A functional group is a small chemical unit consisting of atoms or a group of atoms and their bonds that is part of a larger organic molecule. The functional group is generally the part of the molecule where reactions take place. Organic compounds in the same family have the same functional group. The functional group for alkenes is the carbon-carbon double bond, and for alkynes it is the carbon-carbon triple bond. Aromatic compounds share the hexagonal ring functional group (Figure 15.7). A number of functional groups define several other large families of organic compounds.

Alcohols contain the hydroxyl group, OH. An alcohol can be represented as R-OH, where R is the rest of the molecule attached to the functional group (for instance an alkyl or aryl group). Alcohols take their name from the corresponding alkane molecule and using the ending "ol". For example, methanol (CH_3OH) is methane with the hydroxyl group replacing one of the hydrogen atoms (Figure 15.8). Ethanol, often called ethyl alcohol, has the formula C_2H_5OH. Ethanol results when sugar ferments and is sometimes referred to as grain alcohol. It is the alcohol found in alcoholic beverages. The hydroxyl group gives alcohols polar characteristics, and it is often used as a solvent in many chemical processes. **Denatured** alcohol refers to "spiking" ethanol with a substance making it unfit for human consumption. The substances used to spike ethanol are called denaturing agents. A common denaturing agent for ethanol is methanol. The oxidation

Figure 15.7
Functional Groups Defining Alkenes, Alkynes, and Aromatic Hydrocarbons

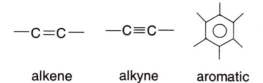

Figure 15.8
Some Common Alcohols

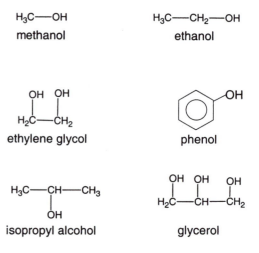

of methanol in humans results in the production of toxic formaldehyde.

Alcohols may contain more than one hydroxyl group. An alcohol containing two hydroxyl groups is called **dihydric**, and one with three is called **trihydric**. Ethylene glycol and glycerol (glycerin) are two examples of common alcohols with more than one hydroxyl group (Figure 15.8). Ethylene glycol is the primary ingredient used in antifreeze, and glycerol is used in soaps, shaving cream, and other cosmetic products.

Aromatic alcohols are called **phenols**. The simplest phenol, also called phenol, forms when a hydroxyl group replaces a hydrogen atom in the benzene ring. Phenol (carbolic acid) was used as an antiseptic in the 1800s. Today other phenol derivatives are used in antiseptic mouthwashes and in cleaning disinfectants such as Lysol. Phenols are easily oxidized, and this makes them ideal substances to use as antioxidants. By adding phenols such as BHT (butylated hydroxy toluene) and BHA (butylated hydroxy anisole) to food, the phenols oxidize rather than the food.

Ethers contain an oxygen atom singly bonded to two hydrocarbon units. The functional group is the oxygen in the general formula R′-O-R. The R and R′ in this general formula may be the same or different. Ethers are identified by naming the attached R groups follow by the word "ether." For example, CH_3—O—CH_2—CH_3 would be methyl ethyl ether. The most common ether is diethyl ether: CH_3—CH_2—O—CH_2—CH_3. This is sometimes referred to as ethyl ether or simply ether. Ether has been used as a general anesthetic since 1846 when William T. G. Morton, a Boston dentist, demonstrated its use at Massachusetts General Hospital. Morton anesthetized a patient who was having a tumor surgically removed from his jaw. The demonstration was so successful that ether was hailed as a miraculous substance that could alleviate pain. Today ether has largely been replaced by modern anesthetics. One of the major problems with ether is its extreme flammability.

An ether called methyl tert butyl ether (MTBE) has traditionally been used as a fuel oxygenate to reduce toxic air emissions from automobiles:

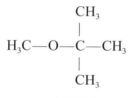

MTBE

Oxygenates are added to gasoline to give more complete combustion. This in turn reduces the amount of carbon monoxide produced. MTBE was widely used throughout the last part of the twentieth century for this purpose, but environmental concerns have curtailed its use in recent years. The problem stems from contamination of groundwater from leaking underground storage tanks and surface water from boats. At first, concerns revolved around odor and taste problems. MTBE has a low concentration threshold (20 to 40 parts per billion) for

odor and taste. More recently, EPA has listed MTBE as a potential carcinogen. While the EPA has not banned MTBE use, many states are regulating and even eliminating its use in gasoline. California, for instance, banned its use after December 31, 2002.

Aldehydes and **ketones** both contain the **carbonyl** functional group:

Aldehydes have at least one of their hydrogen atoms bonded to the carbon in the carbonyl group. In ketones, the carbonyl carbon bonds to other carbon atoms and not hydrogen atoms (Figure 15.9).

Aldehydes use the parent name of their corresponding alkanes and the ending "al." Therefore, the simplest aldehyde is methanal, but it goes by its common name formaldehyde. Aldehydes and ketones are often produced by the oxidation of alcohols. For example, formaldehyde is produced by the oxidation of methanol according to the following reaction:

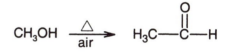

methanol   formaldehyde

Formalin is an aqueous solution of approximately 40% formaldehyde. It was traditionally used to preserve biologic specimens, but its identification as a mild carcinogen has curtailed its use as a preservative. Formaldehyde is widely used to produce synthetic **resins**. Resins are sticky, liquid organic compounds that are insoluble in water. They often harden when exposed to air. Many commercial types of glue are resins. Natural resins are produced by plants as a response to damage. When a plant suffers external damage, natural resins flow to the area and harden to protect the underlying

Figure 15.9
Both aldehydes and ketones contain the carbonyl group, but in aldehydes, the carbonyl carbon bonds to at least one hydrogen atom. In ketones, the carbonyl carbon bonds to two other carbons.

functional group

aldehyde   ketone

structure. An interesting note is that the resin lac, which is used in shellac, is a resin secreted by the insect *Laccifer lacca*. Some familiar natural resins are aloe, amber, and myrrh. Synthetic resins include a variety of familiar materials such as polystyrene, urethane, and epoxies.

The widespread use of formaldehyde to produce resins has led to some concerns about its affect on indoor air quality. Resins are used in plywood, particleboard, synthetic woods, and other building materials. The sweet pungent odor of formaldehyde can often be smelled in a lumber yard. When building materials containing formaldehyde are incorporated into structures, the formaldehyde enters the air through a process called outgassing. Formaldehyde has been identified as a mild carcinogen, and people susceptible to this chemical may experience a variety of reactions including rashes and flu-like symptoms. The use of formaldehyde foam insulation, which was used widely in homes, was discontinued in 1983.

The functional group for ketones shows that the basic ketone contains three carbon atoms. Ketones have the ending "one" so the simplest ketone is called propanone, but it goes by the common name acetone:

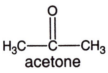
acetone

Acetone is widely used as an organic solvent, as a paint remover (nail polish remover is mostly acetone), and to dissolve synthetic resins.

Acetone and several other ketones are produced in the liver as a result of fat metabolism. Ketone blood levels are typically about 0.001%. Lack of carbohydrates in a person's diet results in greater metabolism of fats, which leads to a greater concentration of ketone in the blood. This condition is called ketosis. People on low carbohydrate diets and diabetics have problems with ketosis because of their increase dependence on fats for metabolism.

Carboxylic acids are organic acids that contain the **carboxyl functional group**, —COOH—. The carboxyl group is a combination of the carbonyl group and the hydroxyl group:

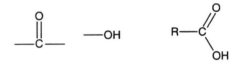

carbonyl + hydroxyl = carboxylic acid

IUPAC names of carboxylic acids come from their parent alkane and use the ending "oic." The simplest carboxylic acid contains two carbons, and therefore, is called ethanoic acid, CH_3COOH. Many carboxylic acids, though, are known by their common names, which reflects their natural origin. Some of these traditional names date back hundreds of years. The simplest carboxylic acid (IUPAC name methanoic acid) is formic acid. "Formica" is the Latin word for "ant." Formic acid, which is responsible for the stinging sensation of red ant bites, was originally isolated from ants. Acetic acid (ethanoic acid) comes from the Latin word "acetum" meaning "sour" and relates to the fact that acetic acid is responsible for the bitter taste of fermented juices. Pure acetic acid is sometimes called glacial acetic acid.

This is because pure acetic acid freezes at 17°C, which is slightly below room temperature. When bottles of pure acetic acid froze in cold labs, snow-like crystals formed on the bottles; the term glacial was thus associated with the pure form. Butyric acid comes from the Latin word "butyrum" for "butter"; butyric acid is responsible for the smell of rancid butter.

As discussed in Chapter 13, organic acids are some of the oldest known compounds. Salicylic acid is found in the plant family *Salix* and is the acid associated with aspirin (see Chapter 13). Several other familiar carboxylic acids are acetic acid, carbonic acid, and citric acid (Figure 15.10) More than one carboxyl group may be present in carboxylic acids, for example, citric acid contains three carboxyls. Oxalic acid, associated with the plant genus *Oxalis* which contains cabbage, rhubarb, and spinach, is a dicarboxylic acid containing just two carboxyl groups bonded together:

Oxalic acid is poisonous to humans, but its concentrations are generally too low in foods to be of concern, although rhubarb leaves are quite poisonous. Lactic acid is produced from the fermentation of lactose, which is the principal sugar found in milk. The taste and smell of sour milk is due to the production of lactic acid from bacterial fermentation. Lactic acid accumulates in our muscles during exercise and strenuous physical activity. It is responsible for the sore, aching feeling often associated with these activities. Benzoic acid is the simplest aromatic carboxylic acid.

Carboxylic acids are prepared by oxidizing alcohols and aldehydes using an oxidizing agent such as potassium perman-

Figure 15.10
Some Common Carboxylic Acids

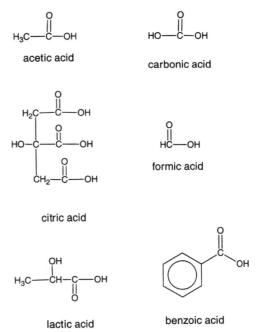

ganate ($KMnO_4$). For example, ethanol can be converted to acetic acid:

$$CH_3CH_2OH \xrightarrow{KMnO_4} CH_3COOH$$

Similar to inorganic acids, the reaction of carboxylic acids and bases produces carboxylic acid salts. Several of these salts are commonly used in foods and beverages as preservatives. The most common are salts from benzoic, propionic, and sorbic acids. The salts of these acids have names ending with "ate," and can often be found in the list of ingredients of baked goods and fruit drinks. Several common preservatives are shown in Figure 15.11.

Fatty acids are carboxylic acids containing an unbranched carbon chain and usually an even number of carbon atoms. Fatty acids do not occur freely in nature, but generally come from **esters** (esters are discussed later). A few common fatty acids and their sources are shown in Figure 15.12. Fatty acids are important in the production

Figure 15.11
Some Common Carboxylic Acid Salts Used
as Preservatives

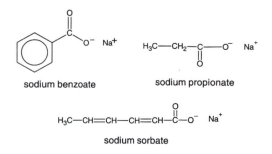

sodium benzoate   sodium propionate

sodium sorbate

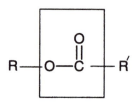

of soap. Before discussing soap production, let's examine another family of organic compounds called esters.

Esters are derived from carboxylic acids when these acids react with an alcohol in the presence of strong acid catalysts. The process of ester formation is called esterification. In esterification, the hydrogen atom in the alcohol's hydroxyl group combines with the hydroxyl portion of the carboxyl group to form water. The remaining segments of the acid and alcohol form the ester. Figure 15.13 represents the esterfication process. The functional group of esters is a carbon bonded to two oxygen atoms, one with a double bond and one with a single bond:

Esters are named by first using the name derived from the alcohol and then the stem of the acid name with the ending "ate." Acetate is derived from the common name acetic acid. The IUPAC name would be methyl ethanoate, with the ethanoate derived from acetic acid's IUPAC name ethanoic acid.

Many of the fragrances and tastes from plants are due to esters. The smells we perceive are generally due to a combination of esters, but often one ester fragrance will dominant. By mimicking these natural esters, the food industry has synthesized hundreds of different flavoring agents. Three of these are shown in Figure 15.14. Many **pheromones** are esters. Pheromones are chemical compounds used by animals for communication. Many medications are also esters. Aspirin is an ester of salicylic acid (see Chapter 13).

Esters can react with water in the presence of acid catalysts to produce an alcohol

Figure 15.12
Some common fatty acids and their sources. Note that the expanded structure is shown for butyric acid and condensed structures for the other acids.

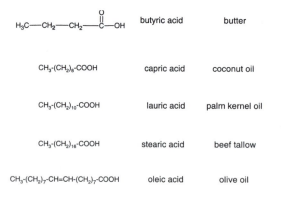

	butyric acid	butter
$CH_3-(CH_2)_8-COOH$	capric acid	coconut oil
$CH_3-(CH_2)_{10}-COOH$	lauric acid	palm kernel oil
$CH_3-(CH_2)_{16}-COOH$	stearic acid	beef tallow
$CH_3-(CH_2)_7-CH=CH-(CH_2)_7-COOH$	oleic acid	olive oil

Figure 15.13
In esterfication, an alcohol reacts with a carboxylic acid to produce an ester and water. The resulting ester contains an alcohol part and an acid part.

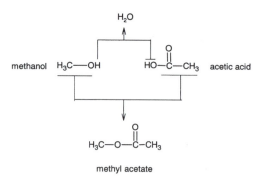

Figure 15.14
Esters are used as flavoring agents. Notice the similarity in the structure of the three flavors.

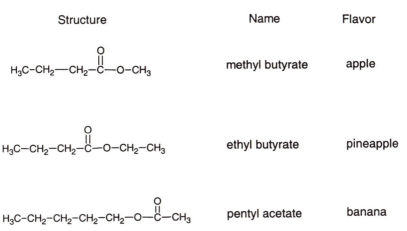

Structure	Name	Flavor
methyl butyrate	apple
ethyl butyrate	pineapple
pentyl acetate	banana

and a carboxylic acid. This process is called **hydrolysis**. Hydrolysis is the reverse of esterification. When hydrolysis takes place in a basic solution, a carboxylic acid salt forms in a process called **saponification**. Saponification is the procedure used to produce soap. Saponification is a two-step process in which an ester is initially hydrolyzed into a carboxylic acid and alcohol. During the second step, the carboxylic acid formed reacts with a base, such as sodium hydroxide or potassium hydroxide, to form a carboxylic acid salt. The traditional production of soap involves the use of animal fats and vegetable oils. The type of soap produced depends on both the type of fat or oil used (beef tallow, lard, coconut oil, palm oil, linseed oil, etc.) and the base. Fats and oils are esters formed from the combination of fatty acids and glycerols. Three fatty acids, either the same or different ones, attach to the three hydroxyl groups in the glycerol to form a triester (Figure 15.15). Numerous triesters may be formed in this process, and the general term for these triesters is triglycerides. During saponification of animal and plant products, the triglycerides are converted back to fatty acids and glycerol. The

fatty acids then react with the base to produce a carboxylic acid salt known commonly as soap (Figure 15.16).

The cleaning power of soaps can be attributed to the polar characteristics of the soap molecule. Long chain hydrocarbons are represented by the "R"s in Figure 15.16. This end of the molecule is nonpolar and tends to dissolve in nonpolar oils and grease (remember the rule from Chapter 11 that "like dissolves like"). The other end of the soap molecule is ionic and dissolves in water. The process of how soap works can be thought of as one end of the soap molecule attracting the grease and oil particles and the other end

Figure 15.15
A triglyceride forms when three fatty acids react with glycerol. When a triglyceride forms, three molecules of water are produced as a byproduct.

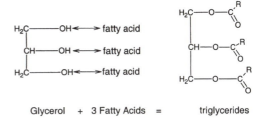

Glycerol + 3 Fatty Acids = triglycerides

Figure 15.16

In saponification, triglycerides are heated in a basic solution to give glycerol and soap. R represents long carbon chains.

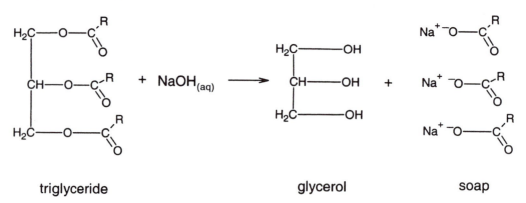

triglyceride   glycerol   soap

dissolving in water. This differential solubility pulls the oil molecules apart and suspends them in water in a process called emulsification. Because the oil and grease are largely responsible for holding the dirt to the fabric, once they have been broken up, the dirt also readily dissolves.

Soaps react with the calcium and magnesium ions in hard water to produce soap curd that greatly reduces its effectiveness. The curds are actually insoluble calcium and magnesium salts. Synthetic laundry detergents have replaced soap for cleaning clothes in the last half century. Synthetic detergents are made from petroleum. They work like soap except they do not react with magnesium and calcium ions to form insoluble precipitates and salts.

Polymerization of esters to produce polyesters is an important commercial process. Polyethylene terephthalate or PET is one of the most common plastics used in food containers (Table 15.4). This ester is formed by the reaction of ethylene glycol and terephthalic acid (Figure 15.17). PET and other polyesters consist of esters linked together. Notice that both terephthalic acid and ethylene glycol have two carboxyls and two hydroxyls, respectively. When a polyester such as PET is formed, a monomer con-

sisting of the ester is sandwiched between the remaining alcohol and carboxyl acid groups. Another acid group can attach to the alcohol and another alcohol group can attach to the acid end (Figure 15.18). Two new acid and alcohol ends are then available to continue the polymerization process. PET goes by a number of commercial names, the most generic of these is polyester. It is also the materials known as Dacron. Mylar is PET in the form of thin films. It is used in recording tapes and floppy disks.

The families of organic compounds discussed thus far have been limited to those containing carbon, hydrogen, and oxygen.

Figure 15.17
PET is a polyester formed from terphthalic acid and ethylene glycol.

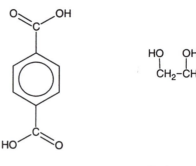

terephthalic acid   ethylene glycol

Figure 15.18
In the formation of PET, a monomer ester forms with a remaining alcohol (OH) end group and carboxyl group (COOH). These represent points to form two more esters and the process continues building the polymer.

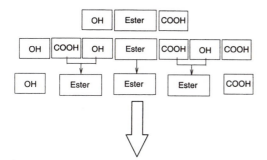

Amines are nitrogen-containing compounds that contain the amino functional group, $-NH_2$. Amines can be viewed as compounds in which an organic group(s), such as an alkyl or aryl group(s), is substituted for a hydrogen(s) in ammonia (NH_3). A **primary amine** is one in which only one of the hydrogen atoms in ammonia has been substituted for, **secondary amines** have two hydrogen atoms substituted, and **tertiary amines** have all three ammonia hydrogen atoms substituted.

Many amines, like other organic compounds, are known by their common names. Their IUPAC names follow that of naming of alcohols, but the ending amine is used. A common practice is to list the names of the groups attached to the nitrogen followed by the ending "amine." Thus, CH_3-NH_2 would be methylamine. The simplest aromatic is called aniline:

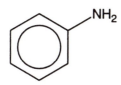

Amines may also be present in other ring structures similar to the cyclohydrocarbons. Ring structures that contain atoms other than carbon in the ring are called **heterocyclic** compounds. Heterocyclic amines are the most common type of heterocyclic compounds.

A number of common amines are familiar to all of us. **Alkaloids** are nitrogen-containing compounds extracted from plants. Caffeine is found in varying quantities in coffee beans (1 to 2% dry weight), tea leaves (2–5%), and cola nuts (3–4%). Caffeine is generally associated with coffee, tea, and soft drinks, but it is also present in many medications such as cough syrups. Nicotine (4–5% dry weight) comes from tobacco leaves, and cocaine is derived from the coca leaves (Figure 15.19). Cocaine was used as a local anesthetic during the early twentieth century, but in the last half cenutry other amine derivatives such as novacaine and benzocaine replaced its use. Morphine, codeine, and heroin are alkaloids that come from opium, which is obtained from poppy plants.

Histamine is an amine found in the cells of all animals (Figure 15.20). The release of histamine in the body is associated with involuntary muscle contraction and dilation of the blood vessels. Allergic reactions release histamine in the body and produce the accompanying side effects associated with hay fever such as coughing, sneezing, and runny nose. These effects can be alleviated with the use of antihistamines.

Antihistamines block the release of histamines, but also tend to produce drowsiness.

Figure 15.19
Caffeine and Nicotine

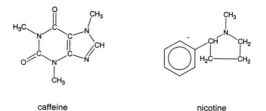

caffeine   nicotine

Figure 15.20
Histamine

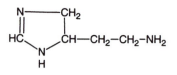

Epinephrine, commonly referred to as adrenaline, is an amine secreted in increased amounts during times of stress (Figure 15.21). Adrenaline increases the heart rate and blood pressure, releases sugar stored in the liver, and constricts blood vessels. It is sometimes administered to people in shock or during periods of acute asthma attacks.

The organic families and compounds presented in this section give only a glimpse of the vast array of organic chemicals present in our world. While each organic family is characterized by a unique functional group, many compounds contain more than one functional group and classifying a compound into a specific family can be difficult. Table 15.5 summarizes those families presented in this chapter. It can be used for a quick reference to the basic organic families.

Figure 15.21
Adrenaline

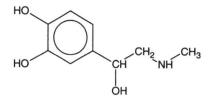

Petroleum

Petroleum is the principal starting material for the vast array of organic chemicals that dominate modern society. Chemicals produced from petroleum are collectively termed **petrochemicals**. The most prevalent use of petrochemicals is as a fuel source, but they are also the basic building blocks of plastics, synthetic fabrics, detergents, pesticides, and many other products. Petroleum is created over a process lasting millions of years as microscopic marine organisms are deposited in sediments and transformed by high pressure and temperature into oil and gas. Over time, the trapped oil and gas molecules migrate through sedimentary rock formations to areas called reservoirs where they become trapped. These reservoirs may be close to or far beneath the surface of the Earth. In certain regions, oil and gas seeps to the surface. Throughout history, surface seeps have been a source of lubricants and tars used for caulking and waterproofing. Medicinal products were distilled from oil obtained from surface sources by the alchemists. The industrial revolution created a greater demand for oil. In the mid-1800s, whale oil was the principal source of oil used in lamps, candles, and industry. The difficulty of obtaining whale oil, and its limited supply, made it and its derivatives quite expensive. This motivated the search for alternative sources of oil. It was known mineral oil could be obtained from the Earth. Besides known surface seeps, oil was sometimes found mixed with water when digging water wells. Individuals in both this country and in Europe dug wells in hope of finding oil that could be distilled into kerosene for lamps. Edwin L. Drake (1819–1880) drilled the first successful commercial well, striking oil on August 27, 1859, in Titusville Pennsylvania. This marked the start of the petrochemical industry.

Table 15.5
Organic Families

Family	Functional Group	Name
alkanes	—C—C—	end with "ane"
alkenes	—C=C—	end with "ene"
alkynes	—C≡C—	end with "yne"
aromatics		may contain either benzene or phenyl in name
alcohols	—OH	end with "ol"
ethers	—O—	named specifically as ether or may contain "oxy" in name
aldehydes	O‖ H—C—	end with "hyde" or "al"
ketones	O‖ —C—	end with "one"
carboxylic acids	OH / —C \\ O	many have common names, use "oic" ending with word acid
esters	O‖ —C—O—	may have "ate" ending, although many common names are used
amines	—NH₂	may contain "amine" in name

Crude oil is a mixture of hundreds of organic chemicals. Through a process of **fractional distillation**, the different components of petroleum can be segregated (Figure 15.22). Fractional distillation, or refining, involves separating the components according to their boiling points as the crude oil is heated. The lightest components, those containing between one and four carbon atoms, are gases at room temperature. Methane, ethane, propane, and butane are used as fuels and are converted into other alkenes such as ethylene and propylene for industrial use. Propane (boiling point –42°C) and butane (boiling point 0°C) can be liquified under pressure, and the term LP

Figure 15.22
Fractionating Column for Oil (Rae Déjur)

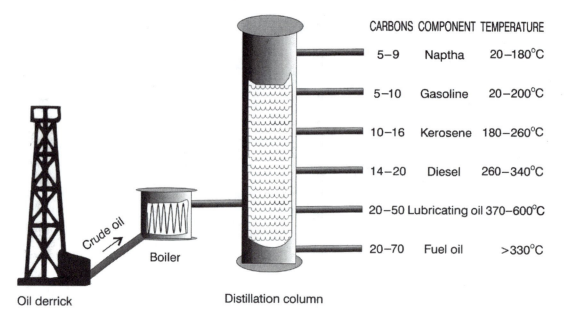

CARBONS	COMPONENT	TEMPERATURE
5–9	Naptha	20–180°C
5–10	Gasoline	20–200°C
10–16	Kerosene	180–260°C
14–20	Diesel	260–340°C
20–50	Lubricating oil	370–600°C
20–70	Fuel oil	>330°C

Oil derrick   Distillation column

is used for liquified petroleum. Propane is readily available at many gas stations for use in barbecues and camp stoves. Butane is used in lighters. Compounds containing between 4 and 10 carbon atoms, C_5 to C_{10}, include those that form petroleum ethers and gasoline. These compounds are obtained at temperatures between approximately 30°C and 200°C. The next component includes compounds ranging from C_{11} to C_{16}. Kerosene, the primary product in this range, is obtained between temperatures of 175°C and 275°C. Heating oils, C_{14} to C_{18} are obtained between 250°C and 350°C, and lubricating oils and paraffin wax are obtained above 300°C. The solid residue remaining after distillation consists of heavy asphalt and tar components.

It is interesting to note that refiners in the 1800s were primarily interested in obtaining kerosene and heavier components. Before the auto industry and the development of the modern petrochemical industry, there was relatively little need for gasoline and its asso-

ciated hydrocarbons. One of the residues obtained from refining oil is petroleum jelly. Petroleum jelly's, or petrolatum's, use as a skin conditioner was developed by an early pioneer in the oil industry named Robert Chesebrough. Chesebrough noticed that oil workers in western Pennsylvania applied a thick oily residue deposited on the drilling rods to their skin as a salve. Chesebrough collected the material and started to market it under the name Vaseline. Evidently, Vaseline was derived from the words "vase" and "line" denoting the fact that Chesebrough stored the product in vases.

Other processes are combined with fractional distillation to obtain the multitude of chemicals used in the petrochemical industry. **Cracking** is a process where large organic molecules are broken up into smaller molecules. Thermal cracking involves the use of heat and pressure. Catalytic cracking employs various catalysts to reduce the amount of heat and pressure required during the cracking process. **Alkylation** is a process in which smaller molecules are combined into larger molecules. Alkylation is used to produce branched hydrocarbons that have much higher octane numbers than straight chain hydrocarbons.

Organic Products

The accelerated use of synthetic chemicals in society, which commenced after World War II, has brought about a concern for their overuse. This has created a market for so-called natural or organic products; in reality, all food can be considered organic. The perception among many consumers of these products is that natural or organic products are safer and healthier, decrease damage to the environment, and are more nutritious. Companies have capitalized on these perceptions by packaging and advertising products as "natural" or "organic." The labels of these products often portray the perceived benefits by using a natural scene or natural colors, or proclaiming the presence or absence of certain substances. The words "includes" or "doesn't include" are often prominently displayed to convey the message that this product is acceptable or better than the competition.

While new markets have been created and new companies founded to meet the demand for natural products, the consumer is often misled or confused regarding these products. When the ingredient labels of many advertised "natural products" are examined, they are often found to include a number of synthetic chemicals. The words "natural" or "organic" should not be interpreted as meaning good or better, and the presence of synthetic chemicals does not mean bad. There is no difference in chemicals derived from natural sources or produced synthetically. A chemical is the same chemical whether it comes from a natural or synthetic source. Synthetic ascorbic acid, vitamin C, and natural ascorbic acid obtained from rose hips are identical. Although synthetic and natural chemicals are identical, there may be an advantage in using one over the other, for example, price, ease of preparation. It should also be noted that individuals may be allergic to certain chemicals, both natural and synthetic, so that it is important in many instances to know the actual ingredients in a product.

The federal government takes an active role in product labeling, and one area of concern during the last decade has been that of organic foods. The organic food market is a $9 billion industry in this country growing at a rate of 20% per year (compare to the average retail growth of 3%). The Organic Foods Production Act (OFPA) of 1990 was passed to establish standards for the growing, handling, processing, and distribution of organic food. A part of the OFPA is intended to address truth in advertising regarding organic food.

During the decade that followed this act, the government has tried to clarify the meaning of the term "organic" in agriculture. The regulations defining this act were set forth on April 21, 2001. The industry has eighteen months to comply with the regulations. The fact that it took the National Organic Standards Board, a panel of fifteen individuals overseeing the implementation of the OFPA, eleven years to develop the standards indicates the difficulty in defining what it means to be "organic." During the 1990s, over forty organizations had their own criteria for organic food. These organizations had diverse interests that included farmers, handlers, processors, and distributors.

According to the 2001 regulations, organic is taken to mean grown and prepared without the use of synthetic chemicals, sewage sludge (as fertilizers), or ionizing radiation. Synthetic pesticides, defoliants, fertilizers, and herbicides are forbidden.

Instead, natural methods such as tillage, mulching, crop rotation, mechanical weeding, and natural pest management must be employed. The regulations would lead one to believe that foods labeled as "organic" contain no synthetic chemicals, but there are exemptions and loopholes that allow synthetic chemicals to be used. Specifically, thirty-five synthetic chemical additives and stabilizers are exempt from regulation. Another aspect of the regulations regards how soon organic crops can be planted on soil to which synthetic chemicals had previously been applied. The new regulations specify a three-year time span has to pass. Opponents of this rule argue that certain synthetic chemical residues do not degrade in that time and remain in the soil after three years. To make sure that organic growers are adhering to the regulations, they must be certified by United States Department of Agriculture (USDA) certified agents.

The most apparent aspect of the OFPA to consumers regards product labeling. By October 2002, USDA labels will specify whether a food is organic. To be labeled organic, the food must be made with at least 95% organic material. Percentages can be included, so some producers may opt to label their foods as 100% organic. When foods are produced with at least 70% organic material, the label "made with organic material" will be used.

The OFPA also regulates animal production. Feed for organic livestock must be 100% organic. No hormones or antibiotics are allowed. Certain vaccinations are permitted. If an animal is treated for a condition with medication, it is no longer considered organic.

The OFPA defines organic food according to the USDA. There are calls by certain groups, such as those producing natural products, that similar regulations are needed in the cosmetic and health care industry. So far the Food and Drug Administration (FDA) has shown no inclination to follow the lead of the USDA to establish organic standards for cosmetics and health care products.

Summary

The material in this chapter traced the history of organic chemistry from Wöhler's synthesis of urea through Kekulé's structure of benzene. The millions of organic chemicals known to exist can be classified into a relatively small number of families, each defined by a common functional group. During the last century, chemists have discovered how to mimic nature to synthesize organic chemicals. A multitude of familiar products are natural or synthesized organic chemicals.

Although organic substances with unique properties tailored to meet the needs of the producer dominate modern society, new organic compounds that are safer, stronger, cheaper, and perform better are continually being sought. Some of the organic substances produced have created unpredictable health and environmental consequences, but that is part of the price for the benefits accrued from these substances. As we search for cures to diseases, seek safer and better products, and try to correct our past mistakes, organic chemicals and organic chemistry will continue to play a vital role in modern society.

16

Biochemistry

Introduction

Biochemistry is the study of chemicals associated with life and how these chemicals interact. Living matter consists primarily of the elements carbon, hydrogen, oxygen, and nitrogen. Many other elements, such as sulfur and phosphorus, are essential to life, but they do not comprise a large portion of living matter. For instance, the elemental mass composition of humans is approximately 65% oxygen, 18% carbon, 10% hydrogen, 3% nitrogen, 1.6% calcium, and 1.2% phosphorus, with the rest of the elements found in the human body totaling the remaining 1.2%. Elements combine to form the molecules that make up living cells. The molecules associated with life are called **biomolecules**, and the main categories of biomolecules include **carbohydrates, proteins, lipids**, and **nucleic acids**. Biomolecules form cells that comprise living matter, while biomolecular reactions provide the energy needed to sustain life. Biomolecules are constantly broken down in a process called **catabolism** and put together through **anabolism** as organisms carry out life processes such as growth, locomotion, and reproduction. The sum total of catabolic and anabolic reactions is known as **metab-**

olism. In this chapter, we examine the main classes of biomolecules and emphasize their role in human metabolism.

Carbohydrates

Carbohydrates are the predominant class of biomolecules. Carbohydrates derive their name from glucose, $C_6H_{12}O_6$, which was considered a hydrate of carbon with the general formula of $C_n(H_2O)_n$, where n is a positive integer. Although the idea of water bonded to carbon to form a hydrate of carbon was wrong, the term "carbohydrate" persists. Carbohydrates consist of carbon, hydrogen, and oxygen atoms, with the carbon atoms generally forming long unbranched chains. Carbohydrates are also known as **saccharides**, derived from the Latin word for sugar "saccharon."

Carbohydrates contain either an aldehyde or a ketone group and may be either **simple** or **complex**. Simple carbohydrates, known as **monosaccharides**, contain a single aldehyde or ketone group and cannot be broken down by hydrolysis reactions. Glucose and fructose are simple carbohydrates (Figure 16.1).° Glucose contains an aldehyde group and fructose a ketone group.

Figure 16.1
Glucose and fructose are simple carbohydrates. Notice that in glucose the carbonyl C is bonded to a hydrogen atom characteristic of an aldehyde, while in fructose the carbonyl C is bonded to two carbon atoms characteristic of a ketone.

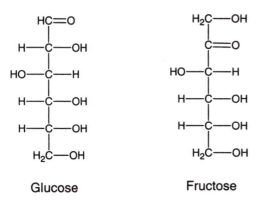

Glucose   Fructose

Although shown as open chain structures in Figure 16.1, most common sugars in nature exist as closed five- and six-carbon ringed structures known as pentoses and hexoses, respectively. Figure 16.2 illustrates the common ring structures for glucose and fructose. The glucose and fructose structures shown in Figure 16.2 are only one of several forms possible depending on the arrangement of atoms in the molecule. The forms shown in Figure 16.2 are the beta forms and are referred to as β-glucose and β-fructose, respectively. There are also alpha forms as shown in Figure 16.3. Notice that the alpha forms are identical to the beta

Figure 16.2
Most carbohydrates exist as closed rings rather than open chains.

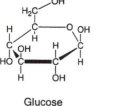

Glucose   Fructose

Figure 16.3
Alpha Forms of Glucose and Fructose

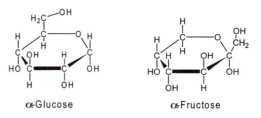

α-Glucose   α-Fructose

forms except that the atoms on the right side of the molecules have been flipped. Additionally, carbohydrates, as well as other biomolecules, may exist as L and D forms. The "L" and "D" come from the Latin *laevus* and *dexter* meaning left and right, respectively. The L and D refer to mirror images of each molecule and are related to the concept of **chirality**. A chiral object is one in which mirror images cannot be superimposed on one another. In achiral objects, mirror images are superimposable. For example, consider the mirror images of the heart and moon shown in Figure 16.4. The heart is achiral because the two images are identical when superimposed on each other. The moon is chiral, because the

Figure 16.4
The heart is achiral and the moon is chiral. The mirror images of chiral objects do not coincide when superimposed on each other.

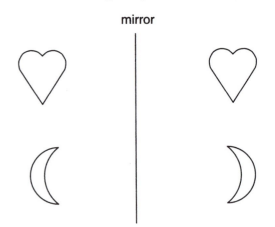

moon's mirror image does not coincide with the original object when superimposed on it.

Chiral molecules are often referred to as left handed and right handed. The concept of chirality can be seen if you hold your two hands in front of you with palms facing away. The left and right hands are similar and can be considered mirror images, but when you place one hand on top of the other, they do not match. In a similar fashion, molecules can have identical formulas but differ according to how they are oriented in space. Molecules that differ according to their arrangement in space are called **stereoisomers**. Stereoisomers can react differently and display different solubilities. This has significance in biochemical reactions in which one form of the isomer may be biologically active and another form may not, for example, enzymes. The sugars shown in Figures 16.2 and 16.3 are D forms. We will not concern ourselves with the rules for determining D and L forms of carbohydrates and other biomolecules, and will generally refer to biomolecules without reference to their specific form. You should be aware, though, that other books might use the designations α, β, D, and L when naming biomolecules.

D-Glucose is the most important and predominant monosaccharide found in nature. It also goes by the names dextrose and blood sugar. The term "blood sugar" indicates that glucose is the primary sugar dissolved in blood. Glucose is the principal energy source derived by cells. D-fructose, as its name implies, is abundant in fruits. Honey also has a lot of fructose. **Invert sugar** is a mixture of glucose and fructose.

Two monosaccharides bonded together comprise a **disaccharide**. Three common disaccharides are sucrose (common table sugar), maltose, and lactose. Sucrose is composed of glucose and fructose. Maltose is composed of two glucose molecules. Lac-tose is made up of glucose and galactose. When sucrose, maltose, and fructose are hydrolyzed, the monosaccharides that comprise these disaccharides are obtained:

sucrose + H_2O → glucose + fructose
maltose + H_2O → glucose + glucose
lactose + H_2O → glucose + galactose

Hydrolysis is very important in biochemical reactions and refers to a reaction in which a substance reacts with water causing the substance to break into two products. The structure of common table sugar, sucrose, is shown in Figure 16.5.

Polysaccharides are composed of many monosaccharides bonded together. Common polysaccharides are cellulose, starch, and glycogen. **Cellulose** forms the structural material of the cell walls of plants. Cellulose is a polymer of glucose and consists of thousands of glucose molecules linked in an unbranched chain (Figure 16.6).

Humans do not produce the enzymes, called **cellulases**, necessary to digest cellulose. Bacteria possessing cellulase inhabit the digestive tracts of animals such as sheep, goats, and cows giving these animals the ability to digest cellulose. Cellulase bacteria also exist in the digestive systems of

Figure 16.5
Common table sugar, sucrose, is a disaccharide composed of a hexose form of glucose bonded to a pentose form of fructose.

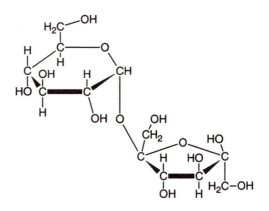

Figure 16.6
Cellulose consists of repeating glucose units bonded together.

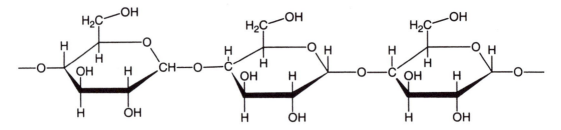

certain insects, such as termites, allowing these insects to use wood as an energy source. Although humans cannot digest cellulose, this carbohydrate plays an important role in the human diet. Carbohydrates that humans cannot digest are called fiber or roughage. Fruit, vegetables, and nuts are primary sources of fiber. These foods seem to have a cleansing effect on the large intestine, and such action is thought to speed the passage of cancer-causing materials through the intestine and reduce the risk of colon cancer. Fiber helps retain water in the digestive system, aiding the overall digestion process. Another benefit of fiber is believed to be its ability to lower blood cholesterol, reducing heart and arterial disease.

Starch is a α-glucose polymer with the general formula $(C_6H_{12}O_5)_n$, where n can be a number from several hundred to several thousand. Starch is the form that plants store carbohydrate energy. Grains, such as rice, corn, and wheat; potatoes; and seeds are rich in starch. Two principal forms of starch exist. One form is **amylose** and it comprises about 20% of all starches. Amylose is a straight chain form of starch containing several hundred glucose units. In the other form, called **amylopectin**, numerous glucose side branching is present. Starch is the primary food source for the human population, comprising nearly 70% of all food consumed. Starch is broken down in the digestive system starting in the mouth where

the enzyme amylase breaks the bonds holding the glucose units together. This process ceases in the stomach because stomach acid creates a low pH environment, but resumes in the small intestine. Glucose, maltose, and other small polysaccharides result from the digestion of starch.

In animals, most glucose is produced from starch and is not immediately needed for energy; therefore, it is stored as glycogen. Because glycogen's function in animals is similar to that of starch in plants, it is sometimes referred to as animal starch. Glycogen is similar in structure to amylopectin, but it is larger and contains more glucose branching. When glucose blood levels drop, for instance during fasting or physical exertion, glycogen stored in the liver and muscles is converted into glucose and used by cells for energy. During periods of stress, the release of the **hormone** epinephrine (adrenaline) also causes glucose to be released from glycogen reserves. When blood glucose is high, such as immediately following a meal, glucose is converted into glycogen and stored in the liver and muscles.

The transformation of glucose to glycogen and glycogen back to glucose enables humans to regulate their energy demands throughout the day. This process is disrupted in individuals who have a form of diabetes known as diabetes mellitus. Diabetes mellitus occurs in individuals whose pancreas produces insufficient amounts of insulin or whose cells have the inability to use insulin. Insulin is a hormone responsible for signaling the liver and muscles to store glucose as glycogen. In Type 1 diabetes, called insulin-dependent diabetes mellitus (IDDM), the body produces insufficient amount of insulin. This type of diabetes occurs in youth under the age of 20 and is the less prevalent form (about 10% of diabetics carry this form). Type 1 diabetes is controlled using insulin injections and regulating the diet. Type II diabetes, known as insulin-independent diabetes mellitus (IIDM), is the most common form of diabetes and is associated with older individuals (generally over 50) who are overweight. In this form of diabetes, individuals produce adequate supplies of insulin, but the cells do not recognize the insulin's signal, and therefore, do not capture glucose from the blood. This type of diabetes is regulated with drugs and a strictly controlled diet.

In recent years, the consumption of carbohydrates in the form of refined sugars has received significant attention from the health professionals. The annual consumption of sugar in the United States is about 50 pounds per person or just under one pound per week. The adverse effects of excessive sugar in the diet include obesity, increase risk of heart disease, diabetes, tooth decay, and disruptive behavior such as hyperactivity in children. Because of these problems, the food industry has employed a number of synthetic or artificial sweeteners in place of sugar. While these artificial sweeteners may reduce the use of sugars, they have also been linked to adverse health problems such as cancer. Saccharin is the oldest synthetic sweetener. It was discovered in 1879 and has been used since 1900. During the 1970s, rats feed large doses of this substance developed bladder tumors, and the federal government proposed banning this substance.

Public resistance prevented its ban, but foods containing saccharin must carry a warning label. Aspartame, known commonly as Nutra-Sweet, is the predominant synthetic sweetener in use today. It consists of two amino acids: aspartic acid and phenylalanine (see section on proteins). It is approximately 160 times sweeter than sucrose. The health issues associated with aspartame are associated with the disease phenylketonuria (PKU). PKU results when excess amounts of phenylalanine occur in individuals due to the lack of the enzyme needed to catalyze the essential amino acid phenylalanine. This genetic condition, if not treated, causes brain damage resulting in mental retardation. Because aspartame contains phenylalanine, it is suggested that infants under two years of age not ingest any foods containing synthetic sweeteners. Cyclamates were once a popular sweetener used in this country, but they were banned in 1969 because of possible carcinogenic effects. Their use is permitted in Canada. A synthetic sweetener approved for use in the United States in 1998 is sucralose. Sucralose is a chlorinated substituted chemical prepared from sucrose itself. It is 600 times sweeter than sucrose, passes unmodified through the digestive system, and unlike aspartame, is heat stable so it can be used in cooking.

Although the use of synthetic sweeteners is increasing annually, the quantity of sucrose is also increasing. This apparent contradiction seems to be related to the increased production of low fat foods in recent years. It seems the fat calories of many low-fat foods have been replaced by sugars. Questions persist regarding the safety of artificial sweeteners, but the consensus is that moderate intake of these substances do not pose a risk to human health. The federal government in approving synthetic sweeteners considers guidelines regarding safe daily limits. The **acceptable daily intake** (ADI) is considered the amount of sweetener that can be consumed daily over a lifetime without adverse effects. For example, the ADI of aspartame is 50 milligrams per kilogram of body weight. No ADI has been established for saccharin, but the FDA limits the amount of saccharin per serving to 30 milligrams.

Lipids

Lipids are biomolecules characterized by their solubility properties. Lipids are molecules that are insoluble in water and soluble in organic solvents. **Triglycerides** are the major class of lipids, comprising 95% of all lipids. Triglycerides include fats and oils. **Fats** and **oils** are classified according to their state at room temperature. Fats are solid at room temperature, while oils are liquid at room temperature. Two other classes of lipids are the **phospholipids** and the **steroids**. Lipids serve a number of functions in the organisms. They provide an energy source to supplement stored glycogen when glucose supplies are low, are principal components of cell membranes, form protective waxy coatings in plants, and serve as the starting material for the synthesis of other biomolecules such as hormones.

Triglycerides are made from three fatty acids attached to a glycerol molecule. The basic structure of fatty acids and triglycerides was discussed in Chapter 15 in the section on organic acids. The fatty acids in a triglyceride determine the triglyceride's nature. One important characteristic of fatty acids is the degree of saturation. **Saturated fatty acids** contain only single bonds and contain the maximum number of hydrogen atoms. A **monounsaturated fatty acid** has a single double bond. The carbon atoms at the double bond are called unsaturated. **Polyunsaturated fatty acids** have more than one double bond per molecule, and at

each of these double bonds, the carbons are unsaturated (Figure 16.7). The melting point of triglycerides is dependent on the degree of saturation. Saturated triglycerides exist as solid fats. As the fat becomes unsaturated, its melting point decreases; therefore, polyunsaturated triglycerides exist as oils rather than fats.

Health professionals concerned about cardiovascular disease recommend monounsaturated and polyunsaturated fatty acids in the diet as opposed to saturated fats. Oils that have high degrees of unsaturation are perceived as healthier compared to those with greater saturation. To rank the degree of unsaturation, a measure called the **iodine number** can be used. The iodine number is the mass of iodine in grams that can be added to 100 grams of a triglyceride. Because the degree of unsaturation is related to the number of double bonds the molecule contains, and the double bonds represent points where iodine atoms can be added, a higher iodine number means a greater degree of unsaturation. Table 16.1 gives iodine numbers and relative percentages of the different types of fatty acids in several common fats and oils.

Figure 16.7
Fatty acids can be saturated, monounsaturated, or polyunsaturated depending on whether the molecule contains double bonds. Where the double bond(s) exists, the carbon atoms are unsaturated.

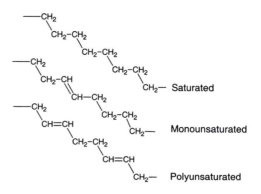

Table 16.1
Fatty Acid Composition and Iodine Numbers

Fat/Oil	Saturated	Monounsaturated	Polyunsaturated	Iodine Number
Coconut Oil	90	8	0	8
Butter	63	36	1	35
Palm Kernel	81	18	1	37
Beef Fat	47	50	3	40
Lard	41	57	2	55
Olive Oil	10	85	5	84
Peanut Oil	11	56	26	95
Corn Oil	14	52	34	110
Canola Oil	4	60	35	115
Safflower Oil	7	17	75	145

Phospholipids are lipids that contain phosphorus. They are similar in structure to triglycerides, except rather than having three fatty acids bonded to a glycerol one of the fatty acids in a triglyceride is replaced by phosphoric acid and an amino alcohol group. Replacing a single fatty acid with the phosphorus group changes the solubility characteristics of the molecule. The fatty acid end of a phospholipid is soluble in fats and oils, while the phosphorus end is soluble in water. The differential solubility character of the opposite ends of phospholipids makes them ideal **emulsifiers**. An emulsifier is a substance that mixes with both fats and water, and allows emulsion. Emulsion is the ability of a lipid and water to mix. The ability of phospholipids to mix in both water and oil makes them important substances for transporting lipids across cell walls in organisms. Phospholipids are a key substance in the cell walls of plants and animals. Common phospholipid groups found in all living tissue are the lecithins. Lecithins are widely employed in the food industry to assist in the mixing of fats and oil into water. Commercial lecithin is extracted from soybeans and egg yolks, and can be found as an ingredient in many food products, for example, margarine, candy bars, cookies, chocolate chips, and so on.

The third class of lipids is steroids. Included in this category of lipids are cholesterol, bile salts, and sex hormones. Steroid structures contain fused rings consisting of three six-carbon rings and a five-carbon ring:

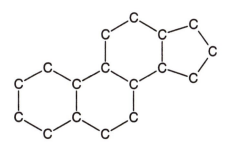

Figure 16.8
Cholesterol

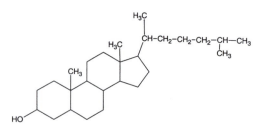

Cholesterol is a steroidal alcohol or **sterol**. (Figure 16.8). It is the most abundant steroid in the human body and is a component of every cell. Cholesterol is produced in the liver, and a number of other substances are synthesized from it including vitamin D, steroid hormones, and bile salts. Cholesterol is commonly associated with cardiovascular disease. High blood serum cholesterol levels are often correlated with excessive plaque deposits in the arteries, a condition known as **arteriosclerosis** or hardening of the arteries. Genetics plays a significant role in blood cholesterol levels, but reducing the amount of saturated fat in the diet can often control elevated cholesterol. Drugs are also employed to reduce cholesterol levels. Bile is produced in the liver from cholesterol and stored in the gall bladder. Bile acts as an emulsifying agent enabling triglycerides to mix with water in the small intestine where enzymes act to digest fats. Steroidal hormones, including the sex hormones, are also steroids. The sex hormones fall into three major groups. These are **estrogens**, which are female sex hormones; androgens, which are male sex hormones; and progestins, which are pregnancy hormones. The basic steroid structure is present in the three main hormones in each of these groups (Figure 16.9).

Biochemists have succeeded in producing synthetic steroids that have found wide use in a number of applications. Oral contraceptives used for birth control are based on structures that resemble progesterone. Oral contraceptives were first marketed in 1960. Synthetic progestins act by producing a state of false pregnancy that prevents the female user from ovulating. Synthetic sex hormones have also been employed, mostly illegally, to boost athletic performance. Anabolic steroids are testosterone derivatives that serve to build muscle mass and speed recovery from strenuous physical activity. The advantage gained by the use of these performance-enhancing drugs is not without cost. Side effects include cancer, heart disease, atrophy of testicles, disruption

Figure 16.9
Estradiol, testosterone, and progesterone are the principal sex hormones from the estrogen, androgen, and progestin groups, respectively. These hormones are synthesized in the human body starting with cholesterol.

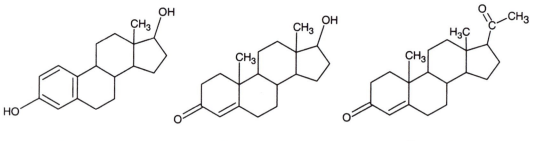

Estradiol   Testosterone   Progesterone

of the menstruation cycle, kidney damage, and the appearance of masculine traits in females.

Proteins

Proteins are the third major class of biomolecules. The word "protein" comes from the Greek word "proteios" meaning primary. Protein is the primary material composing cells. Fifty percent of the human body is composed of proteins, and it is estimated that roughly 100,000 different proteins are found in humans. Protein is a major component of structural and connective tissue found in skin, ligaments, bones, muscles, and tendons. Digestive enzymes, insulin, and other hormones are proteins.

Proteins are polymers made up of **amino acid** units that form long chains. Amino acids are molecules that contain an amino group, NH_2, and a carboxyl group, $COOH$, which characterize carboxylic acids. All amino acids contain a simple chemical backbone consisting of a carbon atom with an amino group, a carboxyl group, and a side chain group attached to the carbon:

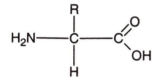

The side chain differentiates one amino acid from another. The simplest amino acid is glycine in which the side group consists of a single hydrogen atom. Proteins found in nature are composed of 20 standard amino acids (Figure 16.10).

Humans have the ability to produce about half of the twenty basic amino acids. The other ten that must be obtained from food are called **essential amino acids**.

A sequence of up to 50 amino acids is called a **peptide**, and the bonds between individual amino acid units are called **peptide bonds**. Peptide bonds form between the amino group of amino acid and the carboxyl group of an adjacent amino acid in a peptide chain. The process of peptide formation is illustrated in Figure 16.11, which shows how two amino acids bond together to form a **dipeptide**. Two amino acids form a dipeptide, three form a tripeptide, several amino acids make an oligopeptide, and more than ten comprise a **polypeptide**. Proteins can be considered polypeptides of more than 50 amino acids. Peptides are named by using the abbreviations for their constituent amino acids separated by dashes. By convention, the peptides start at the nitrogen-containing end and the sequence is read from left to right, for example, Tyr-Gly-Bly-Phe-Met. Numerous hormones secreted by the body's pituitary, hypothalamus, and endocrine glands are polypeptides. Insulin is a hormone that stimulates glucose uptake from the bloodstream. Another polypeptide glucagon stimulates the release of glucose from the liver. Other polypeptides release growth hormones, stimulate production of steroidal hormones, and block nerve messages to control pain.

When the number of amino acids in a polypeptide chain reaches more than fifty, a protein exists. The structure of both polypeptides and proteins dictate how these biomolecules function. There are several levels of structure associated with polypeptides and proteins. The sequence of the amino acids forming the backbone of the protein is referred to as the **primary structure**. A different order or even a minor change in an amino acid sequence creates an entirely different molecule. Just reversing the order of amino acids in a dipeptide changes how the dipeptide functions. An example of this is sickle-cell anemia. Sickle-cell anemia is a genetic disorder that occurs when the amino acid valine replaces

Figure 16.10
The twenty basic amino acids that can be grouped to form proteins. Abbreviations are given in parentheses. Those shown in bold are essential amino acids for humans. Arginine is essential in children, but not adults.

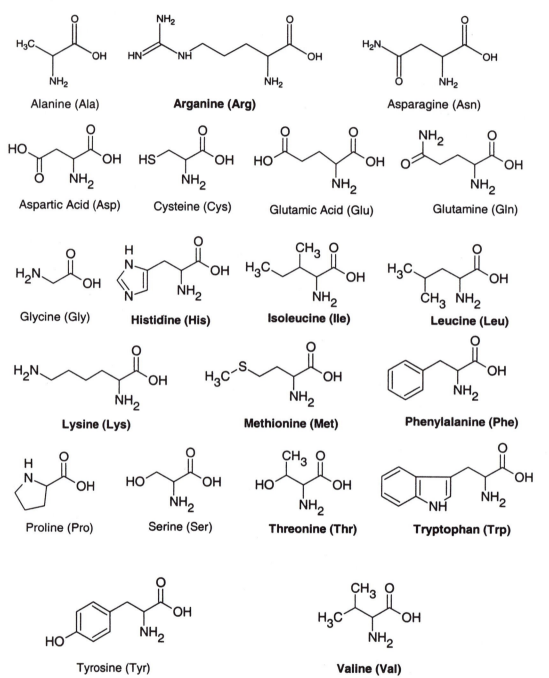

Figure 16.11

The amino and carboxyl ends of two amino acids unite to form a peptide bond. In the process, water is produced. Notice that the dipeptide form has an amino and a carboxyl group at its end allowing other amino acids to bond to the structure.

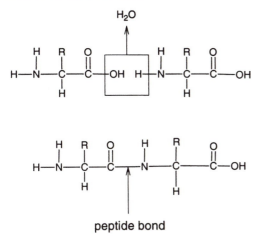

Figure 16.12

In neutral solutions, amino acids acquire a positively charged nitrogen end and a negatively charged carbon end.

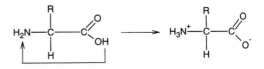

acquires a positive charge and the carboxyl end a negative charge (Figure 16.12). The transfer of the proton within the molecule creates a **zwitterion**, which is a molecule with a positive charge on one atom in the molecule and a negative charge on another atom. The overall charge of the molecule remains neutral. In acidic solutions, a proton can add to the COO^- creating a positively charged ion, while in basic solutions a proton can be donated from the nitrogen end creating a negatively charged ion. In a similar fashion, the side chains of certain amino acids can take on a polar character and have either an acidic or a basic character.

glutamic acid an eight sequence peptide. The presence of valine produces a form of hemoglobin called hemoglobin S.

normal hemoglobin   Val-His-Leu-Thr-Pro-Glu-Glu-Lys

sickle-cell hemoglobin   Val-His-Leu-Thr-Pro-Val-Glu-Lys

Sickle-cell anemia causes red blood cells to assume a sickle shape. The abnormal blood cells disrupt blood circulation, which can lead to infections, organ damage, and chronic pain.

In addition to the primary structure, proteins also exhibit **secondary, tertiary**, and **quaternary** structure. The overall structure of proteins is related to several factors. Primary among these factors is the electrostatic nature of amino acids. The structures displayed in Figure 16.10 do not show the charge distribution displayed by amino acids. In neutral solutions, the carboxyl group tends to donate a proton (hydrogen ion) to the amino group. The transfer of a proton means the amino end of the molecule

The difference in the charge characteristics of different parts of an amino acid molecule results in attractive and repulsive forces within a polypeptide or protein. These forces cause the molecule to coil, fold, and acquire various shapes that produce different structural properties. These structural properties dictate, largely, how proteins function. Secondary structure of a protein refers to how the atoms comprising the backbone are arranged in space around a central axis. Secondary structure results from the hydrogen bonding between hydrogen atoms of the amino group and the oxygen on the carbonyl group of another amino acid. Two common secondary structures exhibited by proteins are alpha helices and beta sheets. An alpha helix structure resembles a coiled spring and slinky beta sheets arise between parallel protein chains arranged along a sheet forming an alternating zigzag pattern (Figure

Figure 16.13
The alpha helix and beta sheets are two types of secondary structures exhibited by polypeptides and proteins. (Rae Déjur)

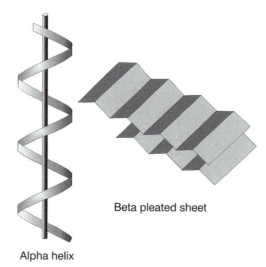

Beta pleated sheet

Alpha helix

16.13). Tertiary structure refers to the three-dimensional shape acquired by the protein due to the forces between amino acid side chains within a single polypeptide chain. The tertiary three-dimensional structure results from the folding of polypeptide chains. Quaternary structure results from the interactivity of different polypeptide chains within a protein. The different levels of protein structure can be compared to a telephone cord. The linear sequence of amino acids comprises the primary structure, the coil the secondary structure, and folding of the coil when the receiver is hung up gives the tertiary structure. The quaternary structure would result when two or more cords became intertwined. The overall structure of proteins leads to two general shapes. Fibrous proteins are long and slender fibers. These proteins are present in muscles, tendons, ligaments, and other structural tissue. The most prevalent protein in the human body is the fibrous protein called collagen. Collagen is a major constituent of tendon, skin, bone, ligaments, and cartilage. Boiling

it in water denatures it, and when allowed to cool, gelatin is produced. Globular proteins have a spherical shape and move through the circulatory system. Examples of globular proteins include insulin, hemoglobin, and myoglobin.

As mentioned previously, structure is critical to how proteins function. Denaturation refers to the process that disrupts protein structure. Heat is a primary cause of protein denaturation. When eggs are cooked, the transformation of clear transparent egg white to a solid white material is due to the denaturation of albumin. Heat causes the hydrogen bonds to break and unravels the globular albumin to produce the fibrous white cooked egg. Denaturation of protein enzymes presents a severe problem when body temperature reaches levels above 105°F. These temperatures inactivate enzymes that control body process, and many systems no longer function. This problem was tragically illustrated during the August 2001 training camp of the Minnesota Vikings with the heat-related death of an offensive lineman whose body temperature rose above 108°F. X-rays or other forms of radiation may also cause denaturation. Damage from radiation may lead to cancer or genetic disorders. Yet, another means of denaturation is due to toxic chemicals such as benzene, heavy metals, or chlorinated hydrocarbons.

Proteins composed solely of amino acid units are referred to as simple proteins. Many proteins contain other chemical groups in addition to amino acids. These proteins are called conjugated proteins. Proteins containing a carbohydrate unit are called glycoproteins, those with a nucleic acid are called nucleoproteins, and those combined with a lipid are called lipoproteins. High-density lipoproteins (HDLs) and low-density lipoproteins (LDLs) are included in a cholesterol test. LDLs, which are mainly lipids, carry fats and large quantities of cholesterol made in the liver to body cells. HDLs, which are primarily protein, transport cholesterol back to the liver. LDL and HDL are often referred to as bad and good cholesterol, respectively. Although both these contribute to the overall blood cholesterol, it is more important for individuals to focus on how cholesterol is distributed between HDL and LDL rather than a total cholesterol level. For this reason, the ratio of either the total cholesterol or LDL to HDL is a better indicator of health risk than the number indicating total cholesterol. For men an acceptable LDL:HDL cholesterol ratio is 4:5 and for women it is 3:5.

Nucleic Acids

Nucleic acids are biomolecules that pass genetic information from one generation to the next. Nucleic acids contained in DNA, deoxyribonucleic acid, and RNA, ribonucleic acid, are responsible for how all higher organism develop into unique species. DNA is present in the nucleus of all cells where segments of DNA comprise genes. DNA carries the information needed to produce RNA, which in turn produces protein molecules. A simplified view of the role of nucleic acids is to produce RNA, and the role of RNA is to produce proteins.

Just as the basic unit of proteins is the amino acid, the basic unit of nucleic acids is the nucleotide. Nucleic acids are polymers of nucleotides. Nucleotides consist of three parts: a sugar, hydrogen phosphate, and a nitrogen or amine base. The sugar unit in DNA is 2-deoxyribose and in RNA it is ribose. The "2-deoxy" means an oxygen atom is absent from the second carbon of ribose (Figure 16.14). Hydrogen phosphate bonds to the sugar at the number 5 carbon, and in the process a water molecule is formed. A nitrogen-containing base bonds

Figure 16.14
Shows that 2-deoxyribose and ribose are identical except for an oxygen atom absent on the second carbon. Carbons are numbered for future reference.

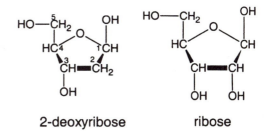

2-deoxyribose   ribose

at the number 1 carbon position also to produce water. The combination of the sugar, hydrogen phosphate, and base produces a nucleotide (Figure 16.15).

The sugar and phosphate groups form the backbone of a nucleic acid, and the amine bases exist as side chains. A nucleic acid can be thought of as alternating sugar-phosphate units with amine base projections:

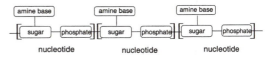

In DNA the sugar molecule in the nucleotide is 2-deoxyribose, and in RNA it is ribose. The amine bases in DNA are adenine, thymine, cytosine, and guanine, symbolized by A, T, C, and G, respectively. RNA contains adenine, cytosine, and guanine, but thymine is replaced by the based uracil (Figure 16.16). The primary structure of nucleic acids is given by the sequence of the amine side chains starting from the phosphate end of the nucleotide. For example, a DNA sequence may be –T-A-A-G-C-T.

Like proteins, DNA and RNA exhibit higher order structure that dictates how these molecules function. Determining the structure of DNA challenged the world's foremost scientists during the middle of the

Figure 16.15

A nucleotide is formed when hydrogen phosphate derived from phosphoric acid combines with a sugar and nitrogen base. Several different nitrogen bases may form a nucleotide, and the hexagon is used as a general symbol for any one of these.

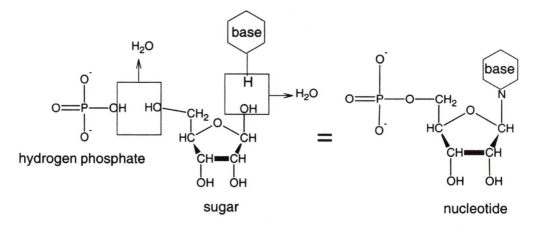

twentieth century. The puzzle of DNA's structure was solved by James Watson (1928–) and Francis Crick (1916–) in 1953. Watson and Crick arrived at a double helix model consisting of two nucleotide strands held together by hydrogen bonds (Figure 16.17). One way to picture the DNA molecule is to consider a ladder twisted into a helical shape. The sides of the ladder represent the sugar-phosphate backbone, while the rungs are the amine base side chains. The amine bases forming the rungs of the ladder exist as complementary pairs. The amine base on one strand hydrogen bonds to its complementary base on the opposite strand. In DNA, cytosine is always hydrogen bonded to guanine, and adenine is always hydrogen bonded to thymine. The fact that cytosine always lies opposite guanine and adenosine always opposite thymine in DNA is referred to as base pairing, and the C-G and A-T pairs are referred to as complementary pairs. Base pairing can be compared to two individuals shaking hands. Both individuals can easily shake hands when both use either their right or left hands. However, when one person extends the left hand and another the right, the hands do not match. In a similar fashion, the base pairing in DNA is necessary to produce a

Figure 16.16

The amine bases making up DNA and RNA. Thymine generally occurs only in DNA and uracil only in RNA.

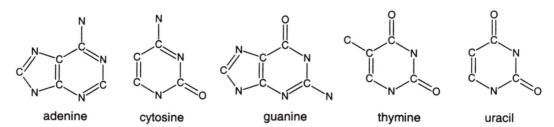

Figure 16.17
DNA Molecule (Rae Déjur)

Figure 16.18
In DNA replication, the two strands making the double helix unravel and nucleotides pair against their complementary nucleotide in the original strands. In this manner, the DNA can replicate itself.

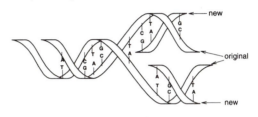

neatly wound double helix. In DNA replication, the two strands unwind and free nucleotides arrange themselves according to the exposed amine bases in a complementary fashion. DNA replication can be compared to an unzipping a zipper. As the DNA strands unzips, complementary nucleotides match up with the unpaired nucleotide. In this manner DNA replicates itself (Figure 16.18).

Genes, residing in the chromosomes, are segments of the DNA molecule. The sequence of nucleotides, represented by their letters, corresponding to a specific gene may be hundreds or even thousands of letters long. Humans have between 50,000 and 100,000 genes contained in their 46 chro-

mosomes, and the genetic code in humans consists of roughly five billion base pairs. The process by which the information in DNA is used to synthesize proteins is called transcription. Transcription involves turning the genetic information contained in DNA into RNA. The process starts just like DNA replication with the unraveling of a section of the two strands of DNA. A special protein identifies a promoter region on a single strand. The promoter region identifies where the transcription region begins. An enzyme called RNA polymerase is critical in the transcription process. This molecule initiates the unwinding of the DNA strands, produces a complementary strand of RNA, and then terminates the process. After a copy of the DNA has been made, the two DNA strands rewind into their standard double helix shape. The RNA strand produced by RNA polymerase follows the same process as in DNA replication except that uracil replaces thymine when an adenosine is encountered on the DNA strand. Therefore, if a DNA sequence consisted of the nucleotides: C-G-T-A-A, the RNA sequence produced would be G-C-A-U-U. The transcription process occurs in the cell's nucleus, and the RNA produced is called messenger RNA or mRNA. Once formed, mRNA moves out of the nucleus into the cytoplasm where the mRNA synthesizes proteins. The transfer of genetic information to produce proteins from mRNA is called translation. In the cytoplasm, the mRNA mixes with ribosomes and encounters another type of RNA called transfer RNA or tRNA. Ribosomes contain tRNA and amino acids. The tRNA translates the mRNA into three letter sequences of nucleotides called codons. Each three letter sequence corresponds to a particular amino acid. Because there are four nucleotides (C, G, A, and U), the number of different codons would be equal to 4^3 or 64. Because there are only twenty standard amino acids, several codons may produce the same amino acid. For example, the codons GGU, GGC, GGA, and GGG all code for glycine. Three of the codons serve as stop signs to signal the end of the gene. These stops also serve in some cases to initiate the start of a gene sequence. The sequential translation of mRNA by tRNA builds the amino acids into the approximately 100,000 proteins in the human body.

Biotechnology

Since Watson and Crick's discovery of the structure of DNA in 1953, advances in biochemistry, molecular biology, and genetics have had tremendous impacts on modern society. No area has been more affected than human health and modern medicine. Scientists continue to develop more precise methods for recognizing genetic diseases, and gene therapies hold promise for curing some of these diseases. In addition to advances in human health, advances in biochemistry and molecular biology affect numerous other fields. These include agriculture with genetically engineered plants and animals, forensics that uses DNA fingerprinting in crime analysis, and bioremediation that uses genetically modified microorganisms to clean contaminated areas. While advances in biotechnology have created new opportunities, they have also presented society with difficult choices concerning the use of this technology. Controversial issues such as human cloning, stem cell research, and alternative forms of reproduction such as artificial insemination receive wide coverage in the popular media. These issues will increasingly capture the public's attention and provide a source of debate as further scientific discoveries are made as the twenty-first century unfolds. To conclude this chapter, we touch upon a few modern biochemical applications, emphasizing the science behind the applications.

The Human Genome Project is an international project to identify all the genes in the nucleus of a human cell. In short, the Human Genome Project seeks to read the biochemical blueprint that is responsible for making a human being. This blueprint, termed the human genome, consists of the sequence of base pairs on all 46 chromosomes. By mapping the human genome, scientists hope to understand the relationship between different human characteristics and diseases and genes. For example, a gene located on the seventh chromosome carries cystic fibrosis.

A technique that has revolutionized the field of molecular biology during the last decade is the polymerase chain reaction (PCR). This method, conceived by Kary B. Mullis (1944–) in 1983, involves making multiple copies of fragments of DNA in a short time. Mullis received the 1993 Nobel Prize in chemistry for his work. PCR mimics DNA's natural ability to replicate itself. It involves three basic steps conducted at different temperatures. In the first step, a mixture of DNA and other basic PCR ingredients is heated in a test tube to approximately 90°C. At this temperature, the DNA strands unwind. The second step involves lowering the temperature to around 55°C, which allows special enzymes called primers to mark the section of DNA to be duplicated. The primers serve as bookends that identify the section of DNA to be copied. Once the primers have attached themselves to the strand, the temperature is raised to 75°C. At this temperature, the polymerase makes a copy of the two DNA segments. The entire process takes only a few minutes, and once completed the process is repeated to produce more copies. In this manner millions of copies of the original DNA material can be produced in a few hours. One of the keys in perfecting this method was the use of polymerase isolated from the bacterium *Thermus aquaticus*. This organism resides in hot springs such as those located in Yellowstone National Park. The polymerase used in the PCR is called Taq polymerase to signify its origin. The ability of Taq polymerase to function over rapid temperature changes and at elevated temperatures made this polymerase ideal for the PCR method. Most polymerases, such as those extracted from humans, were not stable at elevated temperatures.

The ability to produce multiple segments of DNA segments is referred to as DNA amplification. The PCR method gave scientists a quick, simple, and inexpensive means to amplify DNA. Amplification is important because the amount of DNA available often is limited for a particular procedure. Using PCR, the amount of DNA necessary for certain analyses can be produced from very small original samples. One application where this has been used is in genetic fingerprinting in crime cases. A small sample of DNA found in a drop of blood or a piece of hair can be amplified, and then sequenced for comparison to a sample of DNA obtained from a suspect or victim. While most of the DNA sequence in all humans is identical, there are regions in the DNA where individual variation exists. By focusing on these sections of DNA, forensic experts now have the ability to determine whether evidence gathered at a crime scene can be linked to a particular individual. DNA fingerprinting can be used to prove guilt or innocence; in recent years individuals on death row have been freed when PCR and DNA fingerprinting was applied to evidence saved in police files.

PCR amplification has led to more sophisticated and accurate diagnostic techniques regarding diseases. This allows earlier detection of the disease compared to conventional methods, making earlier treatment possible. For example HIV (human immunodeficiency virus) may be detected by searching for the DNA sequence unique to this virus. Amplifying samples and searching for DNA associated with the bacteria responsible for the condition has identified infectious bacterial diseases. Lyme disease, certain stomach ulcers, and middle ear infections have been detected in this manner.

One interesting case of the use of PCR involved analysis of tissue preserved from the eyes of John Dalton. Dalton, who died in 1844, requested that an autopsy be conducted after his death to determine the reason for his color blindness. The modern analysis amplified DNA from Dalton's eye tissue and discovered he lacked a gene necessary for detecting the color green. This explains why Dalton suffered red-green color blindness and could not differentiate between these two colors.

Current PCR technology requires a device about the size of a small television and costs several thousand dollars. Advances are reducing both the size and cost of PCR analyses. This is making PCR screening standard practice in diagnostic medicine and available in numerous other areas. These include such diverse fields as archaeology, ecology, art history, paleontology, and conservation biology.

Genetic engineering involves the alteration of an organism's genetic material in order to introduce desirable effects or eliminate undesirable effects. The most basic form of genetic engineering has been practiced for thousands of years ever since our ancestors used selective breeding to cultivate heartier plants and animals. While the practice of selective breeding continues to flourish, genetic engineering in the twenty-first century involves the modern practice of synthesizing recombinant DNA. Recombinant DNA is DNA in which genetic material from one organism is spliced into the genetic material of another. Using restriction enzymes to cut a section out of plasmid, DNA obtained from an organism, typically a bacterium or yeast, is used in the initial step to create recombinant DNA. Plasmids are small circular DNA molecules that contain a few genes. The same restriction enzyme is then used to cut a matching section of chromosomal DNA (gene) from another organism. The exposed ends of the plasmid and chromosomal DNA contain unpaired bases that are complementary to each other. Using enzymes called DNA ligases, the ends of the DNA from the two different organisms can be joined. Because the ends of chromosomal DNA stick to the ends of the plasmid DNA, the ends are called sticky ends (Figure 16.19). Once chromosomal DNA is inserted into the plasmid, the bacterium containing the recombined plasmid is allowed to reproduce. The reproduction of the bacteria produces clones. Clones are cells with identical DNA that originated from a single cell. In this manner multiple copies of genes coded for specific proteins can be made. The protein produced in the bacterium can then be extracted for human use.

Today thousands of products in medicine, agriculture, and industry are geneti-

Figure 16.19
Recombinant DNA (Rae Déjur)

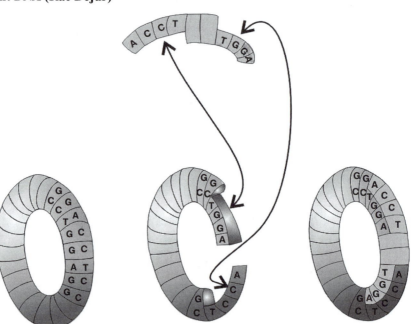

cally engineered using recombinant DNA technology. For example, plants have been engineered to resist viruses, make them more frost-resistant, and delay spoilage. Growth hormones are regularly used to increase dairy and beef yield from cows. In medicine, recombinant DNA is used to genetically engineer proteins that can be used to treat various diseases. The technology is also used to produce vaccines that are used to treat certain diseases, for example, hepatitis and influenza. Insulin was the first recombinant DNA substance approved for human use in 1982. Before this time, diabetics depended on insulin extracted from hogs and cattle. Current methods require the injection of insulin produced using recombinant DNA. The hope for the future is that gene therapy can be used to cure diabetics. Gene therapy involves inserting a healthy gene for an abnormal or missing gene to cure a condition. While this technique is still being perfected, there is hope that gene therapy will eventually provide a cure for many diseases. Currently, gene therapy is being attempted on a limited basis to treat cystic fibrosis and muscular dystrophy.

17

Nuclear Chemistry

Introduction

Chemical reactions involve the interaction of the outer electrons of substances. As one substance changes into another, chemical bonds are broken and created as atoms rearrange. Atomic nuclei are not directly involved in chemical reactions, but they play a critical role in the behavior of matter. Typical chemical reactions involve the interaction of electrons in atoms, but nuclear reactions involve the atom's nucleus. The nucleus contains most of the atom's mass but occupies only a small fraction of its volume. Electrons have only about 1/2000 the mass of a **nucleon**. To put this in perspective, consider that if the nucleus were the size of a baseball, the mean distance to the nearest electrons would be over two miles.

At the start of the twentieth century, it was discovered that the nucleus is composed of positively charged protons and neutral neutrons (see Chapter 4). These particles are collectively called nucleons. During the last half of the same century, scientists learned how to harness the power of the atom. The deployment of two atomic bombs brought a quick and dramatic end to World War II. This was followed by nuclear proliferation, the Cold War, and the current debate over developing a nuclear defense shield. The use of nuclear power as a source of clean, efficient energy continues to be debated. While nuclear power is used by many countries to fulfill their energy needs, accidents such as Chernobyl in Ukrainia and Three Mile Island in this country have dampened the public's enthusiasm for this energy source. Because of nuclear weapons and reactor accidents, people often perceive the term "nuclear" negatively, but there are many positive aspects to nuclear science. Nuclear medicine is used extensively for both diagnostic and therapeutic procedures. Radiometric dating is an invaluable tool to scientists who use this method for applications such as dating relics, determining the age of the Earth, and studying climate change. In this chapter, we extend our examination of the nucleus started in Chapter 4, and examine in detail the scientific principles behind human use of the atom.

Nuclear Stability and Radioactivity

Each element exists as several isotopes. Isotopes of an element have identical atomic numbers, but different mass numbers (see Chapter 4). Therefore, isotopes of an element differ in the number of neutrons in their nuclei. Every element has at least one unstable or radioactive isotope, but most have several. The nuclei of unstable isotopes undergo radioactive decay. Radioactive decay is the process where particles and energy are emitted from the nuclei of unstable isotopes as they become stable.

Nuclear stability is related to the ratio of protons to neutrons. Protons packed into the atom's nucleus carry a positive charge and exert a repulsive force on each other. For the nucleus to remain intact, a force called the strong nuclear force must balance the electrostatic repulsion between protons. The strong nuclear force is the "nuclear glue" responsible for holding the nucleus together. This force is related to the ratio of neutrons to protons in the nucleus. A normal hydrogen atom's nucleus contains a single proton, and therefore, no neutrons are needed because there is no repulsive force between protons. All other elements have more than one proton and require neutrons to supply the strong nuclear force. Helium atoms have two protons and two neutrons in their nuclei. When moving up the periodic table, the atomic number increases as more protons exist in each element's nucleus. When more protons are packed into the nucleus, more neutrons are required to overcome the resultant repulsive force. For elements with an atomic number less than 20, stable nuclei have either an equal number of protons and neutrons or one more neutron than proton. For example, carbon has 6 protons and 6 neutrons and fluorine has 9 protons and 10 neutrons. For elements with atomic numbers greater than 20, the ratio of neutrons to protons becomes increasingly greater than 1. All isotopes above bismuth, atomic number 83, are radioactive. The relationship between nuclear stability and the ratio of neutrons to

Figure 17.1

The ratio of the number of neutrons to protons determines the stability of atomic nuclei. As the number of protons in the nucleus increases, the number of neutrons must increase at a greater rate to be stable.

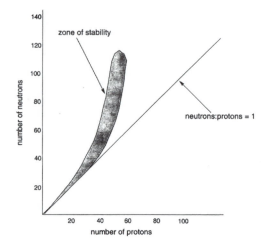

protons is depicted in Figure 17.1. Figure 17.1 shows a zone of stability. Stable nuclei tend to be found in the zone of stability. The neutron-to-proton ratio for isotopes outside the zone of stability characterizes unstable nuclei, and these undergo various forms of radioactive decay.

Stable isotopes prefer a certain combination of neutrons and protons. By far, most stable isotopes have an even number of both protons and neutrons. A smaller number of stable isotopes have either an even number of protons and an odd number of neutrons or vice-versa, and only a few have both an odd number of protons and neutrons (Table 17.1). Stable nuclei are also associated with a specific number of protons or neutrons. These islands of stability occur when the number of protons or neutrons is 2, 8, 20, 28, 50, 82, and 126. To illustrate this, consider tin with an atomic number of 50. Tin has ten stable isotopes, but antimony with an atomic number of 51 has only two.

Table 17.1
Nuclear Stability and Numbers of Neutrons and Protons

number of neutrons	number of protons	number of stable isotopes
even	even	168
even	odd	50
odd	even	52
odd	odd	4

Radioactive Decay and Nuclear Reactions

An unstable nucleus emits particles or electromagnetic radiation, or both, until a stable nucleus results. The main forms of radioactive decay include **alpha** (α), **beta** (β), and **gamma** (γ) **decay**. These three forms of radiation are named for the first three letters of the Greek alphabet. An alpha particle is equivalent to the nucleus of a helium atom and consists of 2 protons and 2 neutrons. A beta particle is equivalent to an electron. Gamma radiation is electromagnetic energy with wavelengths that range from 10^{-14} to 10^{-11} m. When alpha and beta particles are emitted from an atom's nucleus, the atomic number changes. This means the atom changes from one element to another. Alpha particles have a mass number of 4, while beta particles have a mass number of 0. Therefore, the emission of an alpha particle changes both the atomic and mass number, but the emission of a beta particle only changes the atomic number. Because gamma radiation is a form of light, the atomic number and mass number remain constant when gamma radiation is emitted. The emission of gamma radiation does not transform an element. Table 17.2 summarizes the properties and symbols used for the three main types of radioactive emissions.

Conservation of mass and charge are used when writing nuclear reactions. For example, let's consider what happens when uranium-238 undergoes alpha decay. Uranium-238 has 92 protons and 146 neutrons and is symbolized as $^{238}_{92}\text{U}$. After it emits an alpha particle, the nucleus now has a mass number of 234 and an atomic number of 90.

Table 17.2
Main Forms of Radioactive Emissions

Radiation	Form	Mass	Charge	Symbol
alpha	particle = He nucleus	4	+2	$^{4}_{2}\text{He}$ or α
beta	particle = electron	0	-1	$^{0}_{-1}\text{e}$ or β
gamma	electromagnetic radiation	0	0	$^{0}_{0}\gamma$

Because the atomic number of the nucleus is 90, the element becomes thorium, Th. The overall nuclear reaction can be written as

$$^{238}_{92}U \rightarrow \; ^{234}_{90}Th \; + \; ^{4}_{2}He$$

Notice that in this equation mass is conserved, $238 = 234 + 4$, and charge is conserved, $92 = 90 + 2$. In a chemical reaction, reactants are transformed into products. In a nuclear reaction, the terms "parent" and "daughter" correspond to reactants and products, respectively. In our example, the U-238 was the parent that decayed into the Th-234 daughter.

During beta decay, a neutron is transformed into a proton. If Th-234 were to emit a beta particle, it would be transformed into protactinium-234 according to the equation:

$$^{234}_{90}Th \rightarrow \; ^{234}_{91}Pa \; + \; ^{0}_{-1}e$$

Again, both mass and charge are conserved. Gamma emission often accompanies both alpha and beta decay, but because gamma emission does not change the parent element it is often emitted when writing nuclear reactions.

An unstable parent nucleus may decay into either a stable or an unstable daughter. When the daughter is unstable, which is often the case, the daughter will decay. Often the journey from an unstable nucleus to a stable nucleus involves a long series of steps referred to as a radioactive decay series. One example is the decay series for radium (Figure 17.2).

Half-Life and Radiometric Dating

In Chapter 12, the concept of half-life was used in connection with the time it took for reactants to change into products during a chemical reaction. Radioactive decay follows first order kinetics (Chapter 12). First order kinetics means that the decay rate

Figure 17.2

One possible decay series for radium. Ra-226 is transformed into stable Pb-206 through a series of alpha and beta emissions. Horizontal arrows represent beta emissions and diagonal arrows alpha emissions.

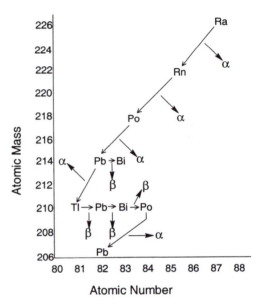

depends on the amount of parent material present at any given time. In nuclear decay, the number of parent nuclei decreases in an exponential fashion (Figure 17.3). The **decay** rate or **activity** of a radioactive substance is equal to the decrease in parent nuclei over time

$$activity \; = \; \frac{\Delta N}{\Delta T}$$

where ΔN equals the change in number of parent nuclei and ΔT is the change in time. The standard unit for activity is the **becquerel**. One becquerel is equal to 1 decay per second.

At any given time, a parent nucleus may decay into a daughter. Although it cannot be known when any individual nucleus may decay into a daughter, the half-life can be used as a collective measure of the time it

Figure 17.3
Exponential Decay (Rae Déjur)

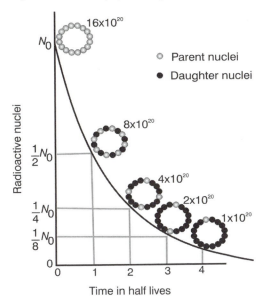

Figure 17.4
The amount of a radioactive substance is cut in half after each consecutive half-life. The amount of the original material remaining after n half-lives is $(\frac{1}{2})^n$.

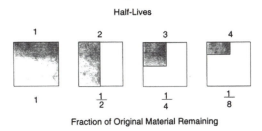

takes for radioactive decay to take place. Half-life, often symbolized using $t_{1/2}$ is the time required for half the parent nuclei in a sample to decay into daughter nuclei. An equivalent definition is that it is the time it takes for the activity of a substance to be cut in half. The concept of radioactive decay and half-life can be compared to popping corn. The kernels represent the parent nuclei and the popped corn the daughters. During the popping process, some kernels pop quickly after heat is applied, while others never pop. Although we cannot say when any one kernel will pop, we could characterize the popping time using half-life. For example, the half-life of a bag of microwave popcorn might be 2 minutes.

During each consecutive half-life, the amount of material remaining is one half of the amount present at the start of the half-life (Figure 17.4). The half-lives of radioactive isotopes vary over a wide range, from a fraction of a second to over billions of years. Table 17.3 lists the half-lives of some common isotopes.

The half-life of isotopes provides scientists with a nuclear clock that can be used to date objects. The concept is based on knowing the fraction of original material that is present in a sample. For instance, if half of the original isotope is present in the sample, then the sample's age is equivalent to the isotope's half-life. If one-fourth of the original material is present, then the sample's age is 2 half-lives. Because the use of radiometric dating involves making accurate measurements of parent and daughter activities in the sample, it is assumed that parent

Table 17.3
Half-Lives of Common Isotopes

Isotope	Half-Life
Polonium-214	1.6×10^{-4} second
Lead-214	26.8 minutes
Radon-222	3.82 days
Strontium-90	28.1 years
Radium-226	1602 years
Carbon-14	5730 years
Uranium-238	4.5×10^9 years

material is not lost from the sample or that the sample has not been contaminated.

Radiometric techniques have been applied in a variety of disciplines including archaeology, paleontology, geology, climatology, and oceanography. The isotopic method used for dating depends on the nature of the object. Geologic samples that are billions of years old would require using an isotope with a relatively long half-life such as potassium-40 ($t\frac{1}{2} = 1.25 \times 10^9$ years). Conversely, the movement of ocean water masses or groundwater might use tritium (tritium is hydrogen containing 2 neutrons, ^3_1H) that has a half-life of 12.3 years. In general, an isotope can be used to date an object up to ten times its half-life; for example, tritium could be used for measurements of up to 125 years.

Carbon-14, $t\frac{1}{2} = 5{,}730$ years, is one of the most prevalent methods used for dating ancient artifacts. Williard Libby (1908–1980) developed the method around 1950 and was awarded the Nobel Prize in chemistry in 1960 for this work. The carbon-14 method is based on the fact that living organisms continually absorb carbon into their tissues during metabolism. Most of the carbon taken up by organisms is in its stable carbon-12 form, but unstable carbon-14 (as well as other carbon isotopes) is **assimilated** along with carbon-12. No distinction is made between the isotopes of carbon assimilated, and so the ratio of carbon-14 to carbon-12 in the organism remains the same as that in the organism's environment, for example, the atmosphere or ocean. The natural occurrence of C-14 in the environment is about one C-14 atom for every 850 billion C-12 atoms. As long as an organism is alive, the carbon-14 to carbon-12 ratio remains constant. Once the organism dies, the carbon-14 stopwatch starts, and the C-14:C-12 ratio begins to decrease due to the decay of C-14. Comparing the activity of C-14 in a sample to that in living material can date the sample. For example, if the activity of C-14 in an excavated bone is half that of bone from a living organism, then the bone must be approximately 5,730 years old. Carbon-14 dating is used extensively with materials such as wood, seeds, fabrics, and bone.

A popular method used to date rocks is the potassium-argon method. Potassium is abundant in rocks such as feldspars, hornblendes, and micas. The K-Ar method has been used to date the Earth and its geologic formations. It has also been applied to determine magnetic reversals that have taken place throughout the Earth's history. Another method used in geologic dating is the rubidium-strontium, Rb-Sr, method. Some of the oldest rocks on Earth have been dated with this method, providing evidence that the Earth is approximately 5 billion years old. The method has also been used to date moon rocks and meteorites.

Nuclear Binding Energy— Fission and Fusion

Because the nucleus consists of a collection of nucleons, it would be expected that the mass of the nucleus is equal to the sum of its constituent nucleons. In fact, the mass of the nucleus is always slightly less than the sum of its parts. This decrease in mass of the nucleus is called the **mass defect**. What happens to this missing mass when the nucleus of an atom is created out of nucleons? The answer is related to the strong nuclear force that holds the nucleus together. The mass defect is converted to energy and released when the nucleons combine to form a nucleus. The amount of energy released is related to the mass defect according to Einstein's famous equation: $E = mc^2$, where m is mass and c is the speed of light. Because the speed of light is so high ($c = 3 \times 10^8$ m/s or 6.7×10^8 mph), it does not take much mass to produce a tremendous amount of energy. The energy calculated from Einstein's equation using the mass defect for m is called the nuclear **binding energy**. To separate the nucleus of an atom into its individual protons and neutrons would take an amount of energy equivalent to the binding energy.

The binding energy provides a measure of the stability of atomic nuclei, the greater the binding energy per nucleon of a nucleus, the greater its stability. When the amount of binding energy per nucleon is calculated for the different elements, it is found that the Fe-56 has the highest binding energy (Figure 17.5). The fact that the highest binding energy curve peaks at the element iron has important consequences for obtaining nuclear energy. When a heavier nucleus splits in a process called **nuclear fission**, lighter nuclei are produced. The lighter nuclei are more stable. Whenever a process involves moving to a more stable state or configuration, energy is released. More stable nuclei can also be obtained when lighter nuclei combine to form a heavier nucleus in a process called **nuclear fusion**. The most common example of a fusion reaction takes place inside the sun, where hydrogen nuclei fuse to form helium. The continual fusion of solar hydrogen provides the energy that makes life on Earth possible.

Modern nuclear power is based on harnessing the energy released in a fission reaction. The development of atomic energy started in the 1930s with the discovery that atoms could be split with neutrons. This discovery laid the foundation for building the first atomic bombs during World War II. A basic reaction representing the fission of uranium can be represented as:

$$ {}_{0}^{1}n + {}_{92}^{235}U \rightarrow {}_{56}^{141}Ba + {}_{36}^{92}Kr + 3\ {}_{0}^{1}n + \text{energy} $$

The equation shows that uranium-235 absorbs a neutron. After absorbing the neutron, the excited uranium nucleus splits and forms barium-141, krypton-92, and three neutrons. Energy is also produced in the reaction. This reaction is only one of a number of different ways that U-235 may split. Several hundred different isotopes have been identified when U-235 undergoes fission.

The three neutrons produced when uranium splits have the ability to split other U-235 nuclei and start a self-sustaining chain reaction. Whether a chain reaction takes place depends on the amount of fissionable material present. The more fissionable material that is present, the greater the probability that a neutron will interact with another U-235 nucleus. The reason for this involves the basic relationship between surface area and volume as mass increases. If a cube with a length of 1 unit is compared to a cube of 2 units, it is found that the surface area to volume ratio of the 1 unit cube is twice that of the 2 unit cube (Figure 17.6). This shows that volume increases at a greater rate than surface area as size increases. The probability that neutrons escape rather than react also depends on the surface area to volume ratio. The higher this ratio is the more likely neutrons escape. When a U-235 nucleus contained in a small mass of fissionable uranium is bombarded by a neutron, the

Figure 17.5
Binding Energy Curve. Those elements such as iron, cobalt, and nickel have the highest binding energy per nucleon, and therefore, are the elements with the most stable nuclei.

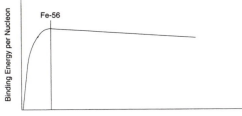

Figure 17.6
As the size increases, volume increases at a greater rate than the surface area.

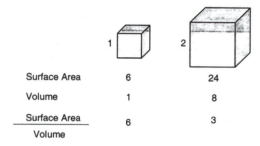

	1	2
Surface Area	6	24
Volume	1	8
$\dfrac{\text{Surface Area}}{\text{Volume}}$	6	3

probability is that less than one additional fission will result from the three product neutrons. The amount of fissionable material at this stage is called a **subcritical mass**. As the mass of fissionable material increases, a point is reached where exactly one product neutron causes another U-235 fission, which in turn produces three more neutrons, and one of these splits another U-235 nucleus. The quantity of fissionable material at the point where the reaction is self-sustaining is termed the **critical mass**. Once the critical mass is reached, a chain reaction can take place. If more than one product neutron reacts with other U-235 nuclei, a **supercritical mass** exists. In this case, the process accelerates, creating an uncontrolled chain reaction. The development of the first atomic bombs involved bringing together simultaneously a number of subcritical masses to form a supercritical mass in a fraction of a second. Once the supercritical mass was assembled, a stream of neutrons detonated the device.

Traditional nuclear power involves using the heat generated in a controlled fission reaction to generate electricity. A schematic of a nuclear reactor is shown in Figure 17.7. The reactor core consists of a heavy-walled reaction vessel several meters thick that contains fuel elements consisting of zirconium rods containing enriched pellets of U-235 in the form of

UO_2. Natural uranium is 0.7% U-235 and 99.3% U-238. Uranium-238 is nonfissionable, and therefore, naturally occurring uranium must be enriched to a concentration of approximately 3% U-235 to be used in common nuclear reactors. The reactor is filled with water. The water serves two different purposes. One purpose is to serve as a coolant to transport the heat generated from the fission reaction to a heat exchanger. In the heat exchanger, the energy can be used to generate steam to turn a turbine for the production of electricity. Water also serves as a **moderator**. A moderator slows down the neutrons, increasing the chances that the neutrons will react with the uranium. Interspersed between the hundreds of fuel rods are **control rods**. Control rods are made of cadmium and boron. These substances absorb neutrons. Raising and lowering the control rods controls the nuclear reaction.

While the basic design of most nuclear reactors is similar, several types of reactors are used throughout the world. In the United States, most reactors use plain water as the coolant. Reactors using ordinary water are called light water reactors (LWR). Light water reactors can be pressurized to approximately 150 atmospheres to keep the primary coolant in the liquid phase at temperatures of around 300°C. The heat from the pressurized water is used to heat secondary water to generate steam. In a boiling water reactor (BWR), water in the core is allowed to boil. The steam produced powers the turbines directly. Heavy water reactors employ water in the form of D_2O as the coolant and moderator. Heavy water gets its name from the fact that the hydrogen in the water molecule is the isotope **deuterium**, D ($_1^2H$), rather than ordinary hydrogen. The use of heavy water allows natural uranium rather than enriched uranium to be used as the fuel.

Figure 17.7
Schematic of Nuclear Power Plant (Rae Déjur)

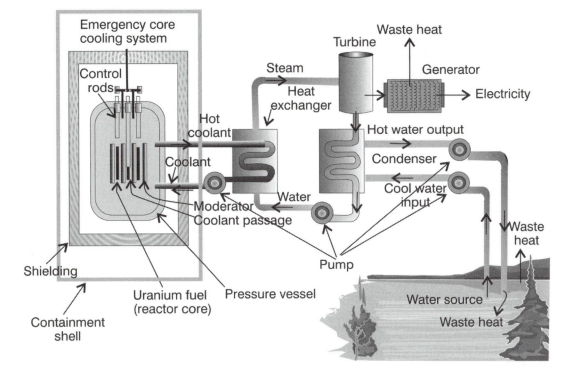

Breeder reactors were developed to utilize the 97% of natural uranium that occurs as nonfissionable U-238. The idea behind a breeder reactor is to convert U-238 into a fissionable fuel material, plutonium. A reaction to breed plutonium is

$$^{238}_{92}U + ^{1}_{0}n \rightarrow ^{239}_{92}U \rightarrow ^{239}_{93}Np + ^{0}_{-1}e$$

$$\rightarrow ^{239}_{94}Pu + ^{0}_{-1}e$$

The plutonium fuel in a breeder reactor behaves differently than uranium. Fast neutrons are required to split plutonium. For this reason, water cannot be used in breeder reactors because it moderates the neutrons. Liquid sodium is typically used in breeder reactors, and the term liquid metal fast breeder reactor (LMFBR) is used to describe it. One of the controversies associated with the breeder reactor is that it results

in the production of weapon grade plutonium and could contribute to nuclear arms proliferation.

The world use of nuclear power to supply a nation's electricity varies widely by country. France, for example, gets around 75% of its electricity from nuclear power, and several other European countries get over half of their energy from this source. Approximately 20% of the electricity in the United States comes from 103 operating nuclear power plants. Nuclear is second only to coal, 50%, and ahead of natural gas, 15%, hydropower, 8%, and oil, 3%, as a source of electrical energy. Although once hailed by President Eisenhower in the 1950s as a safe, clean, and economical source of power, the U.S. nuclear industry has fallen on hard times in the last twenty-five years. Nuclear accidents at Three Mile Island, Pennsylvania, in 1979 and Chernobyl, Ukrainia, in 1986 raised public concerns about the safety of nuclear power. Utility companies, facing more stringent federal controls and burgeoning costs, have abandoned nuclear power in favor of traditional fossil fuel plants. The last nuclear power plant ordered in the United States was built in 1978. Several nuclear plants under construction have been left abandoned, while other operating plants have been shut down and decommissioned.

The lack of emphasis on nuclear power during the last twenty-five years does not necessarily mean it will not stage a comeback in the future. Concerns about greenhouse gases and global warming from the burning of fossil fuels are one of the strongest arguments advanced in favor of a greater use of nuclear power. New developments in reactor design hold the promise that safe reactors can be built. Yet, concerns about the safe handling and transport of nuclear fuels, nuclear waste disposal, nuclear proliferation, and a nimby (not in my back yard) attitude present formidable obstacles to the resurrection of the industry in the United States.

The other type of nuclear reaction is a fusion reaction. The binding energy curve indicates that fusion of light atomic nuclei to form a heavier nucleus is an exothermic reaction. Although in theory many light elements can combine to produce energy, most practical applications of fusion technology focus on the use of hydrogen isotopes. The first application of fusion technology involved development and testing of hydrogen bombs (H-bombs) in the early 1950s. The United States and the former Soviet Union feared that the other would develop the H-bomb. This would pose an unacceptable threat to the country without the bomb. Leaders of each country knew that hydrogen bombs would be much more destructive than atomic (fission) bombs. The reason for

this is related to the small size of hydrogen nuclei. Many more hydrogen nuclei can be packed into an H-bomb compared to the number of uranium nuclei in an atomic bomb. Because the energy produced in a nuclear weapon depends on the number of fission or fusion reactions that take place in a given volume, H-bombs are far more destructive. The quest to produce the first superbomb, as the H-bomb was known, initiated the arms race that characterized the relationship between the United States and the Soviet Union during most of the last half of the twentieth century. It has only been in the last decade that this race has abated with the fall of the Soviet Union, but President Bush's talk of a nuclear defense shield is a reminder that the threat of mutually assured destruction remains.

Temperatures in the range of 20 to 100 million degrees Celsius are required for fusion reactions. For this reason, a hydrogen bomb is triggered by a conventional fission atomic bomb. The atomic bomb produces the tremendous heat necessary to fuse hydrogen nuclei; therefore, fusion bombs are often referred to as thermonuclear. The United States exploded the first hydrogen bomb on Eniwetok Atoll in the Pacific Ocean on November 1, 1952. This bomb was based on the fusion of deuterium:

$$_1^2H + {}_1^2H \rightarrow {}_1^3H + {}_1^1H$$

The fusion of deuterium produces another hydrogen isotope called tritium, $_1^3H$, along with common hydrogen, $_1^1H$. The fusion of deuterium may also produce helium according to the reaction:

$$_1^2H + {}_1^2H \rightarrow {}_2^3He + {}_0^1N$$

There is hope that nuclear fusion may one day prove to be a practical energy source. The U.S. government has spent billions of dollars on fusion energy research, and several prototype reactors have been built. Researchers attempting to produce energy from fusion face several major problems. One problem is how to produce the tremendous temperatures required for fusion. Another is how to confine a fuel, such as deuterium, in a manner to produce a sustained reaction. Yet, another difficulty is obtaining more energy out of the reaction than is required to initiate fusion. One design used to produce temperatures of around 50,000,000°C involves using a donut-shaped coiled electromagnet called a tokamak (Figure 17.8). The electromagnetic field within the cavity of the magnet acts on hydrogen isotopes, stripping them of their electrons and producing a plasma. Increasing the electric current intensifies the electromagnetic field and raises the temperature to a point where the deuterium nuclei can fuse. Confining the fuel in a suspended plasma state within an electromagnetic field isolates the reactor materials from the extreme temperatures required for fusion. In another design called inertial confinement, a series of lasers is focused on glass pellets filled with a mixture of deuterium and tritium at a pressure of several hundred atmospheres. The pellets are then injected into a

reactor. At a precise point, a pellet and the laser beams all intersect to create a controlled thermonuclear explosion. The heat from the explosion can then be used to produce work.

Fusion and Stellar Evolution

The fusion of hydrogen to form helium takes place throughout the universe. Stars are natural thermonuclear machines that are responsible for the formation of all naturally occurring elements. The universe is predominantly composed of hydrogen. Hydrogen is not scattered uniformly throughout the universe. Galaxies are regions of high density, while interstellar space contains little. Some regions of space contain clouds of cool hydrogen gas called **nebula**, and it is in these regions that stars are born. The birth of a star begins when gravitational attraction causes a nebula to contract. As it contracts, the mass of hydrogen generates heat. The nebula continues to contract, and as it does, its temperature rises and its interior pressure increases, forming a **protostar**. Eventually, a point is reached where the protostar's interior temperature reaches several million degrees. At this point, the temperature is sufficient to initiate fusion of hydrogen nuclei. Two primary hydrogen fusion reactions are:

$$_1^1\text{H} + _1^1\text{H} \rightarrow _1^2\text{H} + _1^0\text{e}$$
$$_1^1\text{H} + _1^2\text{H} \rightarrow _2^3\text{He}$$

In the first reaction a positively charged electron is formed. This particle is called a **positron**. In addition to the fusion of hydrogen nuclei, the helium created can also enter into fusion reactions:

$$_2^3\text{He} + _1^1\text{H} \rightarrow _2^4\text{He} + _1^0\text{e}$$
$$_2^3\text{He} + _2^3\text{He} \rightarrow _2^4\text{He} + 2\,_1^1\text{H}$$

Fusion reactions signify the formation of a true star. The fusion reactions balance the gravitational contraction and an equilibrium

Figure 17.8
Tokamak Design (Rae Déjur)

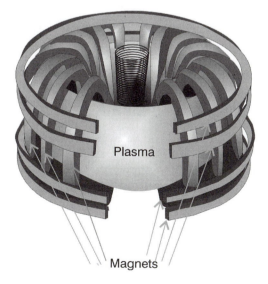

Plasma

Magnets

is reached. Hydrogen burning in the star's interior balances gravitational contraction as long as a sufficient supply of hydrogen is available. Depending on the size of the star, hydrogen burning may last less than a million years or for billions of years. Larger stars expend their hydrogen faster than smaller stars. The sun is considered a normal star, with an expected life expectancy of 10 billion years. Its current estimated age is 5 billion years, so it is about halfway through its hydrogen-burning stage.

Once hydrogen burning stops, there is no longer a balance between gravitational contraction and the nuclear energy released through fusion. The star's interior at this stage is primarily helium. The core will again start to contract and increase in temperature. Eventually, a temperature is reached where helium fusion is sufficient to balance the gravitational contraction. Depending on the size of the star, several contractions may take place. It should be realized that at each stage in a star's life cycle numerous fusion reactions take place. In addition to hydrogen and helium, many other light elements such as lithium, carbon, nitrogen, and oxygen are involved.

Once a star has expended its supply of energy, it will contract to a glowing white ember called a **white dwarf**. The elements produced in the interior of a star depend on the size of the star. Small stars do not have sufficient mass to produce the temperatures required to create the heaviest elements. The most massive stars, though, may go through a series of rapid contractions in their final stages. These massive stars have the ability to generate the temperatures and pressures necessary to produce the heaviest elements such as thorium and uranium. The final fate of these massive stars is a cataclysmic explosion called a **supernova**. It is in this manner that scientists believe all the naturally occurring elements in the universe were created. This would imply that all the elements in our bodies were once part of stars, and the heavier elements came from a supernova that exploded billions of years ago.

Transmutation

Transmutation is the process where one element is artificially changed into another element. Rutherford conducted the first transmutation experiment in 1919 when he bombarded nitrogen atoms with alpha particles. The nitrogen was transmuted into oxygen and hydrogen according to the reaction:

$$^{14}_{7}N \; + \; ^{4}_{2}He \rightarrow \; ^{17}_{8}O \; + \; ^{1}_{1}H$$

The process of transmutation produces most of the known isotopes. In fact, only about 10% of the approximately 3,000 known isotopes occur naturally. The rest are synthesized in large instruments called particle accelerators.

Particle accelerators are devices used to accelerate charged particles and ions to speeds approaching the speed of light. By colliding accelerated particles with other particles, scientists study the ultimate nature of matter. Collisions produced in particle accelerators are like breaking open a piñata. Accelerated particles serve as a club to break open atomic nuclei and reveal their contents. Hundreds of different **elementary particles** have been detected during collision experiments. The term "elementary" was originally used to define a few basic indivisible particles such as the electron, but during the last century hundreds of different elementary particles have been discovered. The term "elementary" has been expanded to include a number of different classes of particles associated with matter. Protons, neutrons, and electrons are the most familiar elementary particles.

Several basic types of particle accelerators are used, but they are all based on the principle that charged particles can be accelerated in an electromagnetic field. The first accelerators built in the 1930s were linear accelerators, or linacs. These consisted of a series of hollow tubes in which the alternating current regulated down the length of the tube was used to accelerate the particle (Figure 17.9). The speed a particle can obtain using a linear accelerator is directly related to the length of the accelerator. The largest linear accelerator is located at Stanford University and is slightly longer than two miles.

Another design for particle accelerators is based on a circular arrangement. A cyclotron is similar to a linear accelerator wound into a spiral. A series of electromagnets causes the particles to move in a circle as they are accelerated by the electric field. According to Einstein's theory of relativity, an object's mass increases as it accelerates. In particle accelerators this is a problem because as the mass increases, the particle slows down and becomes out of sync with the changing electric field. A synchrotron is a cyclotron in which the electric field increases to compensate for the change in

mass of the particle due to relativistic effects.

Particle accelerators are some of the largest scientific instruments. The largest particle accelerator is located at CERN (European Organization for Nuclear Research) outside of Geneva on the border of France and Switzerland. It has a circumference of 25 kilometers and is buried 100 meters below the surface of the Earth. This accelerator is known as LEP for large electron-positron collider. Fermi Labs outside of Chicago contains an accelerator called a Tevatron. The Tevatron has a four-mile circumference and is used to accelerate protons and antiprotons in opposite directions to 99.9999% of the speed of light. These particles are then smashed together and detectors connected to computers are used to analyze the results. The bottom and top quark (Chapter 4) were first detected at Fermi Labs. The charm quark was discovered at the Stanford Linear Accelerator Center. To advance the frontiers of science and learn more about the nature of matter requires building more powerful and larger particle accelerators. The United States started construction on what would have been the largest particle accelerator in 1983. Two billion dollars was spent on the design and initial construction of the Superconducting Super Collider (SSC). This would have been a giant accelerator built in Texas using thousands of superconducting magnets. The SSC would have accelerated protons and antiprotons around an oval track 54 miles in circumference. The proton energies would be twenty times that produced by current accelerators. In 1993, Congress cancelled funding of this project.

Particle accelerators are used to produce most isotopes by transmutation. All elements greater than uranium, known as the transuranium elements, have been produced in particle accelerators. For example,

Figure 17.9
In a linear accelerator, a positive proton is accelerated through a series of tubes by using alternating current. The proton is pushed by the positive charge on the first tube and pulled by the negative charge on the second tube, causing it to accelerate. By timing the change in voltage applied to the tubes, the proton continues to accelerate down the length of the tube.

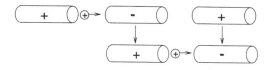

Pu-241 is produced when U-238 collides with an alpha particle:

$$^{238}_{92}U + ^{4}_{2}He \rightarrow ^{241}_{94}Pu + ^{1}_{0}n$$

Tritium is produced in a nuclear reaction between lithium and a neutron:

$$^{6}_{3}Li + ^{1}_{0}n \rightarrow ^{3}_{1}H + ^{4}_{2}He$$

Nuclear Medicine

While our discussion on transmutation, particle accelerators, and elementary particles has focused on basic research, nuclear discoveries have been applied in many areas. Radioactive isotopes produced in particle accelerators have found wide use in nuclear medicine. Radioactive nuclei are used to diagnose and treat a wide range of conditions. **Radioactive tracers** are isotopes that can be administered to a patient and followed through the body to diagnose certain conditions. Most tracers are gamma emitters with short half-lives (on the order of hours to days) that demonstrate similar biochemical behavior as their corresponding stable isotopes. Because tracers are radioactive, radiation detectors can be used to study their behavior in the body. Several examples of common tracer studies in medicine include iodine-131 to study the thyroid gland, technetium-99 to identify brain tumors, gadolinium-153 to diagnose osteoporosis, and sodium-24 to study blood circulation.

A recent development in nuclear medicine that illustrates how advances in basic research are transformed into practical applications is positron emission tomography or PET. PET creates a three-dimensional image of a body part using positron emitting isotopes. Positrons, positively charged electrons, are a form of **antimatter**. Antimatter consists of particles that have the same mass as ordinary matter, but differ in charge or some other property. For example, antiprotons are negatively charged protons, and antineutrons have different magnetic properties than regular neutrons. Antiparticles have been detected in particle accelerators, and antiatoms were detected at CERN in 1995. When matter meets antimatter, mutual annihilation results with the mass converted into short-lived particles and gamma rays. This principle forms the basis of PET.

In PET, positron-emitting isotopes such as C-11, N-13, O-15, and F-18 are used. Each of these unstable isotopes is characterized by lacking a neutron compared to its stable form; for example, C-11 needs one more neutron to become C-12. They undergo positron emission when a proton changes into a neutron:

$$^{1}_{1}H \rightarrow ^{1}_{0}n + ^{0}_{1}e$$

When a PET isotope is administered to tissue, it emits positrons. These positrons encounter electrons found in tissue, the interaction of an electron and positron results in mutual annihilation and the production of beams of gamma rays directly opposite to each other (Figure 17.10). The

Figure 17.10
In positron emission tomography, positrons interact with electrons in body tissue to produce gamma rays that are detected and converted into an image.

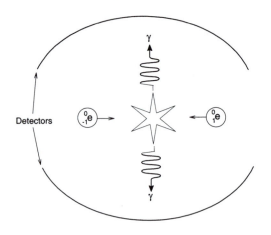

beams strike gamma detectors, and the two signals are translated into an image by a computer. PET has been used to diagnose neurological disorders such as Parkinson's disease, Alzheimer's disease, epilepsy, and schizophrenia. PET is also used to diagnose certain heart conditions and examine glucose metabolism in the body.

Radioisotopes are also used in radiation therapy to treat cancer. The goal in radiation therapy is to kill malignant cells, while protecting healthy tissue from radiation effects. Radioisotopes such as yttrium-90, a beta emitter, may be placed directly in the tumor. Alternatively, the diseased tissue may be subjected to beams of gamma radiation. Cobalt-60 used in radiation therapy is prepared by a series of transmutations:

$$^{58}_{26}Fe + ^{1}_{0}n \rightarrow ^{59}_{26}Fe$$

$$^{59}_{26}Fe \rightarrow ^{59}_{27}Co + ^{0}_{-1}e$$

$$^{59}_{27}Co + ^{1}_{0}n \rightarrow ^{60}_{27}Co$$

Radiation Units and Detection

Radiation measurements are expressed in several different units depending on what is being measured. The strength or activity of a radioactive source is the number of disintegrations that occurs per unit time in a sample of radioactive material. This is most often expressed as the number of disintegrations per second. The SI unit for activity is the becquerel, abbreviated Bq. One becquerel is equivalent to one disintegration per second. The **curie**, Ci, is often used when the activity is high. One curie is equal to 3.7×10^{10} becquerels. Common prefixes such as pico (10^{-12}), micro (10^{-6}), milli (10^{-3}) are often used in conjunction with curies to express activity values. For example, EPA considers radon levels in a home high if the measured activity in the home is greater than 4.0 picocuries per liter of air.

When matter absorbs radiation, a certain quantity of energy is imparted to the matter. The **dose** is the amount of radiation energy absorbed by matter. With respect to human health, the matter of concern is human tissue. The traditional unit of dose is the **rad**. One rad is defined as the 100 ergs of energy absorbed per gram of matter. The SI unit for dose is the **gray**, Gy. One gray is equal to 1.0 joule of energy absorbed per kilogram of matter. The relationship between rads and grays is 1 gray is equal to 100 rads. Similar to activity measurements, prefixes can be attached to the gray unit.

The most important measure of radiation in terms of health effects is the **dose equivalent**. Dose by itself does not consider the type of radiation absorbed by tissue. Radiation effects on human tissue depend on the type of radiation. For instance, a dose of alpha radiation is about twenty times more harmful compared to equivalent doses of beta or gamma radiation. Dose equivalent combines the dose with the type of radiation to give dose equivalent. It is found by multiplying the dose by a quality factor. The quality factors for gamma, beta, and alpha radiations are 1, 1, and 20, respectively. Dose equivalent is the best measure of the radiation effects on human health. The **rem** is the dose equivalent when dose is measured in rads. Rem is the most frequently used unit for measuring dose equivalent. The unit rem is actually an acronym for roentgen equivalent for man. Millirems (mrems or $1/1,000$ rem) are most often used to express everyday exposures (Table 17.4). When expressing dose in grays, the unit for dose equivalent is the sievert, Sv. One rem is equal to 0.01 Sv. The various units associated with radiation measurements are summarized in Table 17.4.

Radiation comes in many different forms. Most of this chapter has dealt with nuclear radiation, those forms of radiation

Table 17.4
Common Radiation Units

Measurement	Unit	Definition	Abbreviation
Activity	becquerel	$1 \dfrac{\text{disintegration}}{\text{second}}$	Bq
	curie	$3.7 \times 10^{10} \dfrac{\text{disintegrations}}{\text{second}}$	Ci
Dose	gray	$1.0 \dfrac{\text{joule}}{\text{kilogram}}$	Gy
	rad	$100 \dfrac{\text{ergs}}{\text{gram}}$	
Dose Equivalent	sievert	gray x quality factor	Sv
	rem	rad x quality factor	

that come from the nuclei of atoms. Nuclear radiation originates from the Earth. When nuclear radiation comes from an extraterrestrial source, it is referred to as **cosmic radiation**. Nuclear and cosmic radiation consist of particles such as protons, electrons, alpha particles, and neutrons, as well as photons such as gamma rays and x-rays. When nuclear and cosmic radiation strike matter and free ions, the term **ionizing radiation** is used. Alpha, beta, and gamma radiation are forms of ionizing radiation. **Nonionizing** radiation includes ultraviolet radiation, responsible for sunburns, and infrared radiation. Humans sense infrared radiation as heat.

Radiation is impossible to detect using our unaided senses. It cannot be felt, heard, smelled, or seen. The fact that radiation ionizes matter as it passes through it provides a basis for measuring radiation. A common device used to measure radiation is the Geiger counter. A Geiger counter consists of a sealed metal tube filled with an inert gas, typically argon. A wire, protruding into the center of the tube, connected to the positive terminal of a power source serves as one electrode. The negative terminal is connected to the wall of the tube and serves as the negative electrode (Figure 17.11). The voltage difference between the wire and tube wall is roughly 1,000 volts. No current flows between the wire and tube wall because the inert gas is a poor conductor of electricity. One end of the sealed tube has a

Figure 17.11
A Geiger Counter detects radiation by ionizing argon atoms. The flow of argon ions and electrons in the tube creates a current that is amplified and detected by the counter.

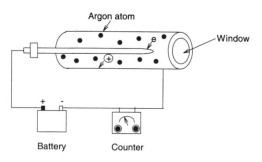

thin window that allows radiation to enter. When radiation enters the tube, it ionizes the atoms of the inert gas. The free electrons produced during ionization move toward the positive electrode, and as they move, they ionize other gas atoms. The positively charged argon ions move toward the negative walls. The flow of electrons and ions produces a small current in the tube. This current is amplified and measured. There is a direct relationship between the current produced and the amount of radiation that enters the window so activity can be measured.

Scintillation counters use materials that produce light when stimulated by radiation. Scintillation materials include sodium iodide crystals and special plastics. Radiation is measured by exposing the scintillation material to radiation and using a photomultiplier tube to count the number of resulting flashes. Photomultiplier tubes are photocells that convert light into electrical signals that can be amplified and measured.

Biological Effects of Radiation

Radiation is ubiquitous in the environment. Life exists in a sea of nuclear, cosmic, and electromagnetic radiation. Our bodies are constantly exposed to radiation and the effect of radiation on cells has several effects. Radiation may 1) pass through the cell and not do any damage; 2) damage the cell, but the cell repairs itself; 3) damage the cell and the damaged cell does not repair itself and reproduces in the damaged state, or 4) kill the cell. Fortunately, most radiation does not cause cellular damage, and when it does, DNA tends to repair itself. Under certain circumstances the amount of radiation damage inflicted on tissue overwhelms the body's ability to repair itself. Additionally, certain conditions disrupt

DNA's ability to repair itself. Figure 17.12 shows several different ways DNA may be damaged by ionizing radiation.

Low levels of ionizing radiation generally do not affect human health, but health effects depend on a number of factors. Among these are the types of radiation, radiation energy absorbed by the tissue, amount of exposure, type of tissue exposed to radiation, and genetic factors. Alpha, beta, and gamma radiation display different characteristics when they interact with matter. Alpha particles are large and slow with very little penetrating power. A sheet of paper or the outer layer of skin will stop alpha particles. Their energy is deposited over short distances, resulting in a relatively large amount of damage. Beta particles are smaller than alpha particles and have a hundred times

Figure 17.12
Radiation may damage DNA in several ways. It may cause breaks in the sugar-phosphate backbone and the base crosslinks. It may also result in incorrect repair. (Rae Déjur)

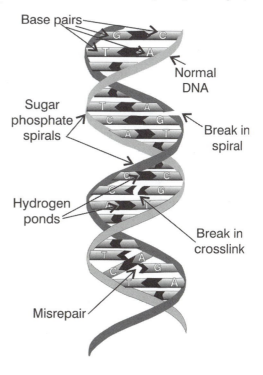

their penetrating power. Beta particles can be stopped by a few millimeters of aluminum or a centimeter or two of tissue. Their energy is deposited over a distance, and hence is less damaging than alpha particles. Gamma particles have a thousand times the penetrating power of alpha particles. Several centimeters of lead are needed to stop gamma radiation. Gamma particles travel significant distances through human tissue, and therefore, deposit their energy over a relatively wide area.

It is important to consider the type of radiation with respect to whether a radiation source is external or internal to the body. External alpha, and to some extent beta radiation, are not as hazardous as external gamma radiation. Clothing or the outer layer of the skin will stop alpha radiation. Beta radiation requires heavy protective clothing, because it can penetrate the skin and burn tissue. The body cannot be protected from external gamma radiation using protective clothing. While external alpha particles are relatively harmless, they cause significant damage internally. When inhaled and ingested, alpha emitters may be deposited directly into tissue, where they can damage DNA directly.

As long as radiation dose equivalent exposures are low, radiation damage is non-detectable. General effects of short-term radiation exposure are summarized in Table 17.5. The dose equivalents in Table 17.5 are listed in rems. These values are several orders of magnitude greater than what humans received in a year. Annual human exposure

Table 17.5
Biological Effects of Short-Term, Whole Body Radiation Exposure

Dose Equivalent in Rems	Effect
Less than 25	None
25-100	Temporary decrease in white blood cells
101-200	Permanent decrease in some white-blood cell, nausea vomiting, lethargy
201-300	Nausea, vomiting, loss of appetite, diarrhea, lethargy.
301-600	Initial vomiting, diarrhea, fever followed by latent period of several days to weeks. Hemorrhaging, inflammation of mouth, and emaciation leads to death in approximately 50% of those receiving this dose equivalent.
Greater than 600	Accelerated effects of those above, with death in nearly 100% of those receiving this dose equivalent

Figure 17.13
Annual human exposure to radiation in millirems (mrems). Note that although radon is radiation from the Earth, it is displayed as a separate category.

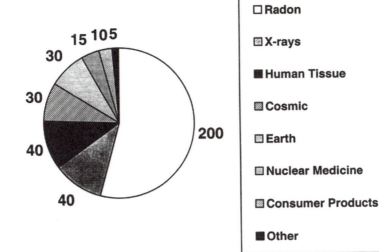

is normally around several hundred mrems. Human exposure to radiation occurs that from both natural and human sources. Natural radiation is referred to as **background radiation**. It consists mainly of radon that emanates from radioactive elements in the Earth's crust and cosmic rays (Figure 17.13). The main human source of radiation exposure is medical x-rays. Humans carry a certain amount of radioactive material inside their bodies. Potassium, as the isotope K-40, is responsible for nearly half of the exposure humans receive from consumed food. Other sources of human exposure include C-14 and radium, which acts like calcium in the body. Radioactive isotopes are incorporated into bone and other tissue along with stable isotopes.

Radiation exposure from both natural and human sources varies widely. Background radiation depends on the local geology and elevation. Areas where radioactive rocks are located close to the surface or where mining has exposed mineral deposits have higher background levels. Higher background levels also exist at higher elevations because there is less atmosphere to screen out cosmic radiation. Patients being treated with radiation therapy may have exposures of several hundred thousand mrems. Exposures of this magnitude involve radiation focused on the diseased tissue, rather than whole-body exposure.

Summary

Chemical change involves the interaction of the valence electrons in atoms as bonds are broken and formed. Even though chemical change creates new substances, atoms of elements are conserved. Nuclear changes result in the transformation of one element into another. These changes also involve a number of elementary particles and energy. Although a sea of radiation surrounds humans, the word "radiation" is often associated with negative images. Humans have used natural and synthesized radioactive isotopes for national defense, medicine, basic research, and energy production, and in numerous commercial products. For example, smoke detectors contain a source of alpha-emitting americium and gas lantern mantles contain radioactive thorium. Luminescent exit signs found in many schools, movie theaters, and airlines contain tubes filled with tritium. As with any technology, the benefits of nuclear applications must be weighed against their costs. This is particularly true with radioactive materials. As the use of these materials continues, scientists will seek to control their deleterious effects.

In the future, nuclear power may be taken for granted, and our generation's burning of fossil fuels may be viewed as primitive. For now, the use of nuclear technology is in its infancy. In Star Trek, the starship *Enterprise* relies on two forms of nuclear power to power it through the universe. Fusion power is used for the impulse engines to achieve speeds below the speed of light. Warp speeds, which are faster than the speed of light, are obtained by using energy obtained from a reaction between matter and antimatter. Protons and antiprotons are used to power the Enterprise's warp drive. While today fusion power use as a practical energy source is science fiction, it may one day be a viable source of energy.

18

Environmental Chemistry

Introduction

Throughout human history, *homo sapiens* have had an impact on our planet. As the planet's human population has grown, accompanied by advances in technology, human impacts have also increased. In the last century, the world population has grown from 3 billion in the year 1900 to over 6 billion today. Each day 2 million more people are added to the planet. The industrial revolution and advances in science throughout the twentieth century have extended the life spans of humans. The Earth's increasing population and use of its resources to support this population have natural consequences for the environment. Human activities continually affect the Earth's **lithosphere** (land), **hydrosphere** (water), **atmosphere** (air), and **biosphere** (organisms). The Earth's natural resources supply the chemicals used by humans for the energy and the raw materials used to sustain life. Throughout history, most of the substances used by humans have consisted of natural products, but in the last fifty years chemists have learned how to synthesize a plethora of new products. While synthetic substances have provided many benefits to society, they have also created additional impacts on the planet.

Environmental chemistry deals with the chemical aspects of the interaction between humans and their environment. Several of the issues of concern to environmental chemists are global in nature and have been well publicized. These include stratospheric ozone destruction, acid precipitation, and the greenhouse effect. Although these issues garner the bulk of the popular media's attention, many common applications of environmental chemistry have a far greater impact on everyday life. For example, disinfecting drinking water with chlorine has eliminated many diseases transmitted through water in many regions. Wastewater is also subjected to chlorination and may undergo further chemical treatments before being discharged into receiving waters. Whenever a vehicle is started, we immediately initiate a number of chemical reactions that transform fossil fuels into other chemicals that are eventually released into the atmosphere. Modern agriculture depends on a variety of synthetic chemicals used as fertilizers, pesticides, and herbicides. Agricultural chemicals have an impact on soil, groundwater, surface water, and the atmosphere. Use of synthetic chemicals is prevalent in many industries. Most of these are presumed safe, although some have been discovered to be harmful after their widespread use. Examples include PCBs in electrical transformers, dioxins in the paper industry, and CFCs used in refrigerants and propellants.

In this chapter, we examine the environmental chemistry associated with both global and local issues. One goal of this chapter is to provide a balanced perspective on the subject. This in turn should help the reader use chemistry to make informed decisions about environmental issues. An axiom of the environmental movement is to "think globally and act locally." While both global and local issues will be treated separately, the nature of having an environmental perspective means knowing that local decisions made by many individuals cumulatively have global consequences.

Stratospheric Ozone Depletion

Ozone, O_3, is a relatively unstable **allotrope** of oxygen consisting of three oxygen atoms. Although it exists in only minute concentrations in the atmosphere, ozone's presence is critical for life on Earth. Ozone acts as a protective shield by absorbing harmful ultraviolet radiation in the stratosphere. The stratosphere is the region of the atmosphere lying between approximately 13 and 20 kilometers (8 to 12 miles) above the Earth's surface. Ozone is constantly created and destroyed in the stratosphere. This occurs when ultraviolet light with sufficient energy splits the double bond of the ordinary molecular oxygen, O_2, producing two oxygen atoms. The reaction can be represented as

$$O_2 + UV \rightarrow 2O$$

Ultraviolet light, like other forms of electromagnetic radiation, can be characterized by its wavelength and frequency. Wavelength and frequency are inversely related, as wavelength gets larger the frequency gets smaller. Figure 18.1 shows the electromagnetic spectrum for common forms of radiation.

The energy of a photon of electromagnetic radiation is given by Planck's equation:

$$E = \frac{hc}{\lambda}$$

In this equation, E is energy of a photon in joules, h is Planck constant = 6.63×10^{-34} J-s, c is the speed of light = 3.0×10^8 m/s, and λ is the wavelength of electromagnetic radiation in meters. Planck's equation shows that the energy of electromagnetic radiation is inversely proportional to its wavelength. The maximum wavelength of light necessary to split a molecule of oxygen is 242 nm. Ultraviolet light includes radiation between about 50 and 400 nm.

Oxygen atoms produced by the ultraviolet dissociation of O_2 can take part in several reactions that create and destroy ozone. One set of reactions involving atomic and molecular oxygen is known as the **Chapman reactions**. The reactions are named after Sydney Chapman (1888–1970) who first proposed them in 1930. In one reaction, an oxygen atom combines with O_2 to form ozone. Alternately, it can recombine with another oxygen atom to produce an oxygen molecule or react with ozone to produce two oxygen molecules. These reactions are summarized as

$$O + O_2 \rightarrow O_3$$
$$O + O \rightarrow O_2$$
$$O_3 + O \rightarrow 2O_2$$

Ozone formed in the stratosphere is unstable. It can be broken down by ultraviolet light shorter than 320 nm (UV-C and UV-B) according to the reaction:

Figure 18.1
The E-M Spectrum (Rae Déjur)

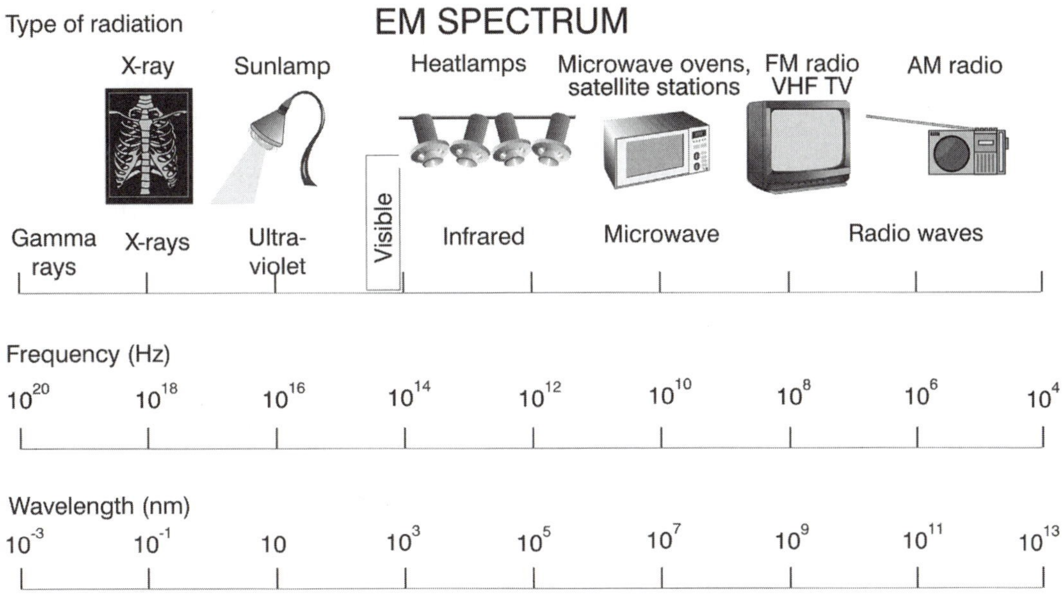

EM SPECTRUM

Type of radiation

| X-ray | Sunlamp | Heatlamps | Microwave ovens, satellite stations | FM radio VHF TV | AM radio |

| Gamma rays | X-rays | Ultra-violet | Visible | Infrared | Microwave | Radio waves |

Frequency (Hz)

10^{20} 10^{18} 10^{16} 10^{14} 10^{12} 10^{10} 10^{8} 10^{6} 10^{4}

Wavelength (nm)

10^{-3} 10^{-1} 10 10^{3} 10^{5} 10^{7} 10^{9} 10^{11} 10^{13}

$$O_3 + UV \rightarrow O_2 + O^*$$

where O* represents an excited oxygen atom. The form of oxygen, which exists in the stratosphere, is related to the altitude. The top of the stratosphere is highly energetic with an abundance of UV-C radiation. In this region oxygen exists predominantly as atomic oxygen. Collisions between oxygen atoms can form molecular oxygen, but the presence of UV-C radiation keeps molecular oxygen concentrations low. This in turn precludes any appreciable accumulation of ozone. In the lower stratosphere, ozone concentrations are low due to the absence of UV-C radiation. Because UV-C has been absorbed in the upper and middle stratosphere by O_2 and O_3, it is not available to produce atomic oxygen. Because atomic oxygen is not present in any appreciable amount in the lower stratosphere, ozone is not formed here. The presence of adequate concentrations of both atomic and molecular oxygen in the middle stratosphere means that stratospheric ozone concentrations peak at about 35 kilometers altitude. Even though ozone reaches its highest concentrations at this level, its concentration is still less than 10 ppm. Thus, the term "ozone layer" is misleading. Ozone does not exist as a protective blanket, but more as a diffuse screen that partially protects the Earth from harmful ultraviolet radiation.

The Chapman reactions show that ozone is continually being created and destroyed in the stratosphere. Natural processes such as volcanoes, climate change, and solar activity change ozone concentrations over long time periods. Seasonal effects in ozone concentrations also exist due to changes in solar radiation. Ozone also reacts with many chemicals in the atmosphere such as nitrogen, chlorine, bromine, and hydrogen. In this process, a substance captures an oxygen from ozone. If X symbolizes a chemical substance that removes oxygen, then a general reaction for the destruction of O_3 is represented as:

$$X + O_3 \rightarrow XO + O_2$$

The XO compound formed in this reaction then reacts with oxygen to regenerate X and molecular oxygen:

$$XO + O \rightarrow X + O_2$$

The regenerated S can destroy another O_3 molecule, and the process can continue indefinitely. Thus a single atom or molecule can destroy many ozone molecules.

Even though stratospheric ozone concentrations are the result of dynamic processes, a natural balance of creation and destruction can be expected to produce a **steady-state** concentration. Steady-state occurs when the concentration of ozone remains constant and exists when the rate of ozone formation equals the rate of ozone destruction. Steady-state is similar to letting water run into a bathtub while it drains out, but the water level in the tub remains the same. Humans have disrupted this equilibrium in recent decades by introducing chemicals into the atmosphere that have increased the rate of destruction. Reduction of ozone levels varies with location, but during the last three decades, there has been about a 5% decrease worldwide. Reduction in ozone means that a greater percentage of UV-B radiation can penetrate the atmosphere. UV-B is ultraviolet radiation with wavelengths between 280 and 320 nm, UV-A includes wavelengths between 320 and 400 nm, and UV-C wavelengths are between 220 and 280 nm. Higher UV-B levels have been associated with increased rates of skin cancer and cataracts. Another consequence of increased UV-B radiation is its ability to suppress the immune system in organisms. UV-B may affect entire ecosystems. Increased UV-B radiation has been mentioned as a contributing factor in the destruction of coral reefs throughout the tropics; it may also destroy or disrupt phytoplankton and reduce productivity in the oceans.

Although ozone depletion is global, the polar regions, especially Antarctica, have experienced the greatest impacts and have received the most attention with respect to this problem. The "ozone hole" appears above Antarctica around September or October. It is called an ozone hole because at this time of the year ozone concentrations drop precipitously by as much as 95%. The reason for this condition is a combination of meteorological conditions and chemistry. During the Southern Hemisphere's winter, a weather condition known as the polar vortex is established over Antarctica. The polar vortex is a wind pattern created as cool air descends over the continent. The descending air is deflected by the **Coriolis force** into a westerly direction around Antarctica. The vortex effectively isolates the atmosphere over Antarctica and creates the region where ozone depletion occurs. The strength and duration of the polar vortex varies annually, but in general it can be thought of as a boundary that defines a giant reaction vessel where ozone-destroying reactions take place.

During the dark, polar winter the temperature drops to extremely low values, on the order of $-80°C$. At these temperatures, water and nitric acid form polar stratospheric clouds. Polar stratospheric clouds are important because chemical reactions in the stratosphere are catalyzed on the surface of the crystals forming these clouds. The chemical primarily responsible for ozone depletion is chlorine. Most of the chlorine in the stratosphere is contained in the compounds hydrogen chloride, HCl, or chlorine nitrate, $ClONO_2$. Hydrogen chloride and chlorine nitrate undergo a number of reactions on the surface of the crystals of polar stratospheric clouds. Two important reactions are:

$$ClONO_{2(g)} + HCl_{(s)} \rightarrow Cl_{2(g)} + HNO_{3(s)}$$
$$ClONO_{2(g)} + H_2O_{(s)} \rightarrow HOCl_{(g)} + HNO_{3(s)}$$

During the long Antarctic night, appreciable amounts of molecular chlorine, Cl_2, and hypochlorous acid, HOCl, accumulate within the polar vortex. When the sun returns during the spring (in September in Antarctica), ultraviolet radiation decomposes the accumulated molecular chlorine and hypochlorous acid to produce atomic chlorine, Cl. Atomic chlorine is a **free radical**. Free radicals are atoms or molecules that contain an unpaired or free electron. The Lewis structures of free radicals contain an odd number of electrons. The unpaired electron in free radicals makes them very reactive. The free radical Cl produced from the decomposition of Cl_2 and HOCl catalyzes the destruction of ozone as represented by the reaction:

$$Cl + O_3 \rightarrow ClO + O_2$$
$$ClO + O \rightarrow Cl + O_2$$

net reaction $O_3 + O \rightarrow 2O_2$

This reaction shows that the original chlorine atom acts like "X" in the general ozone-destroying reaction presented previously. A chlorine atom captures an oxygen atom from ozone in the first reaction. A chlorine atom is regenerated in the second reaction. Thus, a single chlorine atom can lead to the destruction of numerous ozone molecules.

There are numerous natural contributors of chlorine to the stratosphere, for example, volcanic eruptions. The main concern regarding ozone destruction in recent years is associated with human activities that have increased chlorine and other synthetic chemical input into the stratosphere. At the top of the list of such chemicals are chlorofluorocarbons, or CFCs. CFCs are compounds that contain carbon, chlorine, and fluorine; they were first developed in 1928. Common CFCs are called Freons, a trade name coined by the DuPont chemical company. CFC compounds are nonreactive, nontoxic, inflammable gases. Because of their relative safety, CFCs found widespread use as refrigerants, aerosol propellants, solvents, cleaning agents, and foam blowing agents. Their use grew progressively from the 1930s until the 1990s, when their annual production was over a billion pounds.

Chlorofluorocarbons are named using a code that can be used to determine the CFC's formula. Adding 90 to the CFC number gives a three-digit number that indicates the number of carbon, hydrogen, and fluorine atoms in the molecule. Any remaining atoms in the molecule are assumed to be chlorine. For example, the formula for CFC-12 is found by adding 90 to 12 to give 102. Therefore, the CFC-12 molecule has one carbon atom, no hydrogen atoms, and two fluorine atoms. Because carbon is bonded to four other atoms, the remaining two atoms in the molecule are chlorines. The chemical formula for CFC-12 is CF_2Cl_2.

F. Sherwood Rowland (1927–) and Mario Molina (1943–) predicted the destruction of stratospheric ozone in 1974. Rowland and Molina theorized that inert CFCs could drift into the stratosphere, where they would be broken down by ultraviolet radiation. Once in the stratosphere, the CFCs would become a source of ozone-depleting chlorine. The destruction of ozone by CFCs can be represented by the following series of reactions:

$$CF_2Cl_2 + UV \rightarrow Cl + CF_2Cl$$
$$Cl + O_3 \rightarrow ClO + O_2$$
$$ClO + O \rightarrow Cl + O_2$$

net reaction $CF_2Cl_2 + O_3 + O \rightarrow Cl$
$+ CF_2Cl + 2O_2$

This set of reactions shows that ultraviolet radiation strikes a CFC molecule removing a chlorine atom. The chlorine atom collides with an ozone molecule and bonds with one of ozone's oxygen atoms. The result is the formation of chlorine monoxide, ClO, and molecular oxygen. Chlorine monoxide is a free radical and reacts with atomic oxygen, producing another atom of chlorine that is released and available to destroy more ozone.

The actual destruction of ozone in the stratosphere actually involves hundreds of different reactions. Besides the Chapman reactions and destruction by CFCs, many other chemical species can destroy ozone. In 1970, Paul Crutzen (1933–) showed that nitrogen oxides could destroy ozone. Nitric oxide can remove an oxygen atom from ozone and be regenerated according to the following reactions:

$$NO + O_3 \rightarrow NO_2 + O_2$$
$$NO_2 + O \rightarrow NO + O_2$$

Crutzen's work provided arguments against the development of high-flying super sonic planes, such as the Concorde, that would deposit nitrogen oxides directly into the stratosphere.

The pioneering work of Rowland, Molina, and Crutzen raised questions concerning human activity at least a decade before serious depletion of the ozone layer was confirmed. These three scientists shared the 1995 Nobel Prize in chemistry for their work. Ozone measurements taken in the mid-1980s demonstrated that humans were doing serious damage to the ozone layer. This prompted international action and adoption of the Montréal Protocol in 1987. The original agreement called for cutting the use of CFCs in half by the year 2000. Subsequent strengthening of the agreement banned the production of CFCs and halons in developed countries after 1995. Halons are compounds similar to CFCs that contain bromine, for example, CF_2BrCl. Production of other ozone chemicals such as methyl bromide, CH_3Br, is also banned. Methyl bromide later was widely used as a soil fumigant to treat agricultural land.

The phasing out of CFCs and other ozone-depleting chemicals has given rise to new compounds to take their place. Hydrofluorochlorocarbons, or HCFCs, are being used as a short-term replacement for hard CFCs. Hard CFCs are those that do not contain hydrogen, while soft CFCs contain hydrogen. The presence of hydrogen atoms in HCFCs allows these compounds to react with hydroxyl radicals, OH, in the troposphere. Some HCFCs can eventually be converted in the troposphere to compounds such as HF, CO_2, and H_2O. Other HCFCs can be converted into water-soluble compounds that can be washed out by rain in the lower atmosphere. HCFCs provide a short-term solution to reversing the destruction of stratospheric ozone. Their use signaled the first step in repairing the damage that has occurred over the last fifty years. Signs are positive that the ozone layer can repair itself, but it may take another fifty years to bring ozone levels back to mid-twentieth century levels.

Acid Rain

If the pH of natural rain were measured, you might expect a pH of around 7.0. Because a pH of 7.0 indicates neutral conditions, many people assume this to be the pH of rain. The theoretical pH of pure rainwater is actually about 5.6. Pure rain is acidic due to the equilibrium established between water and carbon dioxide in the atmosphere. Carbon dioxide and water combine to give carbonic acid:

$$CO_{2(g)} + H_2O_{(l)} \leftrightarrow H_2CO_{3(aq)} \leftrightarrow H^+_{(aq)} + HCO^-_{3(aq)}$$

Acid rain or precipitation refers to rain, snow, fog, or gaseous particles that have a pH significantly below 5.6. There is no absolute pH that defines acid rain, but a general guideline that can be used is that precipitation below 5.0 can be considered acidic. Although the term "acid rain" is used in our discussion, consider rain to include all forms of precipitation and moisture. Acid rain occurring as rain or snow is referred to as wet deposition. Dry deposition occurs when gaseous acidic particles are deposited directly on surfaces.

Robert Angus Smith (1817–1884), a Scottish chemist, was the first to identify acid rain and study it in depth in 1852. Smith studied the differences in rain chemistry in proximity to Manchester, England. He described how rain changed with respect to geography, fuel combustion, and local conditions. In 1872, he coined the term "acid rain" in his book *Air and Rain: The Beginning of a Chemical Climatology*. The examination of acid rain accelerated around 1920 when Scandinavian scientists started to notice detrimental impacts on lakes in their countries. As more data were collected, it became apparent that acid rain was a global phenomenon that required international attention. It has been a prominent environmental issue in this country since the 1960s.

The principal cause of acid rain is the combustion of fossil fuels that produce sulfur and nitrogen emissions. The primary sources are electrical power plants, automobiles, and smelters. Power plants produce most of the sulfur emissions and automobiles most of the nitrogen emissions. Other sources of acid rain include nitrogen fertilizers, jet aircraft, and industrial emissions. Just as in our discussion of ozone, numerous reactions are involved in the formation of acid rain. The process can be understood by considering the transformation of sulfur and nitrogen oxides into their respective acidic forms: sulfuric acid and nitric acid. Sulfur, present up to a few percent in fuels such as coal, is converted to sulfur dioxide when the fuel is burned. The sulfur dioxide reacts with water to produce sulfurous acid, $H_2SO_{3(aq)}$, that is then oxidized to sulfuric acid. The reactions are

$$S_{8(s)} + 8O_2 \rightarrow 8SO_2$$
$$SO_{2(g)} + H_2O_{(l)} \rightarrow H_2SO_{3(aq)}$$
$$2H_2SO_{3(aq)} + O_{2(g)} \rightarrow 2H_2SO_{4(aq)}$$

Nitric acid, HNO_3, and nitrous acid, HNO_2, form when nitrogen dioxide reacts with water:

$$NO_{2(g)} + H_2O_{(l)} \rightarrow HNO_{3(aq)} + HNO_{2(aq)}$$

Many reactions contribute to the overall formation of acid rain. These reactions often involve a number of complicated steps that depend on the atmospheric conditions. The reactions just shown represent general reactions that form sulfuric and nitric acids, and do not show the numerous reactions that actually occur.

Acid rain has a number of negative effects on humans, plants, animals, building materials, and entire ecosystems. Acid rain directly affects human health by irritating the respiratory passages. Elderly and asthmatic individuals are particularly susceptible to acidic gases. Conditions include persistent coughing, headaches, watery eyes, and difficulty in breathing. Acid rain can have an indirect effect on human health when drinking water is obtained from acidified lakes because acidic water can dissolve toxic metals, such as lead and copper in pipes, introducing these into drinking water.

Acid rain has caused significant damage to forests throughout much of Europe. In some areas of eastern Europe, as much as half of the forests have been damaged by acid rain. Acid rain has both direct and indirect effects on plants. Acid rain dissolves the waxy protective coating of leaves interfering with plant respiration. In areas where acid rain is a problem, plant leaves often become discolored. For example, a yellowing discoloration of conifer needles has been observed at high elevations. Trees may also shed their leaves or needles when subjected to acid rain. Such conditions weaken plants and make them less resistant to disease, insects, and extreme environmental conditions, such as frost and wind. In extreme cases, entire forest areas die. Agricultural plants are subjected to the same impacts as natural plants. Crop damage and disease cause significant economic losses each year throughout the world.

Besides affecting plants directly, acid rain affects plants indirectly by altering the pH of the soil, although some soils are naturally acidic. The pH of some European soils subjected to acid rain has been lowered by one pH unit over the last 60 years. Because the pH scale is a logarithmic scale, this means that the soil is ten times more acidic now than sixty years ago. Lower soil pH interferes with the ability of plants to absorb nutrients. This is why it is important to have the correct pH before fertilizing a lawn. Soil acidity is typically neutralized by applying lime ($CaCO_3$) to the soil. In addition to affecting the nutrient balance in plants, acid rain mobilizes and leaches out soil cations such as Al^{3+}, Mg^{2+}, Ca^{2+}, Na^+, and K^+. These cations dissolve in acidic soil water. This results in the release of toxic metals into the environment that can affect a number of other organisms. This condition is especially pronounced in the spring, when rapid melting of winter snow causes a pulse of dissolved substances in runoff that floods into water bodies. Toxic metals in this runoff can have a pronounced effect on embryos and eggs of amphibians, fish, and other aquatic organisms.

Acid rain has resulted in the loss of life in a number of lakes. The ability of a lake to withstand the impacts of acid rain is related to the geology of the lake's basin. In areas with limestone (calcium carbonate) deposits, a lake has a natural buffering capacity. The buffering capacity refers to the ability to resist changes in pH. In well-buffered lakes, calcium carbonate reacts

with hydrogen ions to produce bicarbonate, HCO_3^-. Bicarbonate reacts with hydrogen ions to produce carbonic acid, H_2CO_3. The reactions are

$$CaCO_{3(s)} + H^+_{(aq)} \rightarrow Ca^{2+}_{(aq)} + HCO_3^-{}_{(aq)}$$
$$HCO_3^-{}_{(aq)} + H^+_{(aq)} \rightarrow H_2CO_{3(aq)} \rightarrow CO_{2(g)} + H_2O$$

In areas where the geology is dominated by granite, lakes have less buffering capacity and are much more susceptible to the impacts of acid rain. Fish and aquatic organisms differ in their ability to adapt to acidic conditions. The natural pH of lakes is approximately 8.0. The pH of poorly buffered lakes is between 6.5 and 7.0. The effects of pH on different aquatic organisms are summarized in Table 18.1.

Another detrimental effect of acid rain is the damage it does to buildings and other structures. Limestone and marble used as building material are readily attacked by acid rain. The calcium carbonate in limestone and marble dissolves in acid. These materials can also react with sulfur dioxide to produce gypsum, $CaSO_4 \cdot 2H_2O$, which is much weaker and more soluble than calcium carbonate. Many cultural treasures have been severely damaged by acid rain, including well-known buildings such as the Taj Mahal, Westminster Abby, and the Parthenon. European cathedrals and statues throughout the world are slowly dissolving as they are subjected to acid rain. In Egypt and Mexico, it is estimated that more damage to ancient ruins has occurred in the last 50 years than in the previous 2,500 years. Acid rain also attacks iron structures, catalyzing the formation of rust. In areas of acid rain, the oxidation of metals is greatly accelerated, greatly increasing maintenance costs.

To control acid rain requires international cooperation and much effort has gone into crafting agreements between countries

Table 18.1
Effect of pH on Aquatic Organisms

pH	Effect
3.0-3.5	Lethal to most fish.
3.5-4.0	Lethal to salmonids (trout, salmon, smelt).
4.0-4.5	Embryos fail to mature, absence of amphibians.
4.5-5.0	Harmful to eggs, low diversity of life.
5.0-6.0	Disappearance of many species of zooplankton, phytoplankton, snails, and aquatic insects.
6.0-7.2	Optimal range for most fish species.
7.5-8.5	Optimal range for algae.
9-10	Slowly lethal to fish.
Above 10	Lethal to most fish

to reduce sulfur and nitrogen emissions. The problem with acid rain is that much of the emissions responsible for this problem are introduced into the atmosphere through tall smokestacks to alleviate local air quality problems. Reactions producing acid rain take place over several days as air masses carrying sulfur and nitrogen emissions move with prevailing winds. When the moisture in the air mass is released as precipitation, it may be several hundred to several thousand miles away from the source of emissions. It is estimated that half of Canada's acid rain is produced in the United States. Similarly, emissions in England and Germany can impact Sweden and Finland. In the United States, the Clean Air Act addresses the problem of acid rain by requiring reductions in sulfur and nitrogen emissions. Much of the legislation is targeted at coal burning electrical power plants. Title IV of Clean Air Act enacted in 1990 mandates reductions in sul-

fur dioxide and nitrogen oxides by the years 2010 and 2000, respectively. The Title IV amendments call for an annual reduction in sulfur dioxide emissions from power plants of 10 million tons from a 1980 level of 19 million tons. Significant progress has been made on reaching the 2010 goal. In 2002, the total emissions from power plants was approximately 11 million tons per year. Progress on reduction of nitrogen oxides has lagged behind that of other pollutants. While power plant emissions have fallen from 6.1 million tons per year in 1980 to approximately 5.0 million tons per year in 2002, total nitrogen oxides emissions have increased by 9% over the past twenty years due to emissions from engines (used for purposes other than transportation) and diesel vehicles.

Industry has adopted both input and output approaches to reducing emissions. An input approach involves using fuels with lower sulfur content. Soft bituminous coal sulfur's content may be as low as a few tenths of a percent compared to hard eastern anthracite coal, which has sulfur contents of several percent. One problem, though, is the transportation of western coal often makes its use economically unfeasible. An alternative input approach is to use a process to remove sulfur from coal before burning it. Coal cleaning involves using a physical process to separate pyrite, FeS, from coal by dispersing coal in oil and allowing the heavier pyrite to settle out. Another input approach includes actually switching from coal to natural gas. Although natural gas emits less sulfur, the higher combustion temperatures involved produces more nitrogen oxides. An output approach involves removing the sulfur from the flue gas after combustion. Emissions can be removed using scrubbers that reduce harmful emissions. In scrubbers, chemicals convert a gas pollutant to less harmful products. For example, a solution of calcium carbonate can be sprayed down stacks emitting sulfur dioxide to produce solid calcium sulfate, $CaSO_4$:

$$CaCO_{3(s)} + SO_{2(g)} \rightarrow CaSO_{3(s)} + CO_{2(g)}$$
$$2CaSO_{3(s)} + O_{2(g)} \rightarrow 2CaSO_{4(s)}$$

The Greenhouse Effect

The greenhouse effect refers to an elevation of the Earth's temperature as a result of changes in the Earth's atmosphere due to human activities. The name comes from the fact that the atmosphere acts like glass in a greenhouse trapping radiation and causing a temperature increase. To understand the greenhouse effect requires knowing how radiation interacts with gases to regulate the Earth's temperature. The Earth's surface temperature is primarily determined by the amount of radiation it receives from the Sun

and how much it reradiates back into space. Any object naturally radiates electromagnetic energy over a wide range of wavelengths. The principal wavelength of radiated energy is determined by the body's surface temperature. **Wien's Law** states that the peak radiation is inversely proportional to the absolute temperature:

$$\lambda_{peak} = \frac{2.9 \times 10^6}{T}$$

In this equation, the peak wavelength is in nanometers (1 nanometer, abbreviated nm, is equal to 10^{-9} meter) and T is the absolute temperature in Kelvin. Wien's equation shows that as the surface temperature increases, the peak wavelength decreases. Assuming the Sun's surface temperature is 6,000 K, the peak wavelength of radiation from the Sun is approximately 483 nm. The peak wavelength for Earth with an average surface temperature of 288 K is about 10,000 nm. The peak electromagnetic energy radiated from the Sun is in the visible range, while that from Earth is primarily infrared radiation. Radiation from the Sun is often termed **shortwave** radiation and radiation from Earth **longwave radiation** (Figure 18.2).

Incoming radiation from the Sun and backradiation emitted by Earth interacts with the atmosphere. Although about half of the Sun's radiation passes directly to Earth's surface, a portion is reflected back directly into space, while another portion is absorbed by atmospheric gases and reradiated. Figure 18.3 shows the fate of radiation intercepting Earth. About half of the incoming solar radiation actually reaches the surface of Earth. The rest is reflected or absorbed by the atmosphere or clouds. Infrared radiation reflected from Earth's surface is partially absorbed and reflected by the atmosphere and clouds. Some of this radiation is reradiated back toward Earth's

Figure 18.2
Peaks of Solar and Earth Radiation (Rae Déjur)

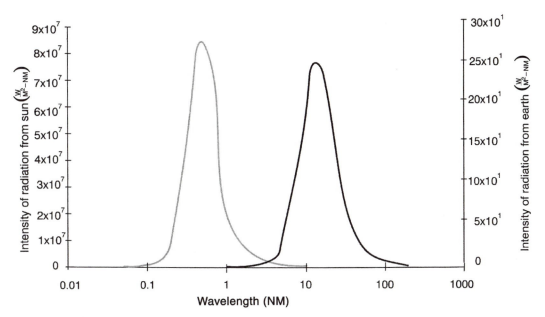

RADIATION FROM SUN AND EARTH

surface, causing a raise in temperature. The atmosphere is responsible for Earth's temperature being about 30°C higher than it would be if the atmosphere was not present. In fact, the atmosphere is responsible for a natural greenhouse effect. Many scientists define the natural temperature increase as the greenhouse effect and the temperature increase due to human activities that modify the atmosphere as the enhanced greenhouse effect. We will define the greenhouse effect as excess warming due to human activities, but keep in mind that a natural greenhouse effect preceded humans on our planet. In summary, Earth's surface temperature is due to the balance between incoming and outgoing radiation. The fact that the atmosphere is more transparent to shortwave solar radiation compared to longwave radiation from Earth leads to a natural greenhouse effect that produces a surface temperature of 288 K (15°C or 59°F).

The ability of certain molecules in the atmosphere to absorb infrared radiation and reradiate back to Earth depends on the molecule's structure. Molecules must possess a **dipole moment** and have bonds capable of vibrating at a frequency in the infrared range. Because the major atmospheric gases N_2, O_2, and Ar do not possess a dipole moment, they do not absorb infrared radiation. Gases capable of absorbing infrared radiation are referred to as greenhouse gases. The most common greenhouse gas is water. Water's ability to absorb infrared radiation is demonstrated by the warming effect of clouds.

Human activities have changed the concentration of a number of greenhouse gases. Unlike water, these gases exist in trace amounts and the relative human impact on their concentrations is significant. The primary greenhouse gases include carbon dioxide, methane (CH_4), CFCs, and nitrous

Figure 18.3
Earth's Radiation Budget (Rae Déjur)

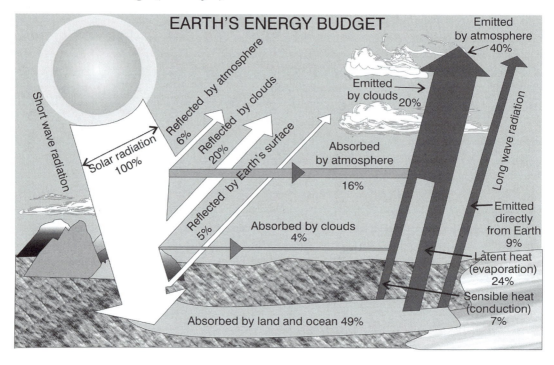

oxide (N_2O). Carbon dioxide is the most important greenhouse gas, being responsible for roughly 60% of the warming. The atmosphere's CO_2 concentration has risen from about 280 ppmv (parts per million by volume) to 360 ppmv in the last 150 years. This increase is due primarily to the burning of fossil fuels. Land clearing has also contributed to increased CO_2 levels. Methane is the next most important greenhouse gas, contributing about 20% to global warming. Methane is produced by methanogenic bacteria under anaerobic conditions. Decaying vegetation from wetlands and rice paddies produces large quantities of methane. It also is produced in the digestive tract of livestock, in landfills, and by biomass combustion. Nitrous oxide contributes about 5% to the greenhouse effect.

The greenhouse effect continues to be a controversial topic as the twenty-first cen-

tury begins. The controversy has shifted from uncertainty concerning whether humans affect Earth's climate to how much humans are affecting the climate and what action, if any, should be taken. A major difficulty surrounding the greenhouse effect is separating human influences from the climate's natural variability. While there are still skeptics, the consensus seems to be that humans are affecting the climate, but exactly how, and to what extent, continues to be debated. The issue is complicated by numerous factors. The oceans play a major role in the carbon cycle because they have a great capacity to absorb carbon dioxide. Vegetation also affects global carbon concentrations through photosynthesis. Some studies have indicated that primary productivity increases with elevated CO_2, while others show no increase. A major human factor is changes in land use, for example, deforestation and agriculture. Other issues include variations in solar output and the role of clouds in Earth's energy balance.

During the last decade national representatives have attempted to craft an agreement to reduce global greenhouse gas emissions. At the Earth Summit held in Rio de Janeiro, Brazil, in 1992, industrialized nations agreed to stabilize emissions at their 1990 levels by the year 2010. This nonbinding agreement was seen as inadequate by many nations, and several subsequent international meetings have attempted to resolve disputes between countries regarding reduction in emissions. The European nations have pushed for more aggressive reductions than the United States. The United States is also dissatisfied that many developing countries, including the two most populous countries on the planet, China and India, are exempt from making immediate emission reductions. The last significant negotiations took place in December of 1997 in Kyoto, Japan. The nations at the Kyoto Summit had trouble agreeing how to curb greenhouse gases. The European Union wanted reductions of 15% below 1990 levels. Several other countries led by the United States resisted such deep cuts, fearing they would cause too much harm to their economies. The last round of talks on the Kyoto Summit took place in November 2001 in Morocco. The meeting in Morocco concluded four years of negotiations and required industrialized countries to reduce greenhouse emissions by 5% below their 1990 levels by the year 2012. The U.S. representative at the meeting announced the Bush administration's dissatisfaction with the Kyoto agreement. The administration does not plan to sign the agreement and is going to put forth its own solution for reducing emissions. This stance has provoked harsh criticism from European leaders and may unravel much of the work done in the

Table 18.2
Carbon Emissions by Country

Country	Rank	1998 Carbon Emissions (Thousand of metric tons)
United States	1	1,486,801
China	2	848,266
Russia	3	391,535
Japan	4	309,353
India	5	289,587
Germany	6	225,208
United Kingdom	7	148,011
Canada	8	127,517
Italy	9	113,238
Mexico	10	102,072

1990s to reduce greenhouse gases. Any workable agreement must include the United States. The United States does not want to jeopardize its economy and believes the overall benefits of entering into an agreement must outweigh the costs. An agreement must also be enforceable. As the world's number one emitter of carbon (Table 18.2), the United States holds the key to global reductions.

Water Quality

Water is essential for human life. Although Earth is often called the "water planet" with nearly three-fourths of Earth's surface covered by water, most of its tremendous supply of water is salty and not directly available for human use. Only 3% of the planet's water is fresh. Approximately 70% of the human body is composed of water, and we can survive only a few days without it. Most of the water in the world, and the largest percentage in this country, is used for agriculture. Large quantities of water are also used in power generation, for domestic use, and in industry. Table 18.3

Table 18.3
Daily Water Use

Use	Gallons per day	%
showers	12.6	17.3
washing clothes	15.1	20.9
dishwashers	1.0	1.3
toilets	20.1	27.7
baths	1.2	2.1
leak	10.0	13.8
faucets	11.1	15.3
other	1.5	2.1

gives the approximate use of water in our daily activities.

Water is often called the universal solvent. Natural water is far from pure and exists as a natural solution containing many dissolved minerals. **Hard water** contains an abundance of divalent metals, primarily Ca^{2+} and Mg^{2+}. Water hardness is measured in parts per million of $CaCO_3$. Soft water has less than 60 ppm $CaCO_3$, moderately hard water has concentrations between 61 and 120 ppm $CaCO_3$, and hard water has hardness values greater than 120 ppm $CaCO_3$. Water hardness does not present a health concern until it reaches several hundred ppm. When the hardness of water is due to dissolved calcium and magnesium bicarbonates, $Ca(HCO_3)_2$ and $Mg(HCO_3)_2$, it is termed **carbonate hardness** or temporary hardness. It is temporary because when water containing carbonate hardness is heated, the carbonates precipitate out. This can lead to scales developing in pipes and boilers used in hot water heating systems. **Permanent hardness** consists of sulfates and chlorides that remain in solution when heated.

Water pollution results from a number of sources including agricultural runoff, erosion, industrial wastes, domestic wastes, and road runoff. Water pollution can be classified as either **point source** or **nonpoint source**. Point source pollution is emitted from a specific, well-defined location such as a pipe. Nonpoint sources refer to pollutants dispersed over a wide area from many different areas. A sewer pipe would be an example of a point source and fertilizer runoff from a field would represent nonpoint source pollution.

A major water quality problem occurs when waterways receive too many nutrients such as phosphate or nitrate. This occurs in areas subjected to increased agricultural runoff, sewage effluent, and erosion. Excess nutrients fertilize water bodies, and this results in **phytoplankton** blooms. Increased algae results in large **zooplankton** blooms. When the heavy plankton populations die and undergo decomposition by bacteria, oxygen levels in the water drop. In extreme cases, the oxygen levels in the lower levels of the lake can drop to zero. This accelerated input of nutrients causes premature aging of the lake. The process is called cultural **eutrophication**. Up until 1980, detergents were a major source of phosphate input to waterways and a primary cause of cultural eutrophication. Many detergents and soap products now carry a "no phosphate" label to let the consumer know that the product is phosphate free.

One of the foremost global issues regarding water quality is access to safe drinking water. Currently, one-sixth of the world's 6.1 billion people do not have access to a source of clean water, and 40% do not have adequate sanitation facilities. Polluted water is responsible for diseases such as typhoid, cholera, and dysentery. Three million people die annually due to the latter. The discovery that diseases were transmitted by bacteria through water in the mid-nineteenth century led to advances in sanitation and water treatment. Several chemicals are used as disinfectants to kill pathogenic organisms in water. The most common disinfectant in this country is chlorine. The first public water supply to be treated with chlorine was in England in 1904 and in the United States in 1908. Throughout the twentieth century, as the practice of water chlorination spread, the incidence of water-borne diseases has decreased.

As mentioned, chlorine, in the form of Cl_2 or the hypochlorite ion, OCl^-, is used as a disinfectant. Free chlorine, Cl_2, reacts with water to produce hypochlorous acid, HOCl, and hydrochloric acid, HCl:

$$Cl_{2(g)} + H_2O_{(l)} \rightarrow HOCl_{(aq)} + HCl_{(aq)}$$

The hypochlorous acid oxidizes the cell walls and kills bacteria. Solid calcium hypochlorite, $Ca(OCl)_2$, and liquid solutions of sodium hypochlorite, NaOCl, can be used to generate hypochlorous acid in place of chlorine gas, for example, in chlorinating swimming pools. The hypochlorite ion generated from $Ca(OCl)_2$ and NaOCl forms an equilibrium with water represented by the equation:

$$OCl^- + H_2O \leftrightarrow HOCl + OH^-$$

The use of hypochlorites tends to raise the pH of water because hydroxide ions are produced, as opposed to chlorine gas, which tends to lower the pH due to the production of hydrochloric acid. Bleach is a dilute solution (approximately 5%) of sodium hypochlorite that is often used as a household disinfectant.

Chlorine has been criticized as a water disinfectant because of undesirable reactions that produce carcinogenic substances.

One drawback of chlorine is that it can react with organic materials present in natural waters to produce harmful substances. **Trihalomethanes** are organic compounds with the general formula CHX_3, where x represents chlorine or bromine. When hypochlorous acid reacts with organic matter, chloroform, $CHCl_3$, can be produced. Chloroform, which was one of the first anesthetics used in 1847, is a suspected carcinogenic. Because of this, the drinking water standard for trihalomethanes has been set at 100 parts per billion. Although chlorine is still the most widely used disinfectant in the United States, some communities have switched to other disinfectants. Chlorine dioxide and ozone are two common alternatives. Ozone is used widely throughout Europe for treating drinking water. It is a much more effective disinfectant and has the ability to kill viruses. Its drawback is that it cannot be stored so it must be generated on site. It also does not provide the residual protection that chlorine provides. **Residual chlorine** is the chlorine dissolved in the water once it leaves the treatment plant. This residual chlorine protects the water from recontamination. Ozone-treated water does not have the residual protection once it leaves the treatment plant.

To protect public health, government agencies have established guidelines for numerous chemicals in drinking water. In 1974, the United States passed the Safe Drinking Water Act. As part of this act, the EPA determined safe levels for chemicals in drinking water. The **maximum contaminant level** (mcl) is the highest acceptable concentration for a chemical in drinking water. Although not an absolute guarantee to ensure safety, the mcl is a guideline that establishes a maximum acceptable value for chemicals in water. Table 18.4 shows the maximum contaminant level for a number of substances in drinking water.

Most of the substances listed in Table 18.4, as well as a number of others, have been associated with serious environmental problems in this country and many others. Arsenic, mercury, and lead are metals that seriously affect human health when found in excessive amounts in drinking water. Arsenic is the quintessential poison, appearing in Shakespeare, Sherlock Holmes, and modern murder scenes. It has been known since ancient times and was traditionally used as an antibiotic; it was commonly used to cure syphilis until the advent of penicillin and modern antibiotics. Arsenic's extensive use in pesticides, primarily before the advent of organic pesticides, and its unintentional release during mining are major sources of water contamination. Arsenic poisoning can lead to several forms of cancer including cancer of the skin, lungs, blad-

Table 18.4
Maximum Contaminant Level

Substance	MCL
arsenic	50 ppb
copper	1.3 ppm
iron	300 ppb
lead	15 ppb
mercury	2 ppb
nitrate	10 ppm
nitrite	1 ppm
benzene	5 ppb
MTBE	5 ppb
PCB	0.5 ppb
total THM	100 ppb

der, and kidney. Initial symptoms include skin discoloration. Arsenic contamination is a serious problem in a number of countries, especially in rural areas that depend on well water as a drinking source. In Bangladesh, it is estimated that millions of people depend on water sources contaminated with arsenic. In the United States, the EPA lowered the arsenic mcl from 50 parts per billion to 10 parts per billion during the waning days of the Clinton administration in 2000. The standard was lowered due to evidence that arsenic in drinking water poses a higher cancer risk than once thought. This ruling was initially suspended by the Bush administration pending further review, but the 10 ppb was finally accepted in late 2001. Water utility companies in areas where arsenic is naturally high fear excessive costs in meeting the stricter standard.

Mercury, known in ancient times as quicksilver, is another metal that has been studied for years. Mercury is widely used in batteries, thermostats, scientific instruments (thermometers and barometers), alloys, fungicides, and precious metal extraction. During the 1800s, mercury (II) nitrate, $Hg(NO_3)_2$, was used in the curing process to manufacture felt hats. Individuals working in the hat industry had a high rate of mercury poisoning, although at the time the effects of mercury on humans were unknown. Because mercury poisoning results in neurological damage, the phrase "mad as a hatter" became associated with hat workers who used mercury compounds. Mercury poisoning symptoms include slurred speech, tremors, and memory loss. Lewis Carroll's character the Mad Hatter from *Alice in Wonderland*, written in 1865, was based on the perception of aging hat workers going mad.

The greatest environmental concern involving mercury involves methylated forms of mercury. These forms result when bacteria convert Hg^{2+} into methyl mercury compounds during **anaerobic** respiration. Dimethyl mercury, CH_3HgCH_3, methyl mercury, CH_3Hg^+, and other methylated mercury compounds appear in the sediments where anaerobic bacteria exist. In this manner, mercury is incorporated into the food chain and ends up in fish consumed by humans. A number of lakes in the United States and Canada contain mercury-contaminated fish. Fish with the highest concentrations are large, carnivorous fish that reside high on the food chain, such as northern pike and largemouth bass. In areas known to be affected, state agencies have posted warnings, and people are advised to regulate their consumption of fish. Mercury poisoning is especially detrimental to native communities whose diet has traditionally consisted of large quantities of fish. Saltwater species such as shark, mackerel, and swordfish are often associated with high mercury levels. The Food and Drug Administration has issued warnings concerning consumption of these species, and pregnant women are especially advised to monitor their fish intake.

Lead is yet another metal that is associated with environmental problems. It has been used for thousands of years and is currently used in lead-acid batteries, nuclear sheathing, pipes, paints, solder, and alloys. Leaden ware is often cited as a primary source of poisoning in Roman times. Studies on the bones of ancient Romans have shown lead levels several orders of magnitude higher than those of modern humans. Lead poisoning can result in anemia, paralysis of joints, fatigue, reproductive problems, and colic. Severe cases of lead poisoning result in damage to the brain, kidney, and nervous system. In recent years, lead has been implicated in the delayed mental development of children, resulting in lower IQ scores. The use of lead has been restricted in recent years as the detrimental effects of lead on human health have been discovered. In the 1970s, leaded gasoline began to be phased out and in the 1980s lead was reduced in paints. Drinking water is generally not a major source of lead in humans, but it can be as high as 60% in infants drinking water-based milk formula. Lead in drinking water is associated with the use of lead pipes and lead solder used in plumbing systems. This is especially prevalent in older water systems built before 1930 where the water is soft. Hard water contains carbonates that form protective deposits on the interior of pipes. In some systems where the possibility of lead contamination from pipes exists, phosphates are added to the water to form protective deposits on the interior of pipes. Ironically, new homes less than five years old may also suffer from lead problems because their pipes have not been coated with deposits. The 1986 amendments to the Safe Drinking Water Act prohibited the use of lead pipe and solder in all public water systems.

Water pollution from benzene, C_6H_6, results from its presence as an octane booster in gasoline. Health effects of benzene include anemia, suppression of the immune system, nerve damage, and leukemia. Groundwater contamination from benzene, and its associated derivates such as toluene and xylene, continues to be a prevalent problem due to leakage from underground storage tanks. Millions of these tanks exist at gas stations and other locales throughout the United States. Starting in the 1990s, the EPA began an intensive effort to put into place a comprehensive series of regulations regarding the monitoring and replacement of underground storage tanks.

PCBs, or polychlorinated biphenyls, consist of a group of chlorinated organic chemicals that found widespread use starting around 1960. The biphenyl molecule consists of two benzene rings joined by a single bond. It is formed by heating benzene in the presence of a catalyst:

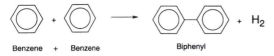

Benzene + Benzene   Biphenyl

PCBs are produced when biphenyl is reacted with chlorine and the chlorine atoms replace the hydrogen atoms. More than two hundred different PCBs can be produced. Because PCBs are nonflammable, inert, and poor conductors, they made ideal electrical insulators in transformers. PCBs also found use in hydraulic fluids, plasticizers, pesticides, copy paper, cutting oils, and fire retardants. Unfortunately, it was discovered that PCBs might be associated with liver, gastrointestinal, and reproductive problems. Some workers in the PCB industry developed a permanent, disfiguring form of acne called chloracne. Animal tests showed cancers developed in test animals. Another concern was that PCBs were fat soluble and **bioaccumulated** in the food chain. Because the structure of PCBs is similar to DDT, it was feared that their continued use could have severe consequences for wildlife (Figure 18.4). Although the acute toxicity of PCBs is low, there was enough evidence that PCBs could have long-range health effects to prompt Monsanto, their sole producer in North America, to discontinue their production.

During the last thirty years the United States has made significant progress in reducing pollution to the nation's waters. The Clean Water Act passed by Congress in 1972 has been the primary legislation responsible for curbing water pollution during this period. This act mandated stricter water standards and regulated those entities that discharged **effluent** into the nation's water. Federal permits were required for many discharge activities, for example, municipal treatment plant effluent. The Clean Water Act also provided grant money for municipalities to improve their wastewater treatment facilities by building new plants and applying modern technology to upgrade existing facilities. Although the media often highlights pollution incidents that tend to capture attention, the overall water quality in the United States has greatly improved since passage of the Clean Water Act.

Air Pollution

Just as water bodies such as rivers, lakes, and the ocean can be considered enormous aqueous solutions, the atmosphere is a gigantic gaseous solution. The atmosphere is composed primarily of nitrogen, 78%, and oxygen, 21%. The third most abundant gas in the atmosphere is argon, which makes up about 0.9%. The remaining 0.1% consists of

Figure 18.4
DDT and PCB. The Xs in PCB denote where chlorine atoms may attach.

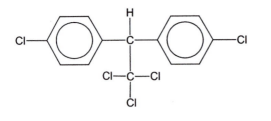

DDT   PCB

a variety of gases including carbon dioxide, water, xenon, helium, methane, and ozone. The previously mentioned problems of ozone depletion, acid rain, and global warming illustrate how human impacts on the chemical composition of **trace gases** in the atmosphere can produce serious problems. On the local level, air quality is a product of many factors such as climate, weather, population density, land use, industry, and transportation.

The Clean Air Act of 1967 forms the legal basis for air pollution control in the United States. This act and its subsequent amendments give the federal government the right to establish national ambient air quality standards (NAAQS) for major pollutants. These standards are intended to protect public health as well as to protect public welfare. The latter includes protecting visibility, nonhuman species, and structures. The NAAQS for the six major categories of air pollutants are given in Table 18.5. Table 18.5 shows that while several of the pollutants are gases, lead and particulate matter consist of solid particles suspended in air. PM-10 refers to solid particles with diameters of less than 10 micrometers. Particles of this size are the most detrimental to human health because they can be inhaled deep into

the respiratory system. The NAAQS pollutants are called criteria pollutants. Criteria pollutants are generic pollutants that all states and local governments need to consider. The Clean Air Act has specific regulations regarding emissions of many other substances such as toxic chemicals, but these substances have unique sources often associated with specific industries.

Chemicals can be labeled as either a primary air pollutant or secondary air pollutant. Primary air pollutants are those such as carbon monoxide and sulfur dioxide that enter the atmosphere directly as a result of human or natural events. Carbon monoxide's primary source in the atmosphere is the incomplete combustion of gasoline. Hundreds of different chemicals are present in gasoline. The combustion of octane, C_8H_{18}, can be used to represent the general reaction of hydrocarbons in an automobile engine to produce energy:

$$2C_8H_{18} + 25O_2 \rightarrow 18H_2O + 16CO_2 + \text{energy}$$

Carbon monoxide results when insufficient oxygen is present during the combustion process. Carbon monoxide's toxic effect results from the fact that hemoglobin's affinity for CO is more than 200 times greater than it is for oxygen. Because carbon monoxide displaces oxygen in the blood, breathing air with high carbon monoxide levels results in a lack of oxygen. The effects of carbon monoxide poisoning depend on the concentration and length of exposure (Table 18.6). If the CO concentration is high enough, people can slowly suffocate unaware (CO is odorless and invisible) that they are suffering from CO poisoning. Several hundred people die each year in the United States due to CO poisoning, mainly due to faulty furnaces or using gas heaters in enclosed areas without proper ventilation. While normal CO levels in areas with heavy

Table 18.5
NAAQS

Pollutant	Standard
Carbon monoxide, CO	9 ppm (8-hour average)
	35 ppm (1-hour average)
Nitrogen dioxide, NO₂	0.053 ppm (annual average)
Ozone, O₃	0.12 ppm (1 hour average)
Lead, Pb	1.5 µg/m³ (Quarterly average)
Particulate Matter, PM-10	50 µg/m³ (annual average)
	150 µg/m³ (24-hour average)
Sulfur dioxide, SO₂	0.03 ppm (annual average)

Table 18.6
Effects of CO Poisoning

Percent of hemoglobin carrying CO	Symptoms
0-9	None
10-19	Headache
20-29	Throbbing headache, nausea, irritability, lack of concentration
30-39	Severe headache, dizziness, fatigue, disorientation
40-49	Increase breathing and pulse rate, fainting.
50-59	Respiratory failure, seizures.
60-69	Low blood pressure, coma
70 and above	Death

traffic congestion are not lethal, they can result in drowsiness, headaches, irritability, and impaired judgment.

Secondary pollutants result from chemical reactions that occur in air. Sulfuric acid and nitric acid responsible for acid rain are examples of secondary pollutants. Another large class of secondary pollutants is **photochemical oxidants**. Photochemical oxidants include ozone, PAN (peroxyacylnitrate), and formaldehyde. A common term for photochemical oxidants is photochemical smog. Smog, a term derived from combining "smoke" and "fog," was originally coined to described polluted air in London. Smog is particularly characteristic in cities where industrial pollution occurs. Photochemical smog is another type of smog that occurs mainly in dry climates where there is ample sunlight and an abundance of cars. Los Angeles and Mexico City are cities where photochemical smog occurs on a reg-

ular basis. Photochemical smog forms through a complex series of reactions involving nitrogen oxides, hydrocarbons, and sunlight in relatively warm climates (Figure 18.5). Automobile exhaust contains nitrogen oxides (NO and NO_2) and hydrocarbons that are the two main ingredients necessary for the formation of photochemical smog. Nitrogen dioxide (NO_2) is a brown gas that produces an irritating odor. When the sun rises in areas with high concentrations of nitrogen dioxide and hydrocarbons, numerous reactions occur to produce a variety of photochemical oxidants. For example, ozone is formed by the following reactions:

$$NO_{2(g)} + \text{sunlight} \rightarrow NO_{(g)} + O_{(g)}$$
$$O_{(g)} + O_{2(g)} \rightarrow O_{3(g)}$$

It is interesting to note that ozone formation in the stratosphere is desirable, but ozone formation in the troposphere is not. PAN results from a complex series of reactions

Figure 18.5
Smog Formation (Rae Déjur)

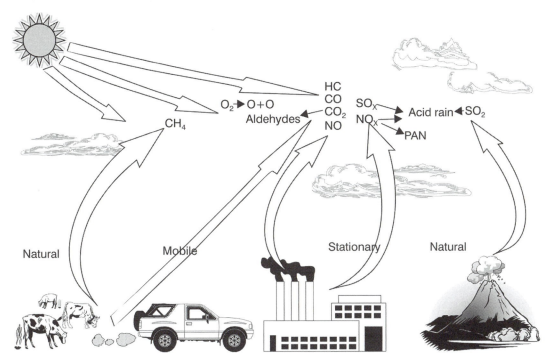

involving nitrogen oxides, oxygen, aldehydes, and hydrocarbons. PAN is a powerful eye irritant and irritates the respiratory system.

Just as the Clean Water Act has improved water quality in the United States, the Clean Air Act has greatly improved air quality since it was passed in 1967 and amended several times after. States and municipalities must meet the NAAQS or face the loss of federal dollars for road projects. Programs such as vehicle emission inspections, reduction in speed limits, emission permitting, alternative fuels, and use of best available technology have served to clean the nation's air.

Pesticides

As long as humans have existed they have been plagued by pests. Today when we observe animal species rolling in mud or primates grooming each other, we can imagine our ancient ancestors performing similar activities to combat pests. Agriculture as a means of obtaining food started approximately 10,000 years ago, and this increased the number of species that could be classified as pests. Several methods can be used to control pests. These include physical, genetic, biological, and chemical methods, but in this section only chemical controls will be considered. The word "pesticide" is a collective term in that it includes numerous other "cides" such as herbicides, insecticides, rodenticides, fungicides, bactericides, and avicides.

The use of chemical pesticides can be traced back several thousand years. The Sumerians, occupying present-day Iraq, burned sulfur compounds to produce fumigants. The Greeks also employed this practice as early as 1000 B.C. Evidence exists that the Chinese extracted pesticides from plants as early as 1200 B.C. The first pesticides used widely in recent history were poisonous metals. In France in the early eighteenth century a mixture of copper sulfate ($CuSO_4$) and lime ($CaCO_3$) was sprayed on grapes to make them unappealing to thieves. It was noted that this mixture also protected the grapes from certain fungi, and thus a pesticide known as Bordeaux mixture came into existence. Paris green, a compound of copper and arsenic, was a common pesticide used throughout the eighteenth century. Lead arsenate, $Pb_3(AsO_4)_2$, and several metal compounds containing antimony and mercury started to replace Paris green at the beginning of the twentieth century. Metal, inorganic pesticides were common throughout the first half of the twentieth century. Sodium fluoride and boric acid, $B(OH)_3$, were commonly used as ant and cockroach poisons; hydrogen cyanide (HCN) was a fumigant. Nicotine solutions made from soaking tobacco leaves in water have been used for several hundred years as insecticides and rodenticides.

The common inorganic compounds used as pesticides have several problems. Although they kill the target pest, they tend to be very lethal to humans and other mammals. Pesticides that kill indiscriminately are classified as broad spectrum pesticides. This means they do not distinguish between target and nontarget species, but kill many harmless and helpful species. Another major problem with the inorganic pesticides is that they are persistent in the environment, lasting years and even decades. A final major detriment to the widespread use of inorganic metal compounds is that they are expensive. To try to alleviate some of these problems and produce large quantities of pesticides that were economically feasible, chemists turned to synthetic organic pesticides in the 1930s. In 1939, while working for the Geigy chemical company, the Swiss chemist Paul

Müller (1899–1965) discovered that the compound dichlorodiphenyltrichloroethane (DDT) was an effective insecticide (Figure 18.6). DDT had first been synthesized in 1873, but it was Müller who discovered its efficacy as an insecticide. DDT was initially marketed in 1941, and found its first widespread use in World War II. During World War I several million deaths, including 150,000 soldiers, were attributed to typhus. There are several forms of typhus, but the most common form is due to bacteria carried by lice. During World War II, fearing a repeat of typhus outbreaks, the Allied forces used DDT to combat not only typhus but also malaria, yellow fever, and other diseases carried by insects. Soldiers liberally applied talcum powder with 10% DDT to clothes and bedding to kill lice. American and European allies were relatively free from typhus and other diseases, while the Germans who did not use DDT had many more noncombat deaths due to infectious diseases. In liquid form, DDT was used in the Pacific Theater to prevent malaria and yellow fever. In addition to its use in the war, DDT was employed in tropical areas as a generic insecticide to prevent infectious diseases, especially malaria. Once the war ended, its use to advance public health in tropical undeveloped countries was expanded for use in agriculture in developed countries. Paul Müller was awarded the

Figure 18.6
DDT

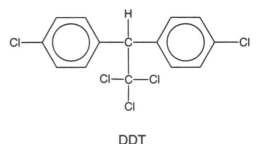

DDT

Nobel Prize in medicine in 1948 for his discovery of the insecticide potential of DDT. By 1950 DDT and several related compounds were viewed as miracle insecticides that were inexpensive and that could be used indiscriminately.

Even though DDT seemed to be a cheap and effective pesticide, enough was known in its early development to raise concerns. DDT is a persistent chemical that lasts a long time in the environment. DDT is fat soluble and not readily metabolized by higher organisms. This meant that DDT accumulated in the fat tissues of higher organisms. As organisms with longer life spans residing higher on the food chain continually fed on organisms lower on the food chain, DDT would accumulate in their tissue. For example, the concentration of DDT in a lake might be measured in parts per trillion, plankton in the lake may contain DDT in parts per billion, fish a few parts per million, and birds feeding on fish from the lake several hundred parts per million. The accu-

mulation of a chemical moving up the food chain is a process known as biological magnification (Figure 18.7). Another concern raised was that certain pests seemed to develop an immunity to DDT and the application rate had to be increased to combat insects. This was because natural selection favored insects that had the genetic characteristics to survive DDT and passed this ability on to their offspring. Direct deaths of bird and fish populations had also been observed in areas with heavy DDT use. The problems associated with DDT and other post-WWII organic pesticides became a national concern with Rachael Carson's (1906–1964) publication of *Silent Spring* in 1962. Carson's book alerted the public to the hazards of insecticides, and while not calling for a ban, challenged the chemical and agricultural industry to curtail its widespread use of chemical pesticides. Most developed countries started to ban the use of DDT and related compounds in the late 1960s. DDT was banned in the United

Figure 18.7
Biological Magnification (Rae Déjur)

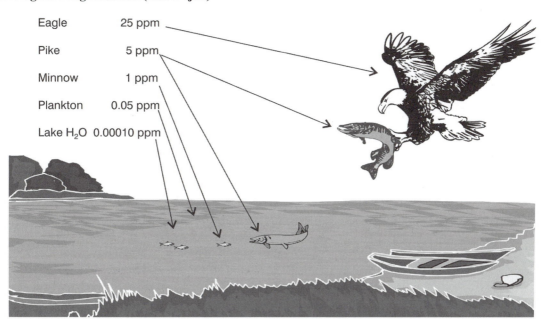

Eagle	25 ppm
Pike	5 ppm
Minnow	1 ppm
Plankton	0.05 ppm
Lake H$_2$O	0.00010 ppm

States in 1973. Although banned in developed countries, its use to improve public health in undeveloped countries continues. The World Health Organization estimates that DDT has saved 25 million lives from malaria and hundreds of millions of other lives from other diseases.

DDT belongs to a group of chemical insecticides known as **organochlorides**. These contain hydrogen, carbon, and chlorine and kill by interfering with nerve transmission; hence, they are neurotoxins. Organochlorines were the dominant type of chemical insecticide used from 1940 to 1970. Some common organochlorines besides DDT are chlordane, heptachlor, aldrin, and dieldrin. Because of their problems, and subsequent ban in many regions, numerous other classes of insecticides have been synthesized to replace organochlorines. Several major groups include **organophosphates, carbamates**, and **pyrethroids**.

Organophosphates were originally discovered in Germany during World War II while researchers were searching for a substitute for nicotine. Nicotine extracts obtained by soaking tobacco in water were commonly used as an insecticide as early as 1700. Organophosphates are derived from phosphoric acid, H_3PO_4, and, like organochlorides, act on the nervous system (Figure 18.8). They work by interfering with specific enzymes in the nervous system. Organophosphates are more lethal to humans and other mammals than organochlorides. Because of this, chemical weapons collectively known as nerve gas are organophosphates. The chief advantage of organophosphates is that they are unstable and break down in days to weeks in the environment. Although this reduces the toxicity to humans on food due to pesticide residues, organophosphates pose a hazard to agricultural workers and nontarget species. Because of this, several countries and different states in the United States ban specific organophosphates. Some common organophosphates are parathion, malathion, and diazinon (Figure 18.8).

Carbamates act similar to organophosphates by interfering with nerve system enzymes. They are derived from carbamic acid, H_2NCOOH, and share many of the properties of organophosphates (Figure 18.9). The carbamate known as carbaryl (commercial trade name is Sevin) was synthesized in 1956 and has been used extensively since that time. Carbaryl's advantages are that its toxicity to mammals is low and it kills a broad spectrum of insects. It is widely used in a number of household lawn and garden products. Other common carbamates include aldicarb (Temik) and carbofuran (Furadan).

Figure 18.8
Organophosphates such as Malathion are derived from phosphoric acid.

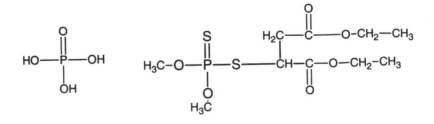

Phosphoric Acid   Malathion

Figure 18.9
Carbamates such as carbaryl are derived from carbamic acid.

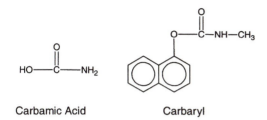

Carbamic Acid   Carbaryl

Pyrethroids are based on mimicking the structure of the natural insecticide pyrethrin. Pyrethins are found in the flowers of chrysanthemums. Ground flowers were traditionally used to obtain pyrethin insecticides and used to kill lice in the early 1800s. Synthetic pyrethins were first produced in the early 1970s. The exact nature of how pyrethroids work is unknown, but because they paralyze insects it is speculated that they affect the nervous or muscular system. Pyrethroids are effective in low dosages and are nonpersistent.

Our discussion of pesticides has focused primarily on insecticides. In the United States the primary use of pesticides is in the form of herbicides, those pesticides used to control weeds. Approximately 70% of the pesticides used in the United States are herbicides and 20% are insecticides. The use and development of herbicides parallels that of insecticides. The first herbicides were inorganic metal compounds and salts. During World War II organic herbicides were synthesized and their use increased dramatically. One of the first major classes of herbicides synthesized in the mid-1940s was phenoxyaliphatic acids. As this name implies, the phenoxyaliphatic acids contain the benzene ring, oxygen, and an aliphatic acid. The two most common phenoxyaliphatic acids are 2,4 dichlorophenoxyacetic acid, called 2,4-D and 2,4,5 trichlorophenoxyacetic acid, known as 2,4,5-T (Figure 18.10). The numbers in these

Figure 18.10
The Two Most Common Phenoxyaliphatic Acids

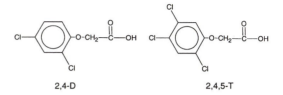

2,4-D   2,4,5-T

compounds refer to the carbon to which the chlorine atom is attached. Dichlorophenoxyacetic acid works by producing excess growth hormones in plants. Plants die due to their inability to acquire sufficient nutrients. The herbicide 2,4,5-T acts as a defoliant that causes plants to shed leaves. During the Vietnam War, 2,4-D and 2,4,5-T were combined into a formulation called Agent Orange. Agent Orange was applied extensively to forests in Vietnam to expose enemy positions. Subsequent birth defects in children of both Vietnamese and American soldiers exposed to Agent Orange became a controversial topic in postwar years. Research has shown that in the manufacture of 2,4,5-T side reactions occur that produce small quantities of **dioxins**. Dioxins are compounds that are characterized by the dioxin structure:

Dioxins have been shown to cause birth defects in tests with laboratory animals. Long-term effects include certain cancers. Because of the health concerns associated with 2,4,5-T, this herbicide was banned in the United States in 1983.

Another class of herbicides is called **triazines**. Triazine compounds consist of a benzene-like structure with alternating nitrogen and carbon atoms in a hexagonal

Figure 18.11
Triazine Structure

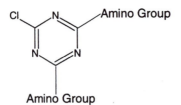

Triazine Structure

ring (Figure 18.11). Amino groups are attached to two of the carbons in the ring and chlorine is attached to the third. The most common triazine is called atrazine. Atrazine disrupts photosynthesis in plants. Atrazine is widely used on corn plants because corn has the ability to deactivate atrazine by removing the chlorine atom.

Another common herbicide is paraquat. Paraquat belongs to a group of herbicides called **bipyridyliums**. The name comes from the combination of two pyridine rings (Figure 18.12). Paraquat works rapidly by breaking down the cells responsible for photosynthesis. It is used as a preemergent herbicide, which means it is applied to soil before plants emerge. Paraquat has been widely used to destroy marijuana plants. Its use to combat marijuana crops has led to speculation that some marijuana users may be susceptible to lung damage from marijuana contaminated with paraquat.

Since their development sixty years ago, synthetic pesticides have increased thirtyfold. Worldwide use of pesticides is approx-

imately 3 million tons, with half of this in the United States and Europe. The use of synthetic chemical pesticides continues to be a controversial topic. There are strong arguments both for and against the use of pesticides. Positive aspects of pesticide use include their ability to save lives and reduce human suffering, increase food supplies, and lower food costs because they are relatively cheap compared to other forms of pest control. On the negative side, pesticides pose some health risks to humans—both short term and long term—kill nontarget species, have uncertain ecological impacts, and produce hardier pests over time. Since the creation of the EPA in 1970, the federal government has heavily regulated the use of synthetic chemical pesticides. Over fifty pesticides have been banned during the last thirty years. Certain pesticides are available only to commercial users who have the credentials to apply them. New pesticides must undergo a rigorous screening that includes various laboratory tests that seek to ascertain the potential human and environmental risks of the pesticide. Currently, approximately 600 different active ingredients are approved by the EPA for pesticide use. Still, even with extensive federal regulations, many people feel oversight is lax and poorly enforced. Most of the chemicals approved by the EPA have never undergone extensive testing. Recent international trade agreements, such as NAFTA, mean that pesticides banned in the United States could be crossing our borders indiscreetly.

Figure 18.12
Paraquat is a combination of two pyridine rings.

Pyridine

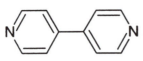

Bipyridine

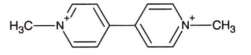

Paraquat

Although a decrease in the use of synthetic chemical pesticides is not foreseen, scientists continue to explore the production of safer pesticides and other methods to augment chemical pesticides. Integrated pest management (IPM) is an approach for controlling pests that utilize multiple techniques in an ecological systematic fashion. In this multifaceted approach, techniques such as cultivation methods, genetic engineering, natural biological controls, and insect sterilization are combined with chemical techniques to control pests. As IPM methods develop during the twenty-first century, the role of synthetic chemical pesticides in modern society will also change.

Summary

As our planet's population continues to grow and undeveloped countries such as China industrialize, an increasing emphasis will be placed on using our **nonrenewable resources** wisely. The planet's land, water, and atmosphere can be considered giant reaction vessels where humans conduct large-scale chemical reactions. In many cases, we know what the results of the reactions will be, but in other instances the results of human activities are unclear. How will an increasing CO_2 concentration in the atmosphere affect global climate? What are the long-term effects of synthetic organic chemicals introduced in the last half century? Can nuclear energy supply our energy needs safely? These are just a few of the many questions in which knowledge of environmental chemistry is critical for determining the answer. Even more important are those areas for which we do not even know a question exists. While we try to anticipate problems and consequences with new discoveries and products, we can never fully anticipate how nature will respond to our actions. Furthermore, the response from nature is not always immediate, but delayed. It is the role of environmental scientists to provide policy makers and the public with their best scientific perspective on issues. In addition to informing the public on what is known, it is also important to tell what is not known. The science of most environmental issues is "fuzzy," and environmental scientists must continually strive to bring the science of these issues into focus. Human-induced changes to the environment must be separated from natural changes, a task that is often difficult. Even when the cause of an environmental problem is known, the course of action is often unclear. In addition to science, the nature of environmental problems involves economics, politics, and culture. While environmental scientists must provide their expertise, it is the role of the individuals to try to understand the fundamental science behind environmental issues and act accordingly. This chapter attempts to help you take a step in that direction.

19

The Chemical Industry

Introduction

All one has to do is take a quick look around to appreciate modern society's dependence on the chemical industry. The paper these words are printed on and the ink used to print them are both products of modern chemical technology. In fact, we would be hard pressed to find materials that are not related in some fashion to the chemical industry. Plastics, medicines, paints, textiles, explosives, fertilizers, cosmetics, fuels, detergents, glass, electroplating, metallurgy, and photography are examples of products and processes that illustrate the impact of the chemical industry on our daily lives. The growth of the modern chemical industry corresponds to the development of chemistry, commencing in the late eighteenth century. The growth of chemical industries during the 1800s was part of the industrial revolution. While Great Britain and the United States led the industrial revolution in areas of agriculture and textiles, it was in Germany where the chemical industry achieved prominence. This situation persisted until World War I, when by necessity, the United States, Great Britain, and other countries at war with Germany were forced to develop their own chemical industries.

Traditionally, chemical technology was more art than science. Processes to produce chemicals and products were crafts that often involved techniques passed between family members and generations. Physicians, alchemists, and apothecaries mixed small quantities of chemicals available for medicines and used as ingredients for products such as soaps, candles, perfumes, foods, and beverages. Regions located close to specific raw materials developed industries that capitalized on their geographic location; for example, coal areas gave raise to the steel industries around Pittsburgh and the Ruhr Valley in Germany. As raw materials became depleted and demand for products increased, chemical technologies were applied to meet demand. Chemical industries developed to produce substances in large batch quantities that were then marketed internationally.

The development of chemistry as a modern science provided a theoretical basis for chemical technology, and knowledge gained in basic chemistry could be applied to meet societal needs. This tradition continues today with the modern chemical industry providing a primary source of research funds. All of the large chemical companies have departments of research and development dedicated to improving current technology and bringing new discoveries to market. In this chapter, we trace the history of the modern chemical industry. Additionally, we look at the basic chemistry used by industry to produce a few of the materials that are part of daily life.

Acids and Alkalis

The first large chemical industries that developed in modern times involved the production of acids and alkalis. The most important industrial chemical used throughout history is sulfuric acid. Each year sulfuric acid tops the list of chemicals used by industry, and it is often said that a country's economic status can be gauged by the amount of sulfuric acid it consumes in a year. In ancient times sulfuric acid was produced by heating the ore green vitiriol, $FeSO_4 \cdot 7H_2O$:

$$FeSO_4 \cdot 7H_2O \rightarrow H_2SO_{4(l)} + FeO_{(s)} + 6H_2O_{(g)}$$

Sulfuric acid produced in this manner was called **oil of vitriol**. During the early 1700s, sulfuric acid was produced in glass jars of several liters using sulfur and potassium nitrate, KNO_3. In 1746, John Roebuck (1718–1794) substituted large lead-lined chambers for glass jars and was able to increase the amount of sulfuric acid produced from a few liters or pounds at a time to tons. The lead-chamber process was used throughout the 1800s to produce sulfuric acid. The reactions representing this process are

$$6KNO_{3(s)} + 7S_{(s)} \xrightarrow{heat} 3K_2S + 6NO_{(g)} + 4SO_{3(g)}$$

$$SO_{3(g)} + H_2O_{(l)} \rightarrow H_2SO_{4(aq)}$$

While the lead-chamber process increased the amount of sulfuric acid that could be produced, it relied on a source of nitrate that usually had to be imported. The process also produced nitric oxide gas, NO, which oxidized to brown nitrogen dioxide in the atmosphere. To reduce the supply of nitrate required and the amount of nitric oxide produced, Gay-Lussac proposed that the nitric oxide be captured in a tower and recycled into the lead chamber. Although Gay-Lussac first proposed this modification to the lead-chamber method around 1830, it was not until the 1860s that John Glover (1801–1872) actually implemented Gay-Lussac's idea with the Glover tower.

The lead-chamber process supplied the world's need for sulfuric acid for a century and a half. In the late nineteenth century, the contact process replaced the lead-chamber process. The contact process utilized sulfur dioxide, SO_2, which was produced as a by-product when sulfur-bearing ores were smelted. The contact process was named because the conversion of sulfur dioxide to sulfur trioxide, SO_3, takes place on "contact" with a vanadium or platinum catalyst during the series of reactions:

$$S_{(s)} \ O_{2(g)} \xrightarrow{heat} SO_{2(g)}$$

$$2SO_{2(g)} \ O_{2(g)} \xrightarrow{heat, \ catalyst} 2SO_{3(g)}$$

$$SO_{3(g)} \ H_2O_{(l)} \rightarrow H_2SO_{4(l)}$$

The production of sulfuric acid was sufficient to meet world demand in the mid-eighteenth century. Sulfuric acid was used in producing dyes, bleaching wools and textiles, and refining metals. Its demand greatly increased at the end of the eighteenth century when a method was discovered for preparing sodium carbonate, also known as soda ash or soda, using H_2SO_4.

Sodium carbonate, Na_2CO_3, and potassium carbonate (potash), K_2CO_3, were two common alkalis used for making soap, glass, and gunpowder. Throughout history these alkalis were obtained from natural sources. Natron imported from Egyptian lakes was one source of soda ash. It was also produced by burning wood and leaching the ashes with water to obtain a solution that yielded soda ash when the water was boiled off. In a similar manner, potash (the name comes from the process used to produce a "pot of ashes") was produced. Potash was used in the manufacture of saltpeter (potassium nitrate), which was a principal component of gunpowder; potash was viewed as a critical chemical by national governments. Another source of soda ash was the barilla plant. The scientific name of this plant is *Salsola soda*, but it goes by the common names of sodawort or glasswort because the soda produced from it was used in making glass. Barilla is a common plant found in saline waters along the Mediterranean Sea in Spain and Italy. Brown kelp from the genus *Fucus* found off Scotland's North Sea coast was also a source of potash. Barilla and brown kelp were dried and burned to produce alkali salts. The depletion of European forests and international disputes made the availability of alkali salts increasingly uncertain during the latter part of the eighteenth century. This prompted the French Academy of Science to offer a reward to anyone who could find a method to produce soda ash from common salt, NaCl. Some of the leading chemists of the day including Scheele and Guyton sought a solution to the soda ash problem, but it was Nicholas LeBlanc (1743–1806) who was credited with solving the problem. LeBlanc proposed a procedure in 1783, and a plant based on LeBlanc's method was opened in 1791. Unfortunately, LeBlanc's association with French royalty led to the confiscation of the plant at the time of the French Revolution. Furthermore, conflicting claims for LeBlanc's method were made by several other chemists, and he never received the reward. LeBlanc, disheartened and destitute, committed suicide in 1806.

LeBlanc's method uses sulfuric acid and common salt to produce sodium sulfate, Na_2SO_4. Sodium sulfate is then reacted with charcoal and lime to produce sodium carbonate and calcium sulfide:

$$H_2SO_{4(l)} + 2NaCl_{(aq)} \rightarrow Na_2SO_{4(s)} + 2HCl_{(g)}$$
$$Na_2SO_{4(s)} + 2C_{(s)} + CaCO_{3(s)} \rightarrow Na_2CO_{3(s)} + CaS_{(s)}$$

The sodium carbonate and calcium sulfide were separated by mixing with water. Because sodium carbonate was soluble in the water and calcium sulfide insoluble, the former would be suspended in solution.

LeBlanc's process increased the demand for sulfuric acid, and the alkali and acid industries were the first large-scale chemical industries. Plants using the LeBlanc process were situated in areas associated with salt mines, and this naturally created locales for other industries that depended on soda ash. The alkali industry using the LeBlanc process created environmental problems near the alkali plants. The hydrogen chloride gas killed vegetation in the immediate vicinity of the plants. To decrease air pollution, the gas was dissolved in water, creating hydrochloric acid that was then discharged to streams, but this just turned the air pollution problem into a water pollution problem. Another problem was created by the solid calcium sulfide product. Calcium sulfide tailings stored around alkali plants reacted with air and water, creating noxious substances such as sulfur dioxide and hydrogen sulfide. Landowners adjacent to alkali plants sought relief from the environmental damage resulting from soda production. The situation got so bad in England that the Parliament's House of Lords enacted the first of the "Alkali Acts" in 1863. These laws regulated the production of soda ash and required producers to reduce their environmental impacts on the surrounding countryside.

The LeBlanc process was the principal method of producing soda ash until 1860 when the Belgian Ernest Solvay (1838–1922) developed the process that bears his name. The **Solvay process**, sometimes called the ammonia method of soda production, utilized ammonia, NH_3, carbon dioxide, and salt to produce sodium bicarbonate (baking soda), $NaHCO_3$. Sodium bicarbonate was then heated to give soda ash. The series of reactions representing the Solvay process are

$$2NH_{3(g)} + CO_{2(g)} + H_2O_{(l)} \rightarrow (NH_4)_2CO_{3(aq)}$$

$$(NH_4)_2CO_{3(aq)} + CO_{2(g)} + H_2O_{(l)}$$
$$\rightarrow 2NH_3HCO_{3(aq)}$$

$$(NH_4)HCO_{3(aq)} + NaCl_{(aq)}$$
$$\rightarrow NaHCO_{3(s)} + NH_4Cl_{(aq)}$$

$$2NaHCO_{3(g)} \xrightarrow{heat} Na_2CO_{3(s)} + H_2O_{(l)} + CO_{2(g)}$$

Explosives

During the beginning of the nineteenth century, the alkali and acid industries provided the model for other chemical industries. One characteristic of the chemical industry is that development in one area often stimulates development in another area. For example, the lead-chamber method produced enough sulfuric acid to make the acid practical for use in the LeBlanc process. Similarly, the Solvay process used ammonia produced when coke was made for steel production. Certain chemical industries were perceived by royalty and national leaders as critical to their nation's welfare. One of these was the manufacture of gunpowder, also known as blackpowder. Gunpowder is a mixture of approximately

75% potassium nitrate (saltpeter), 15% charcoal, and 10% sulfur. Gunpowder was invented by the Chinese in the ninth century and introduced into Europe around the thirteenth century. Gunpowder mills were established in Europe soon after its introduction, and they were heavily controlled by European governments. Lavoisier made extensive studies of the saltpeter (potassium nitrate, KNO_3) used in French gunpowders at the Royal Arsenal in Paris; as a result, French gunpowder improved to the highest quality in Europe.

One of Lavoisier's apprentices at the arsenal was a sixteen-year-old named Eleuthère Irénée DuPont de Nemours (1772–1834). DuPont came from an influential family associated with the French royalty. DuPont's father, Pierre Samuel DuPont (1739–1817), who attached the "de Nemours" to the DuPont family name to distinguish them from numerous other DuPonts, barely escaped being guillotined with Lavoisier. Pierre and his sons were forced to flee to the United States in 1800. The gunpowder industry in the United States had traditionally been a cottage industry; the first gunpowder mill did not appear in the colonies until 1775. Eleuthère Irénée DuPont purchased land along the Brandywine River near Wilmington, Delaware, to establish an American gunpowder mill. Using knowledge gained from working at the French Royal Arsenal, DuPont started his plant in 1802. DuPont began selling gunpowder in 1804, and his business quickly expanded. By the War of 1812, DuPont was the largest manufacturer of gunpowder in the United States producing more than 200,000 pounds annually. One hundred years later the company sold over 1.5 billion pounds of explosives to the Allied forces for World War I. DuPont remained a family business well into the twentieth century. Eleuthère's son Henry DuPont (1812–1889), a graduate of West

Point, ran the company from 1850–1889. Henry's son Lamont DuPont (1831–1884) was a chemist who perfected a gunpowder formula using sodium nitrate in place of potassium nitrate to make an improved, more economical gunpowder. Lamont was killed in a company explosion in 1884. By the end of the nineteenth century, the E.I. DuPont de Nemours Company was experiencing financial difficulties and several elder DuPonts made a decision to sell the company. The young Alfred DuPont and two cousins convinced the older DuPonts to let them run the company. The young DuPonts revitalized the company into the largest chemical company in the United States. While explosives continued to form the core of the business, the company increasingly expanded into other areas during the twentieth century, producing dyes, chemical coatings, synthetic fibers, pigments, and general chemicals. Additionally, E.I. DuPont diversified into other businesses such as automotives (General Motors), agriculture, and petroleum.

Gunpowder was the primary explosive used for almost one thousand years. In 1846, the Italian chemist Ascanio Sobrero (1812–1888) first prepared nitroglycerin, but it was twenty years before Alfred Nobel (1833–1896) developed its use commercially. Nobel was born in Stockholm, Sweden, where his father, Immanuel Nobel (1801–1872), ran a heavy construction company. When Alfred was four, his father's company went bankrupt and Immanuel left for St. Petersburg, Russia, to start over. Immanuel rebuilt a successful business in Russia, in part due to his ability to develop and sell mines to the Russian Navy for use in the Crimean War. Alfred and the rest of his family joined his father in Russia when he was nine, and Alfred received an excellent education with private tutors. He studied in the United States and Paris where he met Sobrero. Nobel studied chemistry, literature, and mechanical engineering as his father groomed Alfred to join him in the construction and defense industry. After another downturn in his fortune, Immanuel and his sons, Alfred and Emil, returned to Stockholm in 1859 to start another business. It was at this time that Alfred began to experiment with nitroglycerin, seeking a safe method for its use as an explosive. Nobel mixed nitroglycerin with other substances searching for a safe way to transport it and make it less sensitive to heat and pressure. Several explosions at Nobel's lab, one of which killed Emil, prompted the city of Stockholm to ban nitroglycerin research inside the city and forced Nobel to move his studies to a barge in one of the city's lakes. By 1864, Nobel started to manufacture nitroglycerin, an explosive that was almost eight times as powerful as gunpowder on a weight basis. In 1864, Nobel mixed nitroglycerin with silica to produce a paste that could be shaped into tube producing dynamite. He also perfected a blasting cap to detonate the nitroglycerin. Nobel received a patent for dynamite in 1867, and his business expanded very rapidly as dynamite was increasingly used for construction and defense purposes. While Nobel's fame and fortune were based on his invention of dynamite, he was a very able inventor and chemist. Over the years he received 355 patents, including ones for synthetic rubber and artificial silk. His will requested that the bulk of his fortune, which approached 10 million dollars, be used to fund annual prizes in the areas of chemistry, physiology and medicine, physics, literature, and peace.

Synthetic Dyes

By the middle of the nineteenth century the chemical industry involved using basic inorganic chemicals to supply basic needs in bulk quantities, for example, soap, glass, and gunpowder. The birth of organic chemistry with Wöhler's synthesis of urea in 1828 provided the promise that chemicals could be synthesized to replace those obtained from natural sources. There were several problems associated with obtaining chemicals from natural sources to supply substances such as medicines, dyes, flavorings, and cosmetics. One problem was that most of these chemicals were available only in limited supplies from specific geographic locations. This meant there was no guarantee that chemicals would be available on a regular basis. Wars, weather conditions, trade barriers, and plant and animal diseases disrupted the supply of chemicals on a regular basis. Even when available in plentiful supplies, the costs of gathering, transporting, and refining chemicals made many of them expensive and unavailable to common citizens. Another problem was that the quality of chemicals extracted from natural sources varied depending on the source, genetic variability of species, local growing conditions, and other factors.

The first **fine chemicals** to be synthesized in large quantities were dyes. Throughout the ages, dyes were extracted from plants, animals, and minerals. For example, the blue dye indigo was obtained from the leaves of various species of the genus *Indigofera* found in India. *Indigofera* was brought to South Carolina, and indigo was the most widely used dye during the eighteenth and nineteenth centuries in association with the cotton industry. *Madder* was another tropical plant family that contained the plant *Rubia tinctorium* used to obtain red dyes. The molecule responsible for the dyeing characteristic is alizarin (Figure 19.1). The *Madder* family also contained coffee and medicines such as quinine, the latter of which played a key role in the development of synthetic dyes as seen shortly.

Wöhler's discovery in 1828 that a natural organic chemical could be synthesized from inorganic chemicals motivated other chemists to search for synthetic substitutes for other natural products. In the years preceding Wöhler's discovery, Faraday had discovered benzene as a liquid residue from the distillation of coal used to produce gas for lighting in 1825. Aniline, $C_6H_5NH_2$ had been prepared from distilling indigo plants (1826). Other developments in Europe also set the stage for the synthetic dye industry. As the eighteenth century progressed, European forests were being depleted, and coke was increasingly substituted for charcoal as a source for carbon in iron smelting. Charcoal and coke are prepared by heating wood

Figure 19.1
Natural and Synthetic Dyes

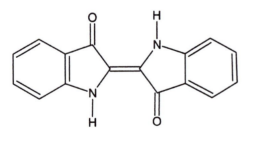

Indigo

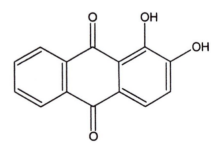

Alizarin

or coal in a low oxygen environment. When coal is heated to make coke, the residue is called coal-tar. By 1845, it was found that coal-tar consisted of a mixture of a numerous organic molecules including benzene, toluene, naphthalene, and xylene. With a ready supply of coal-tar provided by the production of coke, methods were devised to isolate different organic compounds from each other. A leader in early coal-tar work was August Wilhelm von Hofmann (1818–1892) who, although German, was director of the Royal College of Chemistry in London from 1845 to 1864.

William Henry Perkins (1838–1907) enrolled in the Royal College at age 15, and by the time he was 18, he was a laboratory assistant to Hofmann. Hofmann assigned Perkins the problem of synthesizing quinine, $C_{20}H_{24}N_2O_2$ from aniline. Quinine is an alkaloid that was commonly used to treat malaria. It reduced the fever of those suffering from malaria, but it did not provide a complete cure. Quinine's natural source is the bark and roots of cinchonas trees found in the Andean regions of Bolivia and Peru. Jesuits introduced quinine to Europeans who called the drug "Jesuit's powder." During the mid-nineteenth century, the British Empire had colonies surrounding the globe. Many of these colonies were in tropical areas where malaria was prevalent. Because the natural supply of quinine was insufficient to meet its demand for medicinal use, a synthetic source would provide both medicinal and commercial benefits. Perkins attempted to synthesize quinine by oxidizing allyltoluidine, $C_{10}H_{13}N$:

$$2C_{10}H_{13}N + 3O \rightarrow C_{20}H_{24}N_2O_2 + H_2O$$

Perkins did not obtain the white crystals that characterize quinine, but instead got a reddish-black precipitate. Perkins decided to repeat the experiment replacing allyltoluidine with aniline. When he oxidized aniline with potassium dichromate, $K_2Cr_2O_7$ he got a brown precipitate. Upon rinsing out the container containing this residue, he produced a purple substance that turned out to be the first synthetic dye. Instead of quinine, Perkins had produced aniline purple. Rather than continue his studies at the Royal College, to Hofmann's displeasure, Perkins decided to leave school and commercialize his accidental discovery. Backed by his father and a brother, Perkins established a dye plant in London and obtained a patent for aniline purple or mauve. It took several years for Perkins to perfect the manufacture of mauve, and in the process, he had to convince the textile industry of the advantages of his synthetic dye. Mauve first found widespread use in France, but by 1862 Queen Victoria was wearing a dress dyed with mauve. Perkins went on to produce numerous other synthetic dyes at his factory, including alizarin. The synthetic production of alizarin, a red dye, ruined the farming of madder in the Provençe region of France. Other synthetic dyes ruined local economies in areas that had produced natural dyes, for example, indigo in parts of India. Perkins sold his dye factory when he was thirty-seven years old. By this time, he was independently wealthy and devoted the rest of his life to basic chemical research.

Perkins' discovery of mauve and his commercial success should not hide the fact that Perkins worked under the German Hofmann at the Royal College. As mentioned previously, during the last half of the nineteenth century and up until World War I, Germany was the undisputed center for advances in industrial chemistry. Several leading chemists besides Hofmann conducted research on dyes and organic synthesis throughout the country. Justus von Liebig (1803–1873) was the leader of the German group that dominated the chemical industry in the latter half of the nineteenth century. Liebig was professor of chemistry at Giessen where he established a research institution that provided a steady stream of well-educated chemists who advanced chemical theory and technology throughout the country. Liebig's educational philosophy was based on a heavy emphasis in experimentation and chemical analysis. Hofmann was a student at Giessen before being appointed to head the newly created Royal Institute of Chemistry in London in 1845. Hofmann returned to Berlin in 1865. Other illustrious German chemists educated at Giessen included Kekulé, Karl Fresenius (1818–1897), and Robert Bunsen (1811–1899). These individuals in turn established centers for chemical research in other German cities, Fresenius in Wiesbaden, Bunsen in Heidelberg, Hofmann in Berlin, and Kekulé in Bonn. The German government and private industry saw research as the key to the development of the chemical industry. During this time, several major chemical companies were founded to produce synthetic dyes including BASF (Badische Anilin and Soda Fabrik) in 1861, Frederick Bayer & Co. (maker of Bayer aspirin) in 1863, and Hoechst in 1880. Across the German border in Basel, Switzerland, Chemische Industrie Basel (CIBA) was established in 1859. This company merged with the Geigy (founded in 1759) in 1970 to form Ciba-Geigy. While universities, private industries, and the German government worked cooperatively to develop their chemical industry, other countries did not see the benefit of chemical research that did not have immediate financial benefits. Fortunately, many American and European chemists were educated in Germany and returned to their native countries bringing the German chemical philosophy with them.

Although synthetic dyes principally found use in the textile industry, dyes eventually found use in treating disease. Because

of this, the dye industry was directly responsible for the birth of the pharmaceutical industry. Louis Pasteur (1822–1895) established the science of microbiology, but he was educated as a chemist at Paris' École Normale. Pasteur worked closely with Laurent and early in his career worked on crystals and chemicals associated with the French wine industry. His first significant discoveries dealt with **optical isomers**, the ability of solutions to bend polarized light in opposite directions. In 1854, Pasteur's interest in biological problems grew as he studied the different organisms responsible for various chemicals produced in fermentation. Pasteur found that fermentation in wine, milk, butter, and other foods was the result of microorganisms and applied this idea to develop the germ theory of disease. Pasteur developed several vaccines including one for anthrax in livestock. A number of prominent scientists used Pasteur's ideas to advance the study of human disease using dyes. Robert Koch (1843–1910) used aniline dyes to selectively isolate bacterium for study. Paul Ehrlich (1854–1915), assistant to Koch, believed dyes could be used to kill bacteria and prevent disease. For example, he found methylene blue was effective in treating malaria. Another discovery related to Pasteur's work included Joseph Lister's (1827–1912) use of phenol (carbolic acid, C_6H_5OH), a coal-tar derivative, as an antiseptic in 1865 to sterilize wounds and surgical instruments. The Bayer company, originally founded in 1861 to produce the dye fuchsine, acquired its reputation for manufacturing aspirin (see Chapter 13, "Aspirin").

Germany's dominance in the chemical industry ended with the start of World War I in 1914. The war forced the European allies to abandon many German sources and develop their own chemical industries. Furthermore, chemical companies in countries such as neutral Switzerland and the United States, which did not enter the war until 1917, were more than willing to meet the demands of the Allied countries. Specific chemical industries, notably those associated with explosives, particularly benefited from the war. World War I also forced Germany to develop specific technologies such as the ammonia industry as noted at the end of Chapter 12. Germany's defeat in WWI led to a loss of its monopoly of a number of chemicals. Before the war, Germany controlled over 90% of the synthetic dye industry, but by the end of the war, it shared this industry with the United States and Switzerland. For war reparations, German patents, trademarks, and holdings in foreign countries were relinquished to the Allied powers. By the end of WWI, the chemical industry was still dominated by a limited number of industries: synthetic dyes, acid-alkali, fertilizer, and explosives. The war set the stage for a new period in industrial chemistry. International alliances were built between corporations to search for new products and markets using synthetic petrochemicals as the building blocks of the chemical industry.

Synthetic Fibers and Plastics

Throughout human history a limited number of fibers provided the fabric used for clothing and other materials—wool, leather, cotton, flax, and silk. As early as 1664, Robert Hooke speculated that production of artificial silk was possible, but it took another two hundred years before synthetic fibers were produced. The production of synthetic fibers took place in two stages. The first stage, started in the last decades of the nineteenth century, involved chemical formulations employing cellulose as a raw material. Because the cellulose used in these fibers came from cotton or wood, the fibers

were not truly synthetic. Charles Topham was searching for a suitable filament for light bulbs in 1883 when he produced a nitrocellulose fiber. Louis Marie Hilaire Bernigaut (1839–1924) applied Topham's nitrocellulose to make artificial silk in 1884. Bernigaut's artificial silk was the first rayon. Rayon is a generic term that includes several cellulose-derived fibers produced by different methods. Bernigaut's rayon was nitrocellulose in which cellulose from cotton was reacted with a mixture of sulfuric and nitric acid. A general reaction to represent the nitration of cellulose is:

$$[C_6H_7O_2(OH)_3]_n + HNO_3$$
$$\leftrightarrow [C_6H_7O_2(NO_3)_3]_n + H_2O$$

The sulfuric acid acts to take up the water that is formed in this reaction. After converting the cellulose into nitrocellulose, the liquid is removed from the solution, and the nitrocellulose can be forced through a spinneret to produce fibers. Bernigaut's rayon was highly flammable, and he spent several years reducing the flammability of his nitrocellulose rayon before starting commercial production in 1891. Rayon was first produced commercially in the United States in 1910. By this time, several other methods of treating cellulose had been developed to replace nitrocellulose rayon. One involved dissolving cellulose in a copper-ammonium hydroxide solution and is called the cupraammonium process. The third method, known as the viscose process, involved reacting cellulose that had been soaked in an alkali solution with carbon disulfide, CS_2, to produce a cellulose xanthate solution called viscose. The viscose process became the most widely accepted method for producing rayon and accounts for most of the current rayon production.

Cellulose acetate is another form of cellulose that was produced in the late 1800s.

The basic cellulose unit contains three hydroxyl groups. The triester cellulose triacetate forms when cellulose is reacted with glacial acetic acid. Hydrolysis removes some of the acetate groups to form a secondary ester, which averages about 2.4 acetyl groups per unit rather than three. The secondary ester is then dissolved in acetone and the solution ejected through a spinneret to form fibers. Cellulose acetate processed in this manner is referred to as acetate rayon, but it may be more commonly known by its trade name Celanese.

The development of plastics accompanied synthetic fibers. The first synthetic plastic with the trade name Celluloid was made in 1870 from a form of nitrocellulose called pyroxylin, the same substance used to produce the first rayon. Celluloid was developed in part to meet the demand for expensive billiard balls, which at the end of the nineteenth century were produced from ivory obtained from elephant tusks. John Wesley Hyatt (1837–1920) combined pyroxylin with ether and alcohol to produce a hard substance called collodion. Hyatt's collodion, like Bernigaut's original rayon, was unstable and potentially explosive. He solved this problem by adding camphor to the collodion to produce a stable hard plastic he called Celluloid.

Celluloid, like rayon, was derived from cellulose, and therefore, was not a truly synthetic material. The first completely synthetic plastic was Bakelite. This material was produced in 1906 by the Belgian-born (he immigrated to the United States in 1889) chemist Leo Hendrik Baekeland (1863–1944). Baekeland had made a small fortune selling photographic paper to George Eastman (1854–1932). Using this money, Baekeland studied resins produced from phenol and formaldehyde by placing these materials in an autoclave and subjecting them to heat and pressure. Bakelite was a **thermosetting** phenol plastic. A thermoset-

ting plastic hardens into its final shape upon heating, whereas **thermoplastics** can be repeatedly heated and reformed into various shapes. Baekeland established the General Bakelite Company in 1910.

By the 1920s, many companies were producing cellulose-based synthetic materials, and the stage had been set for the production of truly synthetic materials. By this time DuPont had diversified its interest from gunpowder and munitions production, which utilized large quantities of nitrocellulose, into a comprehensive chemical company. In 1920, DuPont started to produce rayon, and at this time DuPont started to invest heavily in research. In 1928, DuPont hired the Harvard organic chemist Wallace Hume Carothers (1896–1937), whose specialty was polymers, to lead a team of highly trained chemists in basic research. Carothers' team quickly started to develop commercially viable products such as the synthetic rubber neoprene in 1931. During the early 1930s the DuPont team produced a number of different fibers, but these had specific problems such as lacking heat resistance. In February 1935, a fiber known in the lab as "fiber 66" was produced that held promise for commercialization. Fiber 66, also called nylon 66, was the first nylon produced. Like rayon, nylon is a generic term used for a group of synthetically produced polyamides. The "66" refers to the number of carbon atoms in the reactants used to produce it. In the case of nylon 66, the six refers to six carbons in adipic acid, $HOOC(CH_2)_4COOH$ and six carbons in hexamethylenediamine, $H_2N(CH_2)_6NH_2$. Nylon 66 is produced when these two reactants are combined under the proper conditions:

$$HOOC(CH_2)_4COOH + H_2N(CH_2)_6NH_2 \rightarrow \text{nylon salt} \xrightarrow{\text{heat}}$$

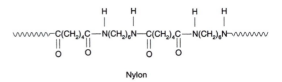

Nylon

The adipic acid used to produce nylon is the product of several reactions starting with benzene. The adipic acid in turn is used to produce hexamethylenediamine.

During the years 1936 to 1939, DuPont developed the production methods necessary to market nylon. Unfortunately, Carothers never lived to reap the rewards as the inventor of nylon. He committed suicide in April of 1937. Nylon's first widespread use was as a replacement for silk in women's hosiery, but by 1941 nylon was being used in neckties, toothbrushes, thread, and some garments. During World War II, the United States government requisitioned the production of nylon solely for the war effort. Nylon replaced silk in military items such as parachutes, tents, rope, and tires. After the war ended, nylon's use in civilian products resumed. The demand for nylon stockings, which DuPont had created before the war, resumed. In addition to its use as a hosiery fabric, it was used in upholstery, carpet, and clothing.

In addition to introducing nylon to the public in 1939, DuPont introduced numerous other synthetic fabrics including acrylics (Orlon, introduced in the year 1950), Dacron (a polyester marketed starting in 1959), Spandex (1959), and Kevlar (1971). Other companies developed countless other fabrics and blends consisting of rayon (1910), polyester (1953), and polypropylene (1961). Gore-Tex, a polytetrafluoroethylene (PTFE) material, was developed by Wilbert Gore (1912–1986) and his wife Genevieve in their basement. Gore was a DuPont engineer who started his business in 1958 after failing to persuade DuPont executives to pursue PTFE fabrics. Gore worked on developing Teflon, which is another PTFE product, at DuPont. The Gores mortgaged their home to start their business, and Gore-Tex outdoor clothing became popular in the late 1980s. Another material invented by individual initiative was Velcro. The Swiss inventor George de Mestral (1907–1990) received a patent for Velcro in 1955 after several years of trial-and-error experimentation. Mestral's idea for Velcro came to him while picking burrs off his clothing and his dog's fur after returning from a hike. He came up with the word "Velcro" from the French terms "velour" (velvet) and "crochet" (hooks).

Synthetic Rubber

Rubber is another natural substance that chemists synthesized during the twentieth century. Spanish explorers discovered native Central and South Americans using rubber in the sixteenth century for waterproofing, balls, and bindings. Natural rubber comes from latex obtained from numerous plants, but the most important of these is *Hevea brasiliensis*. Europeans were intrigued with rubber products brought back by explorers, and the substance remained a curiosity for two centuries. Priestley noted the substance could be used as an eraser if writing was rubbed with it and so the name "rubber" came into use. Rubber factories developed in France and England at the start of the nineteenth century, but it was not until the end of the century that an increased demand led to an expansion of the industry. At this time, rubber tree seedlings (from seeds smuggled out of Brazil) were cultivated in European herbariums and transplanted to the tropical colonies of various countries. Rubber plantations developed in a number of regions, especially in the countries currently known as Malaysia, Indonesia, Sri Lanka, and Thailand. Concurrent with the rise of European plantations in Southeast Asia, the United States obtained rubber from natural sources in Central and South America. The rise of the auto industry and an increased demand for pneumatic tires accelerated the demand for rubber during the twentieth century.

Natural rubber tends to coagulate quickly into a hard tacky substance. To produce different rubbers requires that various additives be combined with the latex obtained from rubber plants. Adding ammonia to the latex prevents coagulation and allows the latex to be shipped and processed as a liquid. Charles Goodyear (1800–1860) was attempting to improve the quality of rubber in 1839 when he accidentally dropped a rubber-sulfur mixture on a hot stove. He discovered the product had superior qualities compared to other natural rubbers of his day. The process Goodyear discovered was **vulcanization**. Vulcanized rubber is more elastic, stronger, and more resistant to light and chemical exposure. Goodyear never reaped the rewards of his discovery and died in poverty. Half a century later the Goodyear Tire Company was named after him and is currently the largest tire producer in the United States and third in the world behind Michelin (France) and Bridgestone (Japan).

Synthetic rubbers include a variety of compounds that mimic the properties of natural rubber. In 1860, Charles Greville Williams (1829–1910) isolated the monomer isoprene from rubber, showing that rubber was a polymer of isoprene:

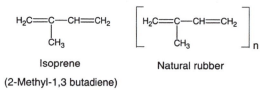

Isoprene
(2-Methyl-1,3 butadiene)

Natural rubber

Numerous attempts were made to synthesize isoprene rubber, but the first successful synthetic rubber was produced by Carothers' group at DuPont. Dupont produced Neoprene (DuPont's name was Duprene) from chloroprene in 1930. DuPont's success was a result of work initially performed by Father Julius Nieuwland (1878–1936), who conducted research on

reactions with acetylene at Notre Dame University. In one series of experiments, Nieuwland passed acetylene into a solution of copper (1) chloride and ammonium chloride to produce several polymers, one of these was monovinylacetylene. It was this compound that the chemists at DuPont were able to convert into chloroprene to make Duprene:

Chloroprene

The double bonds in isoprene and chloroprene allow these compounds to be vulcanized. In vulcanization, sulfur attaches to the doubly bonded carbon to produce cross-linked chains:

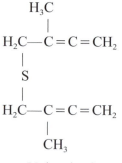

Vulcanization

Other producers were successful in linking single polymers or two polymers, called copolymers, to produce synthetic rubbers. In Germany, a copolymer of butadiene ($CH_2CHCHCH_2$) and acrylonitrile (CH_2CHCN) was produced by using a sodium catalyst. It was called Buna, which came from combining "bu" for butadiene and "Na" for sodium. During the years leading up to World War II, the United States, realizing how dependent it had become on foreign sources of rubber, continued research on synthetic rubbers. The war and Japan's control of the rubber-producing areas in Southeast Asia cut the United States' natural rubber supply by over 90%. In response, several major companies including Standard Oil, Dow, U.S. Rubber, Goodrich, Goodyear, and Firestone signed an agreement to share research to boost synthetic rubber production. The cooperative agreement resulted in improved qualities and grades of rubber and accelerated growth in the synthetic rubber industry. The United States developed a rubber called GR-S (government rubber-styrene) that was a copolymer made from butadiene and styrene. Once the war ended, the development of synthetic rubbers had displaced much of the United States' dependence on natural rubber. Plants controlled by the government to support the war effort retooled to serve civilian needs in transportation, electrical, and consumer products.

Petrochemicals

The use of petrochemicals was introduced in Chapter 15. The modern chemical industry is largely based on the use of petrochemicals, that is, chemicals derived from petroleum and natural gas. Oil, natural gas, and coal have traditionally been used as fuels to provide energy. Coal tar provided the bulk of organic chemicals used for industrial products during the nineteenth century. The production of commercial quantities of oil from wells drilled in western Pennsylvania at the end of the century provided an alternative source of hydrocarbons for synthesizing organic chemicals. Initially, oil production provided kerosene for illumination and oil for heating. The development of the automobile in the twentieth century shifted the emphasis of oil production toward gasoline. Oil refined to produce gasoline also produced a host of other chemicals that could be used as **feedstocks** for other products.

The first petrochemical produced in large quantities was carbon black, which is formed by the incomplete combustion of natural gas. Carbon black consists of fine particles of carbon and was used in the early 1900s for ink, pigments, and tires. The first oil company to produce petrochemicals on a commercial basis was Standard Oil of New Jersey (Exxon). Standard of New Jersey started to produce commercial quantities of isopropyl alcohol from petroleum in 1920. The petrochemical industry grew steadily during the twentieth century as oil companies developed chemical processes for separating and converting petrochemicals from crude oil. Following the lead of Standard of New Jersey, other oil companies developed chemical divisions and started research and design (R & D) departments to process petrochemicals. Additionally, oil companies acquired or merged with established chemical companies to boost their profitability. This trend has continued to the modern day as witnessed by such recent mergers as Exxon-Mobil, Phillips-Atlantic-Richfield (ARCO), and the acquisition of Conoco by DuPont. A number of the largest chemical companies in the world (based on sales) are names traditionally associated with gasoline: Exxon, Chevron, Amoco, Shell, Phillips, Ashland, and British Petroleum (Table 19.1). It should be noted that the figures in Table 19.1 are based on a strict definition of chemical sales. Chemicals such as gasoline, agricultural products ,and pharmaceuticals, although chemicals, are not defined as chemical sales.

A relatively small number of chemicals form the basis of the petrochemical industry. These are methane, ethylene, propylene, butylenes, benzene, toluene, and xylenes. These chemicals are used to derive thousands of other chemicals that are used to produce countless products. Figure 19.2 lists some of the principal chemicals and products derived from these seven basic chemicals.

Table 19.1
Top Ten World Chemical Producers Based on Sales for Year 2000

Rank	Company	Country	Sales in millions
1	BASF	Germany	31
2	DuPont	USA	28
3	Dow	USA	23
4	ExxonMobil	USA	22
5	Bayer	Germany	19
6	TotalElfFina	France	19
7	Degussa	Germany	16
8	Shell	UK/Netherlands	15
9	ICI (Imperial Chemical Industries)	UK	12
10	British Petroleum	UK	11

Source: *Chemical and Engineering News,* July 23, 2001

Industry Histories

Throughout this chapter and in previous chapters, brief histories of several well-known companies started were presented—DuPont, Alcoa, Bayer, and BASF. In this final section on the chemical industry, the histories of several well-known companies using chemical technologies are presented.

DuPont, presented at the beginning of this chapter, and Dow are the largest chemical companies in the United States. Dow Chemical Company was started by Herbert Henry Dow (1866–1930) in Canton, Ohio. Dow was a student at Case Institute in Cleveland who studied the characteristics of salt brines acquired from wells around the Great Lakes. Dow was determined to discover methods to extract chemicals from the salt brine. Rather than use the standard distillation method of his day to obtain chemicals, Dow employed electrolysis to separate

bromides, chlorides, and sodium hydroxide. Bromides were used in medicines, as bleaching agents, and in photographic chemicals. One of Dow's first customers was Kodak. For several years Dow met varying degrees of success both scientifically and financially. He began Midland Chemical Company in 1890, returned to the Cleveland area and formed Dow Process Company in 1895, and finally returned to Midland for good in 1897 where he founded Dow Chemical Company. One obstacle Dow had to overcome to grow his business was fierce competition from the German and English bromide producers. These producers resented Dow marketing his bromide in Europe. The Germans slashed prices in America in an attempt to drive Dow out of business, but Dow shrewdly and secretly managed to purchase the German bromide and resell it at a profit in Europe where prices were not cut. Dow won the bromide

Figure 19.2
A small sampling of some of the chemicals and products obtained from petrochemical sources

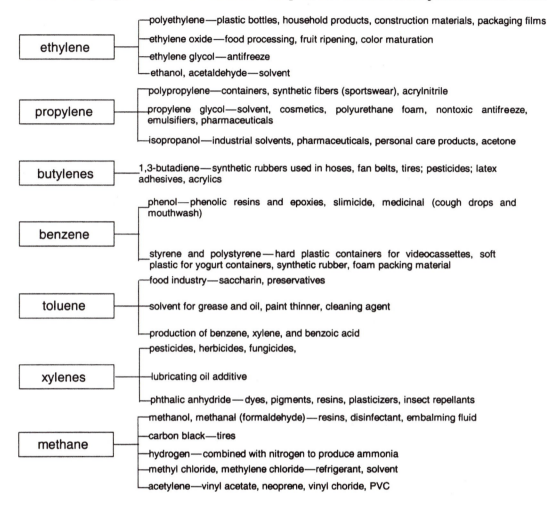

war; he agreed not to sell in Germany, while the Germans would not sell in the United States, but bromide sales elsewhere were allowed. Dow's next assault came from DuPont's attempt to acquire the company in 1915. Dow and his staff threatened to quit if DuPont purchased Dow, and DuPont withdrew its offer. Dow was both an outstanding chemist and businessman. During his career he acquired nearly 100 patents. He expanded his electrolytic methods for separating ions from brine to seawater to produce magnesium and iodine. Today Dow is a comprehensive chemical company producing hundred of different products.

Another familiar name in the chemical industry is Monsanto. Monsanto was founded in St. Louis as a producer of food additives in 1902 by John Francis Queeny (1859–1933). Queeny named the company after his wife's maiden name. Its first product was saccharin. Saccharin was not produced in the United States at that time and had to be imported from Germany. Queeny and his associates sold saccharin to soft drink producers; Coca-Cola was one of his primary customers. Monsanto expanded production into caffeine, vanilla, aspirin, phenolphthalein, and several other chemicals over the next decade. World War I accelerated Monsanto's growth, and it steadily expanded its product line after the war. Its many products include synthetic fibers, rubber, pesticides, detergents, and pharmaceuticals. More recently attempts have been made to merge Monsanto with Dow, which would create the largest chemical company in the United States.

Union Carbide grew out of companies making carbon electrodes and acetylene for city and home lighting. The National Carbon Company founded in 1886 produced carbon electrodes for street lamps. It developed the first commercial dry cell batteries and distinguished its batteries with the Eveready trademark. Union Carbide was founded in 1898 by Canadian Thomas L. Willson (1860–1915) and James T. Morehead (1840–1908). Morehead and Willson were using an electric arc furnace in an attempt to prepare aluminum. In 1892, while searching for a process to make aluminum, Morehead and Willson combined lime and coal tar in their furnace and produced calcium carbide, CaC_2. Acetylene, C_2H_2, is produced when calcium carbide reacts with water according to the following reaction:

$$CaC_{2(s)} + 2H_2O_{(l)} \rightarrow C_2H_{2(g)} + Ca(OH)_{2(aq)}$$

Morehead started promoting the use of acetylene for lighting, and calcium carbide plants were established in Sault Ste. Marie, Michigan, and Niagara Falls, New York. Morehead developed a high carbon ferrochrome in 1897, which was used for armor plating in the Spanish-American War. In 1917, Union Carbide merged with the National Carbon Company, Prest-O-Lite (another calcium carbide producer), Linde Air Products, and Electro Metallurgical to form the Union Carbide & Carbon Company. In 1957, the company's name was shortened to Union Carbide. In addition to batteries, some of its major products include Prestone antifreeze, Glad wrap, and Champion spark plugs. Union Carbide's reputation was severely hurt in 1984 when its pesticide plant in Bhopal, India, had an accident resulting in the death of 2,500 Indians. In February of 2001, a merger between Union Carbide and Dow was completed, making Union Carbide a subsidiary of Dow.

In the section on rubbers it was noted that the Goodyear Tire and Rubber Company was not founded by Charles Goodyear. Frank A. Seiberling (1859–1955) and his brother Charles (1861–1946) founded Goodyear in 1898, thirty-eight years after Goodyear's death. Originally, the Seiberling brothers produced carriage and bicycle tires, but in 1901 they started producing automobile tires for the infant automobile industry. Although Goodyear built its reputation on rubber products, it expanded interests into textiles, chemicals, and a number of petrochemical products related to the rubber industry. The B.F. Goodrich rubber company was started by Benjamin Franklin Goodrich (1841–1888). Goodrich, along with several partners, founded a rubber company on the Hudson River in 1869. Goodrich, as president of the company, believed a location further west would be more suitable for his company and explored Akron, Ohio, as a possible site for relocation. With backing from Akron business interests, Goodrich moved his operation to Akron, incorporating as Goodrich, Tew & Company on December 31, 1871 (Tew was Goodrich's brother-in-law). Goodrich's company went through several reorganizations and grew slowly during its first decade. It was not until after Goodrich's death that the company established itself as a viable producer of bicycle and carriage tires. Goodrich's most significant innovation during this period was the development of a pneumatic tire suitable for the speeds and load of automobiles. Goodrich expanded into the chemical business after World War I and developed a method for producing polyvinyl chloride. This development continued, and in 1943 B.F. Goodrich opened a chemical division producing plastics, vinyl flooring, and chemicals. A third major American rubber and tire company was founded by Harvey Samuel Firestone (1868–1938). Firestone started Firestone Rubber Company while working for his uncle's carriage shop in Chicago in 1896. In 1900, he relocated to Akron, Ohio, and founded Firestone Tire & Rubber Company. Firestone built his company by supplying tires to Ford Motor Company. He was a close friend of Henry Ford (1863–1947) throughout his life. In 1988, Firestone Tire and Rubber was acquired by the Japanese firm Bridgestone Tires.

As noted previously, Exxon Corporation is one of the leading chemical companies in the world. It regularly ranks among the top five world chemical producers. Exxon was one of a number of oil companies formed from the breakup of John D. Rockefeller's (1839–1937) Standard Oil Trust. Rockefeller started his oil business with Samuel Andrews in 1863 in Cleveland and quickly developed into the major refiner of crude coming out of the fields of Pennsylvania. In 1870, Rockefeller and his associates formed the Standard Oil Company. Rockefeller built an oil monopoly by obtaining rebates from the railroads and buying out competitors. His practices were challenged in state and federal courts forcing Rockefeller to relocate from Ohio to New Jersey in 1892, where state laws were more favorable to his attempts to form a trust and preserve his monopoly. At the time, Rocke-

feller controlled more than 90% of the refining and distribution of oil in the United States. In 1911 following years of litigation, the Standard Oil Trust was finally dissolved under the Sherman Antitrust Act. Standard of New Jersey became Exxon and continues as a major force in the oil and petrochemical industry. Other major petroleum companies that resulted from Standard Oil include Amoco (Standard of Indiana), SOHIO (Standard of Ohio and subsequently taken over by British Petroleum), Chevron (Standard of California), and Mobil (Standard of New York). It is ironic that Standard Oil trust, which was dissolved by the courts at the beginning of the twentieth century, has been somewhat rebuilt in recent years with mergers such as Exxon and Mobil.

Procter & Gamble is a familiar company that has supplied household products for over 150 years. William Procter (1801–1884), an immigrant candle maker from England, and James Gamble (1803–1891), an Irish soapmaker, met in Cincinnati when they married sisters. Procter and Gamble were encouraged by their father-in-law to start a candle- and soap-making business together. They started their business in 1837 in Cincinnati. At the time, Cincinnati was a center for hog butchering, and the slaughtering houses provided an abundance of animal fat that was used to produce lye (NaOH), a key ingredient in both soap and candles. Procter & Gamble's business grew steadily. The Ohio River and the railroads provided easily accessible transportation to ship goods both east and west. Procter & Gamble was a major supplier to the Union Army during the Civil War. During this period, due to the shortage of raw material created by the war, Procter & Gamble developed new methods for candle and soap making. In the mid-1870s, Procter & Gamble, under the direction of Gamble's son William, sought to develop a low-cost high-quality soap to rival expensive castile soaps made from olive oil. One early batch of their soap had inadvertently been mixed excessively when a worker failed to shut down a machine during the lunch break. Excessive air was introduced into the batch, causing the hardened cakes to float. Rather than scrap the batch, the soap was sold and customers demanded more of the floating soap. The soap's ability to float was advantageous to people who often bathed in rivers and lakes at the time. If the soap was dropped, it was easier to find than other soaps that sunk when dropped. Harley Procter, son of William, named the soap Ivory inspired by Psalm 45 in the Bible: "The perfume of myrrh and aloes is on your clothes; musicians entertain you in palaces decorated with ivory." Harley Procter organized a large advertising campaign to promote Ivory soap. As part of the advertising campaign, Procter & Gamble, which had always stressed quality control in its products and used its moon and stars trademark to distinguish P & G products from others, used independent chemical analyses of Ivory soap, which showed 0.56% impurities. Hence, the slogan 99 $^{44}/_{100}$% pure was born. Soaps continued to be Procter & Gamble's main product throughout the nineteenth century. The company pioneered the process of hydrogenation of unsaturated fats and introduced Crisco to the public in 1912. In 1957, Procter & Gamble acquired Clorox, the largest producer of household bleach in the country. Clorox was founded in 1913 by five businessmen in Oakland, California. The original Clorox bleach solution, 21% sodium hypochlorite (NaOCl), was delivered in five-gallon returnable buckets to dairies, breweries, and laundries that used the solution for disinfecting equipment. A 5.25% solution was sold door-to-door and in local groceries starting in 1916, and Clorox bleach remained the company's most sold product up until it was acquired by Procter & Gamble.

Procter & Gamble's number one competitor is another well-known producer of household items: Colgate-Palmolive. William Colgate (1783–1857) was a candle maker from New York City who opened his business in 1806. Colgate sold candles, starch, and soap out of his Manhattan shop. These products formed the core of Colgate's business for its first fifty years. In 1873, Colgate marketed toothpaste in jars, and it introduced the familiar tube dispenser in 1896. Today it is the world's largest producer of toothpaste. Colgate merged with The Palmolive Company in 1928, forming Colgate-Palmolive. The Palmolive Company was originally founded as B.J. Johnson Soap Company in Milwaukee in 1864. In 1898, Johnson introduced Palmolive soap. Palmolive soap was made with palm and olive oil, hence the name Palmolive, rather than with animal fat and became so popular that the company eventually changed the company's name to that of the soap.

This section has focused on a few well-known chemical companies started in the United States. Remember that Germany led the world in industrial chemistry for half a century before World War I forced other countries to develop their own chemical infrastructure. Germany continues to be a leader in the modern chemical industry. BASF (Badische Analin und Soda Fabrik) and Bayer (of aspirin fame), traditionally two of the largest chemical companies in the world, have exerted a major influence on the world chemical industry since the mid-nineteenth century. Before World War II, BASF partnered with the two other German giants: Bayer and Hoechst, and with hundreds of smaller German companies to form the chemical cartel known as IG Farben. IG Farben was a major supporter of Adolf Hitler and instrumental in rearming Germany in the period between World War I and World War II. IG Farben factories were heavily damaged during WWII. After the war, many of the directors of IG Farben were tried as war criminals, and allies argued about the fate of the German chemical cartel. It was eventually decided to break IG Farben into BASF, Bayer, and Hoechst and nine smaller companies. While BASF and Bayer continue to be major chemical producers, in 1998 Hoechst merged with the French company Rhône-Poulenc to form the pharmaceutical and agrochemical company Aventis.

Summary

This chapter has given a brief overview of the chemical industry, with emphasis on chemical production in the United States. The importance of the chemical industry in the United States cannot be overstated. The United States is the largest producer of chemicals in the world, accounting for 27% of the world's production in the year 2000. The total value of chemicals produced in the United States during 2000 was $460 billion. Of this total, $80 billion was exported and the United States' trade surplus in chemicals was $6.3 billion. Thirty-one billion dollars was spent on research and design in the year 2000. There are over 12,000 chemical plants in the United States. One hundred and seventy U.S. chemical companies operate an additional 3,000 chemical plants in foreign countries. The chemical industry employs over 1,000,000 workers and accounts for 1.2% of our gross domestic product. The top chemical producing states are Texas, New Jersey, Louisiana, North Carolina, and Illinois.

In recent decades, much of the chemical industry has been scrutinized with respect to its products and practices. National stories on hazardous wastes generated by the industry (Love Canal, Times Beach) paint a picture of a toxic cornucopia.

The popular movie *Erin Brockovich*, although focusing on the power industry, is an example of how the popular media can simplify the public's perception of chemical exposure. In *Erin Brockovich* the chemical of concern was hexavalent chromium, Cr^{6+}. Hexavalent chromium is listed as a carcinogen by EPA, but complicated questions dealing with exposure (drinking, inhalation, absorption through skin), toxicity levels, and specific health effects were lost in Hollywood's version.

The modern chemical industry is a highly regulated industry. National legislation passed in the last few decades such as the Resource Conservation and Recovery Act (RCRA), Toxic Substances Control Act (TSCA), and Comprehensive Environment Response, Compensation, and Liability Act (Superfund) have required chemical companies to take responsibility for their wastes. Furthermore, chemical industries have discovered that minimizing waste and being energy-efficient are sound business practices that increase profits. While accidents in the chemical industry will continue to happen and a small proportion of companies will try to "cut corners," overall the chemical industry is safe and responsible. Society has become dependent on many of the products marketed by the chemical industry. Everyday life would be significantly changed without many of these products.

This chapter has touched upon only a few of the core chemical industries: petroleum refining, rubber, plastics, and textiles. A comprehensive review of the chemical industry would include agrochemicals, glass, pharmaceuticals, paper, metals, photography, brewing, cement, ceramics, printing, and paints. As the twenty-first century unfolds, the chemical industry will continue to play a dominant role in dictating our quality of life. The core industries such as petroleum, chlor-alkali, metal smelting, and rubber will continue to provide the products of daily life. However, new industries based on biotechnology, newly created materials, and exotic processes (products created in space) will attract entrepreneurs to form new industries. Additionally, the old guard will not stand still. The large chemical companies, which have developed over the last century, will continue to develop new products while at the same time will search the globe for acquisitions, partnerships, and mergers to increase their value.

Chemistry Experiments

Introduction

Science begins with careful observation using all of our senses. Most people associate observation with the sense of vision, but the senses of smell, taste, hearing, and touch are also used to observe. All five of these senses are important in working with chemicals. Observations accompanying chemical reactions are often made by smelling an odor. For example, a distinctive odor is produced in reactions where ammonia is formed. Inadvertently coming into contact with a strong acid or base causes a burning sensation. One way to characterize acids and bases is by taste; acids taste sour and bases bitter. Nearly all of our observations made every day are accepted at face value, but frequently there are observations that do not fit our expectations. These observations gain our attention. Many are quickly dismissed, but some tend to peak our curiosity. We might question whether we can believe our senses and then attempt to verify an initial strange observation. If we trust our senses, we might begin to ask questions to explain our observations. Many paths can be taken when trying to explain observations, but the one used by scientists is termed the scientific method. In this chapter, we take a brief look at the scientific method, discuss some elements of experimentation, and conclude with a few chemistry activities and experiments.

The Scientific Method

As stated in the previous section, questions naturally arise from observations made in daily activities. Questions may also arise from something read, a discussion, or an experience. There are various methods used to seek the answers to questions. Asking another person, consulting a book, or searching the Internet is often sufficient. Scientific questions are questions to which the scientific method can be employed. The scientific method is a systematic procedure used to answer questions in order to develop logical explanations for the world (universe) around us. The general scientific method follows a common pattern for answering a question. The procedure consists of developing a question into a testable **hypothesis**, developing an experiment to test the hypothesis, performing the experiment, and using the results to come to some conclusion about the original question. As described in this fashion, the scientific method seems like a straightforward procedure for tackling a question. In reality, the scientific method often follows a circuitous path. Applying the scientific method is like driving in new locations. In familiar locations, there is little problem of getting from one place to another. Several routes may be used, but steady progress is made in arriving at a destination. Conversely, driving in an unfamiliar area, such as a large city, often consists of wrong turns, backtracking, and running into dead ends. Even with a road map and directions, the path taken is less than direct. In the rest of this section, some of the key elements of the scientific method are briefly discussed.

When using the scientific method it is important to formulate a scientific question about a subject and that this question points the researcher toward further knowledge on the subject. Posing a question may seem trivial, but great scientists are often characterized by having asked the right question. This often means looking past the obvious and focusing on questions that have not been asked. Alternatively, perhaps everyone else has ignored the obvious and overlooked a very basic question. Scientific questions must be clearly defined. Students starting projects often use terms loosely in posing questions. For example, a question such as "which battery is better?" is unclear. Questions arise concerning what is meant by better (cheaper, longer life, more powerful), under what conditions, and the type of battery. An alternate question might be "how does the life of rechargeable batteries on a single charge compare to nonrechargeable alkaline batteries in audio equipment?" Formulating an appropriate question leads the researcher to developing a testable hypothesis.

A hypothesis is a possible explanation that answers the question. It is often pre-

sented as a concise statement of what the outcome of an experiment will be. A hypothesis is often defined as being an educated guess. Before a hypothesis is stated, as much background information about the subject should be gathered. In fact, it may be possible that an answer to the question will come from the preliminary research. Books, journal articles, the Internet, and experts on the subjects can be used to guide the researcher in refining the question and developing a hypothesis. Researching the subject allows the researcher to draw tentative conclusions based on the work of others. This is why the hypothesis is only a possible explanation. Researching the subject can help identify important **variables** in formulating the hypothesis. Variables are conditions that change and may affect the results of an experiment. For example, in a study comparing the life of batteries it may be important to know how batteries operate at different temperatures. If the purpose were to compare battery use of flashlights used both indoors and outdoors, then temperature would be an important variable to consider. This would affect how the hypothesis was formulated, and ultimately, how the experiment was designed.

Once a hypothesis is formulated, an experiment is conducted to test the hypothesis. Experimentation is what distinguishes chemistry, and other experimental science, from other disciplines. In simple experiments, the researcher designs the experiment to examine one **independent variable** and attempts to hold all other variables constant. Again, using the battery example, if the experiment was designed to test battery life in flashlights, then "battery life" would be the dependent variable. It would be im-portant to define in exact terms the variable "battery life" and exactly how it would be measured. The researcher would then attempt to **control** all other variables, except for one independent variable that is allowed to change, thereby conducting a controlled experiment. Temperature might be controlled by conducting the test in a controlled environment, and the type of flashlight could be controlled by using the same or identical flashlights. Although the researcher makes every attempt to control for every variable except the independent variable, it is impossible to control for all variables. In the flashlight example, even using the same flashlight means that one set of batteries would be placed in a flashlight that was slightly more used. On the other hand, the researcher might use two identical flashlights and conduct the experiment using rechargeable batteries in one and alkaline batteries in the other during a first trial. Then a second trial would be conducted with a new set of batteries and switching flashlights. The results of the two trials could then be averaged to control for order of batteries placed in flashlights. Although this controls for order, the fact that two identical but different flashlights were used means that the variable "not exactly the same flashlight" is introduced. There are methods that could be used to control for order using a single flashlight, but the point of this example is to illustrate how difficult it is to totally control for variables. Another variable important to control in this experiment is "age of the batteries." If this was a student experiment and the student randomly picked batteries off the store shelf without considering the expiration date, this could seriously bring into question any conclusions reached. Even when the expiration date is used, this is no guarantee that the batteries are the same age.

This example also demonstrates why it is important to identify the important variables when formulating the hypothesis and designing the experiment. A few variables will generally be identified as important to control; most variables will be ignored. For example, variables such as air pressure, rel-

ative humidity, position of flashlight during the experiment, color of batteries, and so on, would seemingly have no effect on battery life. It is important to remember that an experiment without proper controls is generally useless in reaching any valid conclusions concerning the hypothesis.

In conducting an experiment, it is important to keep careful and accurate records. The standard method for this is using a laboratory notebook. Official laboratory notebooks are sold at bookstores or a notebook of graph paper may be used. A simple laboratory notebook can be made by placing sheets of graph paper in any three-ring binder. The first several pages of the notebook should be left blank to allow for a table of contents. Your name and phone number should be on the inside cover along with any emergency information. The latter is important in situations where an accident might occur and authorities may have to be contacted. The laboratory notebook is used to write a comprehensive record of the experiment. Entries should be dated and important relevant information on the experiment should be recorded. Information such as reactions, notes, times, observations, questions, calculations, and so on should be written in the laboratory notebook. Concentrations of solutions, forms of chemicals (examples include hydrated form or company that produced), and quantities used should be recorded. By going back through the laboratory notebook, it should be possible to retrace the steps taken during the experiment. This is important in analyzing why things worked or did not work as expected. It is also important to use the laboratory notebook to identify areas where the experiment can be improved or modified. Results recorded in the laboratory notebook should provide the basis for future experiments. The notebook should be written neatly in ink. Changes should be crossed out with a single line, enabling the crossed-out material to be read.

Once the experiment has been conducted and data collected and analyzed, a conclusion can be drawn concerning the original hypothesis. The experiment does not establish absolute truth concerning the hypothesis. A researcher should resist the temptation to state that the hypothesis was proven or not proven. In science, absolute truth can never be established. Results of an experiment may either support or not support the hypothesis. Ambiguous results can cause the researcher to modify the hypothesis and redesign the experiment. Even when the results are not ambiguous, an experiment is often repeated. One of the benchmarks of science is reproducibility. When an experiment is conducted and certain conclusions reached, another researcher should be able to repeat the same experiment and obtain similar results.

When analyzing the results of an experiment, it is important to consider sources of errors and how these sources affected the results. Two primary types of errors occur in experiments. These are random error and systematic error. The researcher has no control over random error. Random error involves the variability inherent in the natural world and in making any measurement. As its name implies, random error varies in a random manner. An attempt is made to control for random error by taking multiple measurements. For example, say we wanted to know the mass of an average penny. Naturally, older pennies might be expected to have smaller masses due to wear and newer pennies more mass. (We will ignore the fact that penny composition changed in 1982.) If a large random sample of pennies was used, it would be expected that there would be a good mixture of pennies. A large sample of pennies should include pennies of many ages. When measured on a scale, some pen-

nies would have more than the true mass of a penny, others less, but it is assumed that these would tend to balance out and give a fairly accurate measure of a penny's mass.

Systematic error is error that occurs in the same direction each time. For example, if a scale was used that had a dirty pan, each measurement would be increased by the mass of dirt on the pan. Another example of introducing systematic error in an experiment might be reading the meniscus level of a graduated cylinder at an angle. Looking down when measuring a liquid's volume would systematically underestimate the volume. Systematic error can be eliminated using proper techniques, calibrating equipment, and employing standards. An example of accounting for systematic error is checking a thermometer to see if it measures 0°C in a freshwater-ice mixture and 100°C in boiling freshwater at 1 atmosphere pressure. If both measurements were 2°C high, then it can be assumed that a systematic error of +2°C exists in the temperature measurements made with this thermometer.

Chemistry Experimentation

The most important aspect of conducting chemistry activities and experiments is safety. The excitement of chemistry is often portrayed to the public through fire, explosions, and large noises. Both Jay Leno and Dave Letterman regularly have scientists perform attention-grabbing demonstrations on their shows. The caveat "don't try this at home" is often spoken in jest, but should be taken seriously. Individuals performing public demonstrations have practiced them and honed their skills in science classrooms and museums. They know what conditions (amount of reagents, safety precautions) are needed to conduct the demonstration safely. Chemical demonstrations, like magic tricks, are often "hyped" for the audience. Nevertheless, if you are not thoroughly familiar with a chemistry activity and its possible hazards, do not attempt it.

Even when doing simple experiments it is important to practice safety. It is a good habit to wear safety glasses, avoid loose-fitting clothing, wash your hands after handling chemicals, dispose of chemicals properly, have extinguishers and other safety equipment available, and when in doubt ask questions. The simple activities listed in the following section can be performed safely at home. Before doing any activity or experiment written by someone else, it is important to read and think through it before actually conducting it. Safety begins and ends with common sense. If a procedure is unclear, advice should be sought from an adult.

One of the difficulties in doing chemistry at home is obtaining materials. While many chemicals available in even a modest chemistry lab are not available at home, a surprising number of basic chemicals can be obtained at local groceries and pharmacies (see Table 5.6 for some common chemicals). Building supply stores, aquarium shops, garden supply stores, and hardware stores carry a number of chemicals. Substitutions of commercial products can often be made, for example, nail polish remover for acetone. A small collection of inexpensive instruments may be assembled by searching the shelves of various stores, secondhand shops, and garage sales. Eyedroppers, tongs, tweezers, syringes, funnels, and measuring cups can be used without modification. Rubber or plastic tubing can be obtained from hardware or automotive stores. Alcohol thermometers can be found for less than $2; they are usually encased in a housing from which they can be removed. Mercury thermometers should not be used. A postage scale can substitute for a balance, a small camping stove for a Bunsen burner, Pyrex cookware for beakers, and a warming plate

or iron can be used to make a hot plate. If it is safe, gas or electric ranges, microwave and regular ovens, and refrigerators may be used to conduct experiments.

Chemical Activities

This section contains a few chemical investigations that may be conducted at home. They are not full experiments, but they should provide useful activities that could be developed into a full-fledged experiment. Each illustrates basic concepts explained in the previous chapters. Most involve a few simple steps, while a few are more complicated. Again, the most important aspects of these activities are to be safe, have fun, and learn. Before actually trying to do the activity, read and understand the activity. Assemble all materials before starting and do the activity in a safe place. If chemicals are stored, label and mark the contents properly. Pay close attention to safety and dispose of used chemicals and materials safely.

Quantities in this section are given using cooking measures. Although metric units are preferred in science, most measuring units found in the home consist of traditional cooking units such as cup and teaspoon. For this reason, these units are used. Approximate equivalent units are given in Table 20.1.

Table 20.1
Approximate Metric and English Equivalents

English	Metric
1 ounce (mass)	28 grams
1 inch	2.5 centimeters
1 pound	454 grams
1 teaspoon	5 milliliters
1 tablespoon	15 milliliters
1 ounce (fluid)	30 milliliters
1 cup	250 milliliters

Activity 1   Dissolving an Eggshell

Materials: egg, vinegar, jar or container

In this activity, an eggshell is dissolved with vinegar, illustrating a simple chemical reaction. To dissolve the eggshell, fill a jar or other suitable container with enough vinegar to cover an egg. Place an egg in the vinegar. Vinegar is diluted acetic acid, $HC_2H_3O_2$. In two or three days, the shell will be gone. As the egg dissolves, bubbles will be observed. These bubbles are carbon dioxide, CO_2. Handle the egg carefully after the shell is gone. The egg's membrane remains with the albumen and yolk inside. Acetic acid reacts with the calcium carbonate in the eggshell. The reaction can be represented as

$$CaCO_{3(s)} + 2HC_2H_3O_{2(aq)} \rightarrow$$
$$Ca(C_2H_3O_2)_{2(aq)} + H_2O_{(l)} + CO_{2(g)}$$

Dissolving an eggshell in vinegar demonstrates a modern environmental problem. Calcium carbonate is present in Earth's crust as marble, limestone, and chalk. Many modern buildings, statues, and stone structures contain calcium. Acid rain is slowly dissolving these structures.

Do not throw away the egg; it can be used for the next activity.

Activity 2   Osmosis

Materials: egg with shell removed from Activity 1, corn syrup, water, food coloring, drinking glass or other container to hold egg, spoon

Osmosis takes place when a solvent moves across a semipermeable membrane. The solvent moves from an area of high solvent concentration to an area of low solvent concentration. Osmosis can be demonstrated using an egg with its shell removed. Remember to handle the egg carefully so the membrane does not break. Place the egg in a container with enough corn syrup to surround the egg. Observe what happens to the egg over the next day. The egg should shrivel up. This occurs because water passes from inside the egg (area of higher concentration of water) across the egg's membrane into the corn syrup (lower concentration of water). The membrane allows the water molecules to pass through; the sugar molecules are too large to fit through the openings in the membrane. The water molecules move in both directions across the membrane, but overall more molecules move from the water inside the egg into the corn syrup than vice versa.

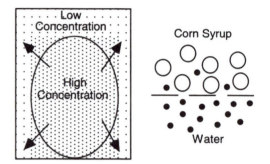

To reverse the process, use a spoon to take the egg out of the corn syrup. Carefully blot off the syrup. Now place the egg in water to which several drops of food color have been added. Again, observe what happens over the next day or two. The egg will expand and the inside of it should be tinted by the food color.

Activity 3   Surface Tension

Materials: small drinking glass such as juice glass, water, small metal objects such as pennies, paper clips, small nails, medicine dropper (optional)

Surface tension is the energy needed to increase the surface area of a liquid per unit area. It is a measure of the intermolecular attractions of a liquid. Because of hydrogen bonding, water has a high surface tension. To demonstrate this, take a small glass and fill it to the very top with water. Once you believe the glass is completely full, start to place the small metal objects into the glass one at a time. Count how many of the objects you can place in the glass before water spills out.

Another variation of this activity is to guess the number of water drops that can be placed on a penny using a medicine dropper.

In both variations of this activity, a much larger number of objects or drops than expected will probably be found. The reason for this is that the intermolecular forces and surface tension of water cause the water molecules to form into a spherical, tightly bound shape.

Activity 4   Electrolysis

Materials: water, 9 V battery, vinegar, distilled water, salt, clear drinking glass

The electrolysis of water can be seen by taking a 9 V battery and placing it in enough distilled water to cover the entire battery. Make sure the electrodes are several centimeters below the water's surface. After placing the battery in the distilled water, note any evidence of a reaction. There are not enough free ions in distilled water to conduct electricity and no evidence of a reaction should be observed. Now add a teaspoon of vinegar to the water and note what happens at the battery terminals. Bubbles form around the terminals and then a steady stream of tiny bubbles emerge from both terminals of the cell.

When vinegar is added to the water, it partially dissociates into free hydrogen and acetate ions:

$$HC_2H_3O_{2(aq)} \rightarrow H^+ + CH_3COO^-$$

The ions enable a current to flow through the water-vinegar solution. At the anode (negative terminal), oxidation of water occurs:

$$2H_2O_{(l)} \rightarrow O_{2(g)} + \div H^+_{(aq)} + 4e^-$$

While at the cathode (+ terminal), reduction of hydrogen ions takes place:

$$2H_3O^+ + 2e^- \rightarrow H_{2(g)} + H_2O_{(l)}$$

Try this activity with tap water. Evidence of bubbles would indicate the presence of free ions from dissolved minerals in the water.

Instead of vinegar, add salt to the distilled water. Bubbles will again appear at each terminal. If the concentration of salt is high enough, chlorine gas, Cl_2, is produced at the anode from the oxidation of chlorine ions:

$$2Cl^- + 2e^- \rightarrow Cl_{2(g)}$$

Hydrogen is produced at the cathode by the reaction:

$$2H_2O + 2e^- \rightarrow 2H_2 + 2OH^-$$

Activity 5   Solubility of Packing Materials

Materials: Styrofoam packing peanuts, starch packing peanuts, acetone, water, glass

Polystyrene is a common polymer used to make Styrofoam containers, packing peanuts, and insulation. Another kind of packing peanut used is made of starch. Place about a cup of acetone in a glass container. Acetone is available wherever paint is sold. Do this in a well-ventilated area. Be careful with the acetone, it can remove paint and ruin finishes. Add a starch packing peanut to the acetone and then a Styrofoam packing peanut. The starch peanut will not dissolve, but the polystyrene peanut will. Dilute the acetone with ample amounts of water before disposing down the sink.

Repeat the activity using warm water instead of acetone. When water is used, the starch peanut dissolves and the polystyrene peanut does not. The acetone and polystyrene do not exhibit hydrogen bonding, while starch peanuts and water do exhibit hydrogen bonding. This illustrates the general principle of "like dissolves like." Not all the polystyrene dissolves because of the presence of cross-linked units that form the polystyrene.

Activity 6   Cabbage pH Indicator

Materials: red cabbage, pot, strainer, household liquids to test (ammonia, vinegar, tap water, lemon juice, etc.)

Acid-base indicators can be extracted from different plant materials. An easy way to prepare an indicator is to cut about half of a small red cabbage into small pieces. Place these pieces in a pot, cover with water, and boil until the water turns a deep purple. It should take about 5 to 10 minutes of boiling. Let the solution cool and then either pour off the liquid or strain to separate the liquid from the cabbage leaves.

The cabbage can now be used as a pH indicator to test different substances. An approximate scale is shown below.

Color	red	pink	purple	blue	green	yellow
pH	1	4	6	8	10	13

Different household liquids can be tested such as ammonia, vinegar, tap water, lemon juice, and Sprite. Items such as orange juice that possess a characteristic color should be diluted and then tested. Diluted solutions of solids like dishwasher soap, cream of tartar, and baking soda can also be made. The indicator can be frozen or mixed with alcohol to preserve for later use.

The chemical compounds responsible for the indicator properties of red cabbage are called anthocyanins. These compounds lose hydroxide ions at low pH, creating a color change. Other plants can be tried as indicators including beets and red onions beets.

Activity 7   Putty

Materials: small glass jars, measuring cup, Elmer's glue, Borax, food coloring

A simple putty can be made with Elmer's glue by mixing 2 tablespoons (30 mL) of glue with 4 teaspoons (20 mL) of water in a jar. In another jar, mix about $\frac{1}{2}$ cup (120 mL) of water, $\frac{1}{2}$ teaspoon of Borax, and 2 or 3 drops of food color. Take a tablespoon (15 mL) of the Borax solution and add it to the glue solution. Mix the two solutions until a viscous putty forms. Remove the putty from the jar and note its ability to stretch, bounce, and so on.

Activity 8   Copper to Gold

Caution: This activity should be done with adult supervision in a well-ventilated area. By using only small amounts of chemicals, it can be done safely. Wear safety glasses.

Materials: safety glasses, several shiny new pennies, metal file, hot-dipped galvanized zinc nails 8 cm (3 inches) or longer are preferred, paper or cardboard, a hot plate, Red Devil lye, a small ceramic pot (capacity of around $\frac{1}{2}$ cup), tweezers, several plastic containers such as yogurt cups or butter tubs

In this activity, galvanized nails are used as a source of zinc. Galvanization is a process by which metals such as steel are dipped in zinc to protect them from rusting. One type of nail is dipped in a hot bath of molten zinc to form a protective coating. Zinc is more easily corroded than iron, so it oxidizes rather than the steel.

Before preparing the zinc, mix the lye solution. **Be careful with lye! Read the precautions on the package!** Do this part outside or in a well-ventilated area. Place about $\frac{1}{2}$ cup of cold water in a yogurt or other plastic container. Slowly add a tablespoon of lye and mix gently. The solution will warm considerably. Place the solution in a safe place where it won't spill. It should sit for at least 30 minutes. To prepare the zinc, file the outside of the nails on a piece of paper or cardboard. Your goal is to obtain a small pile of minute zinc chips. Enough zinc can be obtained using as little as four nails. Only a little zinc is needed, perhaps a pile large enough to cover this circle:

Turn on the hot plate and fill a plastic container with about a cup of tap water. Set the container aside for now. Place 3 tablespoons of the previously prepared lye solution in your small pot. Add the zinc shavings. The zinc should sink to the bottom. A little swirl of the pot will cause the zinc to form a small pile on the bottom of the pot. Place a penny on top of the zinc pile with the tweezers. Now put the pot on the hot plate. Bring the pot just to its boiling point; if you see bubbles that indicate boiling, turn down the heat. As the solution heats, zinc will plate on the penny. The penny will obtain a shiny silver luster. Use the tweezers to remove the penny and rinse it in the tapwater container. Blot the penny dry. Now place the penny on the hot plate. The penny will turn a dull gold color. It may only take a few seconds or a half a minute or so depending on the hot plate's temperature. Use the tweezers to remove the penny and drop it in the water. The penny will turn into a shiny "gold" coin. If a gold color change is not observed on the hot plate, the hot plate's temperature is too low. In this case, the silver coin can be held with tongs and slowly waved through a gas stove flame until it turns color. Immerse the hot coin in water immediately after flaming.

Once the cooking process is perfected, several pennies can be prepared at a time. The lye-zinc solution may have to be swirled during the coating process to get good coverage. You may also need to add more lye solution. Add small amounts of solution carefully.

Silver-colored coins do not have to be transformed into gold coins immediately. The coins can be plated with zinc and saved for heating at a later time. Excess lye solution can be poured down the drain after diluting with plenty of cold water. Plate enough pennies until the zinc is used. Old pennies can be plated, but will produce a duller, dirty finish.

The silver color is due to the plating of zinc on the penny. The final heating process causes the copper and zinc to fuse, producing a brass alloy. Brass is an alloy of zinc and copper. Typical brass alloys contain between 3% and 30% zinc.

Activity 9   Polar Water Molecule

Materials: comb

Water is a polar molecule. The oxygen atom in the water molecule carries a partial negative charge and the hydrogen atoms carry a partial positive charge. This polarity can be observed by using a charged comb. Turn on a faucet so that a very small continuous stream of water is coming out of the faucet. Open the faucet the minimum amount to get a continuous stream. Take a comb and run it several times through your hair. Your hair should be dry, and you should be able to hear little discharge sparks as you comb your hair. Alternately, the comb can be rubbed on a piece of fur or even a pet's fur. Bring the charged comb up to the stream of water. Notice how the water is attracted to the comb. Move the comb around and observe how the water "dances" in response to the comb's movement. The positive hydrogen end of the polar water molecules is attracted to the negatively charged comb, while the negative oxygen is repelled.

Activity 10   Immiscible Liquids

Materials: old fruit juice glass bottle (8–16 oz.) with screw-on lid, ethyl alcohol, charcoal lighter fluid, olive oil, water, dishwashing detergent

Immiscible liquids can be combined to give a colorful display in a bottle. Take a clear glass bottle and remove the label. A bottle used for fruit juices works well. Soaking the bottle in soapy water for several hours makes it easier to remove the label. Some scraping may be necessary. Make sure the bottle has a screw-on cap. Once the bottle is clean fill it halfway with water, then add ethyl alcohol until the bottle is about 95% full. Ethyl alcohol is available in paint and hardware stores. Finally, add several drops of charcoal lighter fluid until the total level of the fluids in the bottle is just below the lip. Tightly screw the cap on. Turning the bottle on the side produces a rolling display of lighter fluid suspended in the alcohol. Adding a drop or two of food coloring makes a more colorful display.

As another example of immiscible liquids, take a jar with a lid and place water and olive oil in the jar. The water and oil will form two distinct layers. Shake the jar to mix the oil and water, and they will separate. You can take an identical jar with the same amount of water and olive oil and add a little dishwashing detergent. Shake both jars and observe how long it takes for the water and oil to separate in the jars. The jar with detergent should keep the olive oil suspended in water longer. Detergents weaken the intermolecular forces between water molecules and the detergent molecules surround the oil molecules with water molecules. Detergents' action on oil in water gives them their cleaning ability.

Activity 11   Creation of Carbon Dioxide and Oxygen

Materials: baking soda, vinegar, 3% hydrogen peroxide, yeast, wood splint, plastic containers such as butter or cottage cheese tubs, lids, matches

Gases are produced in many common reactions. The reaction of baking soda, $NaHCO_3$, and vinegar, $HC_2H_3O_2$, produces carbon dioxide according to the reactions:

$$NaHCO_{3(s)} + HC_2H_3O_{2(aq)} \rightarrow NaC_2H_3O_{2(aq)} + H_2CO_{3(aq)}$$
$$H_2CO_{3(aq)} \rightarrow H_2O_{(l)} + CO_{2(g)}$$

Place about a tablespoon of baking soda in a 16 oz. plastic container. To the container, add about a $\frac{1}{2}$ cup vinegar and immediately cover the container with a loose-fitting lid. Do not snap the lid on tight, but simply lay the lid across the top of the container. A piece of cardboard or any suitable covering can be used for a lid. Let the reaction proceed for 15–30 seconds, light a match, and then lower it into the tub. The match will immediately be extinguished by the CO_2. Carbon dioxide is used in certain types of fire extinguishers.

To produce oxygen, place about $\frac{1}{2}$ cup of 3% hydrogen peroxide, H_2O_2, in the bottom of a tub. To the tub, add a teaspoon of yeast and immediately cover the container. After the reaction has proceeded for a minute or two, light one end of a wooden splint. A shaved piece of wood or Popsicle stick can be used for a splint. Blow out the flame on the splint and immediately lift the lid and place the still glowing end of the splint inside the tub. The splint should reignite in the rich oxygen environment. Oxygen gas is produced from the decomposition of hydrogen peroxide according to the reaction:

$$2H_2O_2 \rightarrow O_{2(g)} + 2H_2O_{(l)}$$

The yeast acts as a catalyst to speed up the decomposition reaction.

Activity 12   Air Pressure and Gas Laws

Caution: This experiment should be done with adult supervision. Wear safety glasses or goggles.

Materials: safety glasses, hard-boiled egg, large glass juice bottle with a mouth slightly narrower than an egg, rubbing alcohol, cotton, aluminum pie pan, tongs

On Earth partial vacuums are used for all kinds of applications. The most common is the vacuum cleaner. Another one is drinking out of a straw. The word "suck" is used when dealing with vacuums or straws, but this term is misleading. What happens when a liquid moves up a straw is that a partial vacuum is created and atmospheric pressure pushes liquid into this partial vacuum, forcing it up the straw.

To do this activity, first boil and peel an egg. The egg should be slightly wider than the mouth of the bottle. Wet the cotton ball with a little alcohol. You don't need a lot. **Put the cap back on the rubbing alcohol container and remove it from the area.** Now place the cotton ball in the pie pan and light it. Using the tongs, grab the flaming cotton ball and drop it in the bottle. Place the egg, narrow end up, in the opening. In a few seconds the flame should go out and atmospheric pressure should push the egg into the bottle. To get the egg out, position the egg with its narrow end in the opening. Tilt the bottle back and blow into it as if you're blowing a trumpet. The egg should slide out.

When asked why the egg is forced into the bottle, most people respond that a partial vacuum is created because the oxygen is removed during combustion. While this is true, it also must be remembered that carbon dioxide is created and replaces the oxygen. The egg is forced into the bottle

Source: Rae Déjur

because of the relationship between temperature and pressure. The flame heats up the air in the bottle. Because the bottle is initially open to the atmosphere, the pressure inside the bottle is the same as the atmospheric pressure. Once the opening of the bottle is blocked with the egg, the flame goes out due to lack of oxygen. The gas molecules comprising the air inside the bottle are no longer heated and are less energetic. This translates into a drop in pressure inside the bottle and in pops the egg.

To get the egg out, it wouldn't make sense to suck on the bottle's mouth. The pressure inside the bottle is the same as atmospheric pressure. Extracting more air would just lower the pressure inside the bottle even further. To get the egg out, you have to increase the pressure inside the bottle so that it is higher than the atmospheric pressure. This is accomplished by blowing into the bottle. Give the bottle a good blow and out slides the egg. It may take a little practice, but you should be able to get the egg out with a good blow.

Activity 13   Balloon in the Bottle

Materials: safety glasses, gloves, baby bottle with screw-on lid, 15″ balloon, microwave oven, water, potholder

Wear safety glasses for this activity. As an alternative or addition to Activity 11, this activity can be performed. Prepare a baby bottle for this activity by removing the nipple from the top lid. The screw on lid should now have a hole in place of the nipple. Take a balloon and inflate several times to stretch it. Stretch the balloon over the opening of the lid. Stretch the balloon evenly over the hole in the lid so that the balloon's opening is centered. Put the lid close to the microwave oven. In the bottle, place several tablespoons of water. About 3 cm (1 inch) of water on the bottom of the bottle is sufficient. Put on gloves. Place the bottle in a microwave for a minute or however long it takes to boil the water in the bottle for at least 30 seconds. After the water has boiled for at least 30 seconds, using a potholder or mitten, quickly remove the bottle from the microwave oven and screw on the lid. Once the lid is screwed on, make sure the balloon is upright and not folded over to one side. You may have to move it slightly to the upright position.

In a short period, usually less than a minute, the balloon will be pushed into the bottle similar to the egg being pushed into the bottle. In this case, the baby bottle fills with water vapor when it is boiling. When the bottle is removed from the oven, it is filled with steam and it immediately begins to cool. Quickly placing the lid on the bottle traps the steam. As the bottle continues to cool, the steam condenses. This decreases the pressure inside the bottle, and the balloon is forced into it. The same principle is used to seal jars when making jelly or canning.

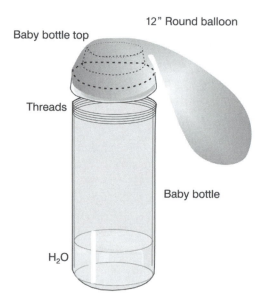

Source: Rae Déjur

Activity 14   Heat Capacity of Water

This experiment should be done with adult supervision. Do this activity outdoors or in a well-ventilated area.

Materials: old cotton socks, container, 70% rubbing alcohol, measuring cup, aluminum pie pan, tongs, bucket of water, long fireplace matches

Water has a very high heat capacity. Substances with high heat capacities experience smaller temperature changes when heat is exchanged between them and their surroundings as compared to substances with smaller heat capacities. In this activity, the high heat capacity of water enables a piece of fabric to stay below its kindling temperature and in the process the fabric dries. Prepare a mixture of $\frac{1}{3}$ cup water and $\frac{2}{3}$ cup 70% rubbing alcohol. Once you are done with the alcohol, remove it from the area. Cut a small, approximately 10 cm (3 inch), square out of an old cotton sock or another piece of clothing. Soak the fabric in the water-alcohol mixture. Ring out the fabric, dry your hands, and place the fabric in the pie pan. Light the fabric with a long fireplace match. Alternately, hold the fabric with tongs and light it, then place it in the pie pan. The alcohol in the fabric burns, but the fabric does not ignite. The water absorbs much of the heat and keeps the temperature below the ignition temperature of the fabric.

This activity can be done with ethyl alcohol or 95% rubbing alcohol. In this case, use $\frac{1}{2}$ cup each of water and alcohol. This activity can also be done with paper such as a dollar bill.

Activity 15   Paper Chromatography

Materials: white construction paper, alcohol, nail polish remover, felt tip pens and markers, food coloring, small dish, small juice glasses or jar, toothpicks

Chromatography is a method used to separate mixtures of substances into their constituents. It is based on the principle of differential attraction of the components of a mixture to mobile and stationary phases as the mobile phase moves through the stationary phase. In this activity, acetone will be the mobile phase and the construction paper will be the stationary phase. The components in inks will be separated.

Cut a sheet of white construction paper into several strips approximately 15 cm (6″) long and 2 cm (1″) wide. About 2 cm from the bottom of the strip, make several small marks using the different felt pens.

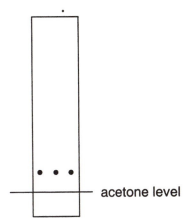

acetone level

Pour enough nail polisher remover (the primary ingredient of nail polish remover is acetone) into a jar or small glass to cover the bottom. Only a small amount of acetone is needed. The depth of the acetone in the jar does not have to be more than 1 cm ($\frac{1}{2}$ inch). Place the prepared construction paper in the acetone. The dots should be above the acetone. The acetone will move up the construction paper and the ink mixture will be separated. Compare several different colored inks and note the components present. Individual food colors or mixtures of food colors can be made and separated. To apply food colors, place a drop of food color on a dish or piece of paper, use a toothpick, and dab the color onto the paper. When finished, the acetone can be diluted with water and flushed down the sink.

Chemists use a number of very sophisticated chromatographic methods for analysis of drugs, DNA, petroleum products, and air samples. Chromatography is used to separate and identify different substances.

Activity 16   Solution Saturation

Materials: safety glasses, quart jar with lid, tall drinking glass, slaked lime, [$Ca(OH)_2$], straw, permanent marker

Solvents have the ability to hold only a certain amount of solute, and then they become saturated. Once saturation is reached, a solute generally precipitates out of solution. A solution can become saturated in a number of ways. Some of these include a change in the solution's temperature or pressure, a chemical reaction occurring in the solution, and adding more solute to the solution. In this activity, the addition of CO_2 to limewater causes a precipitate of $CaCO_3$ to form.

Limewater is a saturated aqueous calcium hydroxide, $Ca(OH)_2$, solution. To make limewater, a small amount of calcium hydroxide is needed. Calcium hydroxide is marketed commercially as slaked lime or hydrated lime. It is used for cement, increasing the pH in soils, and water treatment. Lime may be obtained from building material stores in the cement section and in agricultural stores. The smallest quantities sold are generally 5- or 10-pound bags, which cost a few dollars. Because only a teaspoon of lime is needed (the solubility of calcium hydroxide in water is 0.1g per 100 mL), ask the sales clerk if there are any broken bags from which you can take a tablespoon of lime. Often there will be enough lime dust where it is stored to obtain an ample amount for this activity.

Take a jar and using a permanent marker label the jar "Limewater DO NOT DRINK!" Fill the labeled jar almost full with water and add $\frac{1}{4}$ teaspoon of lime. Cap the jar with the lid and shake the solution. The water will become cloudy. Set the jar in a safe place and let the $Ca(OH)_2$ settle. It will probably take several hours. A good idea is to mix the solution and let it settle overnight. After the settling period, the limewater should be clear. Solid calcium hydroxide will be present on the bottom of the jar. Handling the jar gently so as not to disturb the solid $Ca(OH)_2$ carefully pour about $\frac{1}{2}$ cup of the clear solution into a tall drinking glass or jar. Put on safety glasses and using the straw blow gently into the limewater. Continue to blow bubbles into the limewater and notice how it changes. Be careful not to ingest any limewater. The safety glasses protect your eyes in case any limewater splashes out of the jar.

The limewater should eventually turn cloudy as you blow into it. It may take a couple of minutes of blowing, so be patient. Blowing into the limewater introduces CO_2 from your breath into the solution. Carbon dioxide reacts with the $Ca(OH)_2$ in the limewater to produce calcium carbonate, $CaCO_3$:

$$CO_{2(g)} + Ca(OH)_{2(aq)} \rightarrow CaCO_{3(s)} + H_2O_{(l)}$$

Calcium carbonate is insoluble in water, and this produces the cloudy appearance of the limewater.

Make sure to dispose of the used limewater immediately upon completion of this activity and wash out the drinking glass. Water can be added to the unused limewater in the labeled jar and stored until used again. A pH indicator can be added to the limewater before blowing into it to see if a change in pH can be detected. Blowing into the limewater causes carbonic acid, H_2CO_3, to form as CO_2 dissolves in water. This lowers the pH of the limewater:

$$CO_2 + H_2O \rightarrow H_2CO_3$$

Activity 17   Self-Sealing Polymer

Materials: 9–15″ round balloon, bamboo skewer, Vaseline or baby oil

Inflate a round balloon to a moderate size and tie off the end. Take a bamboo skewer and rub some Vaseline or baby oil on it. Using the skewer, pierce the balloon near the knot. You may need to use a twisting motion to break through the latex. The balloon should not pop when it is pierced. Continue to thread the skewer through the balloon and pierce the end opposite the knot. The balloon will now have a skewer through it, and it may slowly deflate. The skewer can be taken out of the balloon and used to pop the balloon by piercing through the side.

Latex balloons are made of polymers. The latex near the knot and top of the balloon are not stretched as tightly as latex on the side. This can be seen by observing the transparency of the different parts of the balloon. The knot and top areas of the balloon have a greater polymer density. Therefore, the balloon in these areas has a greater ability to stretch and partially seal itself around the skewer. On the sides of the balloon, the tightly stretched latex cannot seal around the needle and the balloon pops.

Activity 18   Cathodic Protection

Materials: nongalvanized steel nails, penny, metal file, small dish

Cathodic protection involves connecting a metal to be protected to another metal that is more easily oxidized. The more easily oxidized metal serves as the anode and the metal to be protected is the cathode in an electrochemical cell. The metal that is oxidized is called the sacrificial anode because it is sacrificed to protect another metal. Metals such as zinc and magnesium are often used to protect iron. Cathodic protection is demonstrated in this activity by using two steel nails. The nails are placed on a shallow dish. Using a white or light-colored dish displays the oxidation better [iron (III) oxide, Fe_2O_3, referred to commonly as rust is more visible on a light-colored surface]. To obtain zinc, take a penny and file off the thin copper coating. Pennies dated before 1982 are 95% copper and 5% zinc. In 1982, the composition of pennies was changed to 2.5% copper and 97.5% zinc. Place the filed zinc penny on one side of the dish. Now place the two nails in the dish. Place one nail so its tip is in contact with the penny and place the other nail by itself. Make sure the nails are separated an appreciable distance. Pour just enough water in the dish to cover the sides of the nails, keeping the top exposed to air.

Over the next couple of days observe the pennies. The nail in contact with the zinc should accumulate much less rust compared to the unprotected nail. The nail not in contact with zinc will show signs of oxidation in just a few hours. Appreciable rust will accumulate on this nail. The zinc oxidizes on the other nail, but the oxidation results in a protective coating of zinc oxide forming on the zinc.

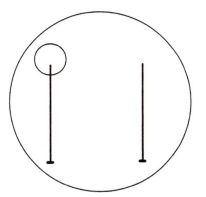

Activity 19   Water Purification

Materials: activated charcoal, coffee filter, funnel, jars, food coloring, dirt, water

Activated charcoal is a pure form of charcoal made by heating wood to a high temperature in the absence of oxygen. Activated charcoal is used in granular form to remove color, odor, and impurities from gases and liquids. The charcoal granules provide a large surface area to contact the fluids being purified. In this activity, a char-

coal filter is used to purify water. Take a round coffee filter and flatten it out. Fold it in half and then fold it in half once more to form a quarter of a circle. Place the pointed portion of the folded filter down into the funnel and open it up to form a cone. Fill the inside of the cone with activated charcoal. Activated charcoal can be found wherever aquarium supplies are sold. A small bag can be purchased for a few dollars. Place the funnel's neck inside a jar.

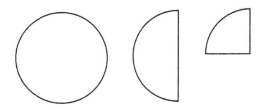

Next, place some water in a jar and produce your "polluted" water by adding a couple of drops of food coloring, a little dirt, and other items to give the water an odor. To produce an odor, a little vinegar, ammonia, onion juice, garlic juice, or other pungent items can be used. Filter the polluted water through the charcoal and observe the filtrate. Its color should be clear and much of the odor should be gone.

Activated charcoal filters are used as a form of treating wastewater. Small filters can also be installed on home faucets to remove impurities. This improves the appearance and taste of drinking water.

Activity 20   Alka-Seltzer Rocket

Materials: safety glasses, Alka-Seltzer tablets, 35 mm film canisters (the white Fuji brand ones work better than the black ones), aluminum pie pan, water

Alka-Seltzer contains several ingredients including citric acid, $C_6H_8O_7$, and sodium bicarbonate, $NaHCO_3$. When an Alka-Seltzer tablet is placed in water, the citric acid and sodium bicarbonate react to produce carbon dioxide. This reaction can be used to propel a film canister into the air.

Contaminated liquid

Funnel lined with coffee filter

Activated charcoal

Clean liquid

Source: Rae Déjur

Caution: Wear safety glasses and do this activity in an open area. Do not do it indoors with a low ceiling under light fixtures.

Put on your safety glasses. To make a rocket, fill a film canister about $\frac{1}{3}$ full of water. Place the canister in the center of the pie pan. Break an Alka-Seltzer tablet into three piece of about equal size. Drop one of the pieces into the canister, snap on its top, and quickly invert the canister so its top is resting on the pie pan. Step back and wait. The pressure inside the canister increases due to the accumulation of carbon dioxide gas inside it. This pressure builds up until the canister body suddenly explodes from the pie pan launching pad. If the top of the canister is quickly replaced and the canister repositioned, another blast can occur. In fact, several blast-offs can be obtained with one Alka-Seltzer tablet.

Alka-Seltzer is a good substance to use to explore how temperature, surface area, and quantity affect the rate of a chemical reaction. The degree of fizzing is a direct indication of the amount of carbon dioxide being produced and how fast the reaction is proceeding. Bayer Company, the producer of Alka-Seltzer, has several experiments on its Web site: *http://www.alka-seltzer.com/as/experiment/student*.

Activity 21   Dry Ice

Materials: film canisters, pie pan, cabbage indicator, plastic drinking cup, tongs, insulated gloves, dry ice. **Caution:** Dry ice exists at approximately −100°C. Use tongs and insulated gloves when handling dry ice. Because it sublimates directly from solid to gas and is denser than air, always use and transport dry ice with adequate ventilation. When transporting dry ice in a vehicle, keep windows cracked open. Use an insulated container with a loose-fitting top to hold dry ice. Do not place dry ice in containers with a tight lid, it will explode. Never take dry ice in an elevator.

Dry ice is solid carbon dioxide. It is sold in some grocery stores, fish markets, and specialty stores. The Alka-Seltzer rocket can be repeated using dry ice in place of Alka-Seltzer. Do not use water, but place a small piece of dry ice into the canister.

Dry ice can also be used with cabbage leaf pH indicator. Place several drops of cabbage leaf indicator into a plastic tub containing water. Drop in several cubes of dry ice. The color should change indicating a drop in pH as the carbonic acid, H_2CO_3, concentration increases in the water.

Carbon dioxide is colorless. The vapor seen in the presence of dry ice is water vapor condensing as the dry ice cools the surrounding air.

Activity 22   Cooling Curve of Water

Materials: Styrofoam cups with lids, thermometer, graph paper

A cooling curve for water can be graphed by taking heated water and placing it in a Styrofoam cup and taking temperature reading at regular time intervals. A graph can then be plotted with time on the x-axis and temperature on the y-axis. This activity can be developed into an interesting investigation. Take one cup of room-temperature water and place it in a Styrofoam cup with a lid. Punch a hole in the lid with a sharp pencil. Place a thermometer through the hole in the lid. Put the cup assembly in the freezer and monitor its temperature every five minutes for a given period—30 minutes or one hour. Plot the cooling curve for this first trial. Now repeat the activity, but this time use one cup of hot water. Boil water in a pan and use this water to repeat the experiment. Compare the cooling curves. Common sense says that the room-temperature water should be colder than the hot water after the same time period. Its temperature starts roughly 70°C lower than the hot water.

The results of this activity may be surprising. Try to determine some reasons for your results.

The activities described in this section should be used as a starting point for scientific investigation. There is no end to the number of scientific investigations that can be performed in any given area. Many professional scientists spend their whole life researching one specific area. They continually probe, ask questions, design experiments, collect data, and analyze the results. Throughout the process new questions and surprises appear that require further investigation. The process of science is a never-ending attempt to understand nature; the more it is understood, the more it seems that there is to understand.

A Future in Chemistry

Introduction

This book began by portraying a chemist as the quintessential absent-minded professor. A bespectacled male concocting strange reactions in the lab is the stereotypical picture of a chemist perpetuated in movies and cartoons. In reality, a relatively small proportion of chemists spend their days dressed in white lab coats conducting experiments, many are women, and they work in a variety of settings. At the end of Chapter 19, it was stated that over 1,000,000 workers are employed in the chemical industry. Many of these workers are not chemists. Conversely, many chemists are employed in a myriad of professions not directly related to the chemical industry, including banking, law, sales, teaching, and public service. Approximately 100,000 chemists are employed in chemical fields in the United States, with roughly 60% of these involved in chemical manufacturing. This final chapter presents information for exploring careers in chemistry and chemistry-related fields.

Chemist's Job Description and Training

Because chemistry is such a broad field, it is difficult to give a concise job description of a chemist. In general, chemists study the composition, structure, and characteristics of matter in an attempt to improve substances and products; discover new substances and products; and improve and develop chemical processes. Chemists work in a multitude of industries including agriculture, medicine, pharmaceuticals, textiles, petrochemicals, mining, plastics, hazardous materials, electronics, adhesives, paints, household products, and cosmetics.

Chemists are frequently grouped into subcategories to distinguish their area of expertise. Five large chemical specialties into which many chemists may be classified are inorganic chemist, organic chemist, physical chemist, biochemist, and analytical chemist. Inorganic chemists work with noncarbon-based compounds and frequently work in areas such as solid-state electronics, the metal industry, and mining. Organic chemists focus on carbon-based compounds working with petrochemicals, pharmaceuticals, plastics, synthetic fiber, and rubber. Many organic chemists use organic synthesis to create new compounds. Physical chemists focus on the relationship between the chemical and physical properties of matter. Employment opportunities exist for physical chemists in alternative energy fields, nuclear power, and material science. Biochemists work on compounds associated with life and are frequently employed by the drug industry, in medical research, and in genetic engineering. Analytical chemists determine and quantify the composition of substances. Analytical chemists are heavily employed in the pharmaceutical industry to determine the nature and purity of drugs. They also are heavily employed in the environmental field where they sample the environment for pollutants. Analytical chemists also develop new methods for characterizing matter.

A job in chemistry generally requires a four-year undergraduate degree in chemistry, although chemical technician positions are available with a two-year degree. In areas such as industrial research or college teaching, an advanced degree, typically a Ph.D., is required. Students preparing for a career in chemistry should focus on the sciences and math in high school. In addition to having the proper academic background to enter a college chemistry program, students should be naturally inquisitive, enjoy laboratory work, be able to work both individually and in groups, and be comfortable working with computers and instruments.

Hundreds of colleges and universities offer bachelor degrees in chemistry in the United States. Around 600 bachelor programs, 300 master's programs, and 200 doctoral programs are approved by the American Chemical Society. With so many

programs at all levels, it is expected that a variety of curricula and programs are in place. Chemistry programs at large universities offer a wide range of courses that allow students to focus in one of several specialty areas in chemistry. Small schools have more focused programs, but may have developed a unique niche in one specialty area. Although there is a wide diversity of bachelor programs, an undergraduate student can expect to complete a core of basic chemistry, science, and math courses. While colleges differ in how they organize and name their courses, there is consistency in the content. For example, some schools may offer laboratory work as independent courses, while others may include laboratory work within the framework of a main course.

A model undergraduate curriculum for a bachelor degree in chemistry is presented in Table 21.1. This model shows the core technical courses for a traditional four-year undergraduate degree in chemistry.

The model program is based on a semester system and generic names are used to identify courses. Most undergraduate chemistry programs reflect the courses listed in the model program. Embedded within the chemistry curriculum is content that may not be apparent in the course titles. Topics such as lab safety, chemical information retrieval, and record keeping fit into this category. Depending on the school, actual coursework in chemistry may only comprise about one-third of the total amount of credit hours to obtain the bachelor degree. The remaining coursework is filled with additional science courses and basic requirements such as English, humanities, and social sciences.

Elective courses can be used by the student to enhance their technical education. These courses are often neglected by students who assume that mastering the tech-

Table 21.1
Model Undergraduate Chemistry Curriculum

1st Semester	2nd Semester	3rd Semester	4th Semester
General Chemistry 1 with lab	General Chemistry 2 with lab	Organic Chemistry 1 with lab	Organic Chemistry 2 with lab
Calculus 1	Calculus 2	Physics 1 with lab	Physics 2 with lab
	Computer Science		Quantitative Analysis
5th Semester	**6th Semester**	**7th Semester**	**8th Semester**
Physical Chemistry 1 with lab	Physical Chemistry 2 with lab	Biochemistry with lab	Senior Thesis or Research Course
Analytical Chemistry	Instrumental Methods	Advanced chemistry elective	Advanced chemistry elective

nical material is sufficient to become a successful chemist. For example, one of the most highly valued skills sought in graduates by employers is the ability to communicate clearly in writing and speech. A technical writing or speech course helps students develop these skills. Another benefit of elective courses is to help the student develop career goals. For example, if a student wanted to work in chemical sales or work in a non-English speaking country, courses in marketing or foreign language would be beneficial.

Another option that the student should consider is completing an internship as part of the undergraduate program or summer employment in the chemical industry. Some colleges require an internship as part of their programs. There are many advantages to completing an internship. An internship provides a professional experience that exposes the student to practical applications of chemistry. The internship can be used to network with professionals to assist in obtaining employment upon graduation or being admitted into a graduate program. An internship exposes students to advanced equipment and processes. For instance, work as an intern in a refinery would acquaint the student with distillation towers, cracking columns, heat exchangers, reactors, and other specialized equipment. Students completing research internships often have access to modern instrumentation not available at the college. Internships can be found through the college's chemistry or career service department. Students may also contact the human resource department of large chemical companies, the American Chemical Society, or individual chemists or faculty members to learn about internships. The Internet is an invaluable tool for exploring internship opportunities.

Careers in Chemistry

In addition to the five general specialties of inorganic, organic, analytical, physical and biochemist, there are numerous other specialties that are named by placing a descriptive word in front of the word "chemist." A small sampling includes environmental chemist, food chemist, cosmetic chemist, soil chemist, atmospheric chemist, medicinal chemist, polymer chemist, petroleum chemist, geochemist, and industrial chemist. Other specialties combine different fields of study with chemistry and use the term "chemical" as a descriptive term. Examples include chemical engineer, chemical oceanographer, and chemical technician. There are other professions that require a thorough knowledge of chemistry, for example, pharmacy and hazardous material management. To fully describe the numerous jobs dependent on chemistry would be impossible. In this section, a brief description of selected chemical careers is given. Although only a limited number of careers are presented, the descriptions serve to demonstrate the importance of chemistry in a number of fields and the breadth of opportunities available to people with backgrounds in chemistry.

Chemical engineers use engineering principles to solve problems of a chemical nature. Nearly three-fourths of the approximately 33,000 chemical engineers in the United States work in chemical manufacturing. They design chemical plants, develop chemical processes, and optimize production methods. A primary concern of chemical engineers working in industry is to develop economic methods to lower production costs. Such measures might include reducing waste, improving energy efficiency, or modifying reaction conditions. Chemical engineers work in research and development departments to develop new products, improve existing products, and determine the best way to produce a product. Undergraduate programs in chemical engineering include a basic core of chemistry courses in inorganic, organic, and physical chemistry along with a core of general engineering courses such as thermodynamics, computer-aided design, fluid dynamics, and economics. Specialized chemical engineering courses in an undergraduate curriculum include plant design, process control, unit operations, and transport processes.

Chemical technicians assist chemists and chemical engineers. Much of their time is spent in the laboratory, where they perform chemical analyses, test products, calibrate equipment, prepare solutions and reagents, and monitor **quality control**. In addition to laboratory work, chemical technicians may collect samples from the field. For example, technicians in environmental work may collect water samples from lakes to analyze in the field or transport back to the lab. Petroleum technicians monitor conditions at oil and gas wells. Chemical technician positions do not require a four-year degree, although many individuals working as technicians have or are in the process of obtaining a bachelor's degree in chemistry. A job as a chemical technician is an excellent entry-level position for chemists. There are a limited number of two-year chemical technician programs offered. These are generally found in community colleges and offer a study in basic chemistry, science, and math with a heavy emphasis on laboratory work. Although not technically considered as chemical technicians, many other technician jobs involve work in chemistry, for example, medical technician, pharmacy technician.

Hazardous materials managers oversee the removal, transportation, and disposal of harmful materials and substances. The handling of hazardous materials requires an understanding of the chemistry of these materials. Especially important is knowledge on the compatibility of different chemicals, how they move through the environment (air, soil, groundwater), and their effects on humans and other organisms. In addition to courses in chemistry, the academic preparation of hazardous materials normally includes courses in environmental regulations, toxicology, and **industrial hygiene**. Because hazardous waste management is a relatively new field, only a few colleges offer four-year or graduate degrees specifically in this area. Many schools offer a certificate program requiring the completion of a limited number of courses in hazardous material management. These certificate programs are often completed by individuals already possessing a degree in science or engineering who are already working or desire to work as hazardous material managers.

Pharmacists provide patient care that involves dispensing drugs, monitoring their use, and advising patient and health care workers about their effects. Although the most obvious role of the pharmacist is as a dispenser of drugs, this role is changing to that of a general health care provider involved in education and advising patients. Actually less than half of all pharmacists work at local pharmacies. The majority work at hospitals, universities or research institutes, government agencies, or industry. Currently there are around 200,000 pharmacists in the United States. In the year 2000, people in the United States spent $75 billion on prescription and nonprescription

drugs. Because the population of the United States is aging, pharmacy is a growing profession. The education required to become a pharmacist has changed in recent years; the bachelor of pharmacy degree is being phased out and students seeking to become pharmacists will need to complete a doctorate in pharmacy program. This is generally a four-year competitive program that requires two years of prepharmacy coursework; therefore, a total of six years of postsecondary education will generally be required to become a pharmacist. Preparatory courses in general chemistry, organic chemistry, math, physics, and the humanities are needed for entrance into a college of pharmacy. The pharmacy program builds on this education with courses in biochemistry, medicinal chemistry, pharmaceutical chemistry, and pharmaceutics. Pharmacists must be licensed to practice in all of the fifty states. To obtain a license, a prospective pharmacist must pass a qualifying exam.

Many chemists find employment as secondary or college teachers. Students can prepare themselves as secondary teachers by majoring in chemical education or combining a chemistry major with the appropriate education courses to become certified to teach. A number of colleges have master's programs that enable a person with an undergraduate degree in chemistry to become a certified teacher. Many areas of the country are experiencing a shortage of qualified chemistry teachers. Large cities such as New York City currently recruit chemistry teachers from foreign countries to help fill their shortage. Teaching in a community college requires at least a master's degree, but most college positions require a doctorate in chemistry.

Chemical sales and marketing is another area where chemists are widely employed. Knowledge in chemistry allows a chemical sales representative to better communicate product information to customers. It also helps the sale representative to provide information back to chemical manufacturers about customer needs, helping to shape product improvements and development. In addition to their chemical training, chemical sales representatives benefit from courses in business, for example, marketing. A popular career path taken by many chemists and chemical engineers is to obtain a master's of business administration (MBA). This degree allows technically trained scientists and engineers to move into numerous management positions with chemical manufacturers.

A field that combines chemistry with crime investigation is forensic chemistry. Forensic chemists use the tools of analytical chemistry to analyze samples collected from crime scenes. They often work with small samples such as blood stains, glass shards, paint chips, hairs, and fabric threads in an attempt to analyze the evidence of a case. Forensic chemists are employed primarily by government crime labs at the federal, state, and local levels. There are a limited number of college programs offering degrees in forensic science. Many people enter the profession with degrees in chemistry or other sciences and receive specialized training from agencies such as the FBI or Drug Enforcement Agency (DEA).

A few of the many employment opportunities available to those with an interest in chemistry have been summarized in this section. The few examples should illustrate the breadth of possibilities available to individuals who obtain degrees in chemistry and related fields. Chemistry majors also branch out into other diverse fields that make use of their technical training. Examples include law (patent law, environmental law), insurance inspectors, journalists, technical writers, and librarians or chemical information specialists. Pay for chemical jobs vary widely depending on geographic location, specialty field, and education required. In the year 2000, median salaries for newly graduated chemistry majors was approximately $33,000 and for chemical engineers was close to $50,000. Chemical technicians tend to be at the low end of salaries starting in the low to mid-$20,000 range. At the other end of the salary range, pharmacists' starting salaries average about $65,000.

Professional Development and the American Chemical Society

All scientific disciplines have professional organizations that promote their disciplines and assist their members in their professions. The American Chemical Society (ACS) is the predominant chemical organization in the United States. It is unique because it is the largest scientific society in the world with 150,000 members. ACS's headquarters are in Washington, D.C., where elected officers and a large professional staff coordinate a myriad of activities. These activities include publishing journals, conducting meetings, offering courses for members, assisting in member job placement, providing outreach to the public, producing chemistry curriculum materials, approving college programs, distributing chemical information, and interacting with government and industry officials. ACS has student memberships and approximately 100 campuses have student affiliate organizations.

The American Chemical Society publishes a weekly magazine called *Chemical & Engineering News*. This publication informs ACS members and others on current news and significant events in chemistry and related areas. Topics include new advances in research, awards in chemistry, book reviews, government legislation, perspectives on major global issues related to chemistry, and a classified section listing current job openings in education, private industry, and government.

One of the roles of ACS is to promote chemical science among young people. In addition to ACS's college program, it coordinates a number of activities targeted at elementary and secondary students. Especially pertinent to this chapter are two publications written for high school students interested in chemistry: *Chemistry and Your Career: Question and Answers* and *I Know You're a Chemist, But What Do You Do*. Information on how to obtain them can be found under "Student Programs" on the "Education Page" accessed through the ACS Web site (*http://www.acs.org*). In this same section of the ACS Web page, summaries of different job descriptions are given. Look for "Chemical Careers in Brief."

As a professional organization, one of the roles of the American Chemical Society is to present chemistry in a positive light. To promote the highest standards among its members, the ACS has adopted a code of conduct. The code presents basic principles to guide the action and responsibilities of chemists.

Chemists acknowledge responsibilities to:

The Public
Chemists have a professional responsibility to serve the public interest and welfare and to further knowledge of science. Chemists should actively be concerned with the health and welfare of coworkers, consumers, and the community. Public comments on scientific matters should be made with care and precision, without unsubstantiated exaggeration, or premature statements.

The Science of Chemistry
Chemists should seek to advance chemical science, understand the limitations of their knowledge, and respect the truth. Chemists should ensure that their scientific contribution, and those of their collaborators, are thorough, accurate, and unbiased in design, implementation, and presentation.

The Profession
Chemists should remain current with developments in their field, share ideas and information, and keep accurate and complete laboratory records, maintain integrity in all conduct and publications, and give due credit to the contributions of others. Conflicts of interest and scientific misconduct, such as fabrication and plagarism, are incompatible with this Code.

The Employer
Chemists should promote and protect legitimate interests of their employers, perform work honestly and completely, fulfill obligations, and safeguard proprietary information.

Employees
Chemists as employers should treat subordinates with respect for their professionalism and concern for their well-being, and provide them with a safe, congenial working environment, fair compensation, and proper acknowledgment of their scientific contribution.

Students
Chemists should regard the tutelage of students as trust conferred by society for the promotion of the student's learning and professional development. Each student should be treated respectfully and without exploitation.

Associates
Chemists should treat associates with respect, regardless of their level of formal education, learn with them, share ideas honestly, and give credit for their contributions.

Clients
Chemists should serve clients faithfully and incorruptibly, respect confidentiality, advise honestly, and charge fairly.

The Environment
Chemists should understand and anticipate the environmental consequences of their work. Chemists have a responsibility to avoid pollution and to protect the environment.

Modern chemistry was born in the late eighteenth century. Building on three hundred years of tradition, chemists continue to build models attempting to understand the mysteries of matter. This knowledge is not collected in secrecy and coveted like the ancient alchemists. Guided by the principles embodied in the Code of Conduct, chemists operate in the world community and collectively have a major influence on all humans. Chemists have developed the ability to modify matter in an amazing number of ways. This has led to improved materials, new compounds, and even new chemical elements, the basic building blocks of matter. New knowledge is helping us to understand the chemical basis of life itself, enabling scientists to modify life and even create new life forms. This in turn has raised questions on the role of chemists in modern society. Many benefits have accrued from chemistry in the last three hundred years, and chem-

istry continues to improve the quality of life. As chemists continue to advance the frontiers of science, an educated public is required to understand these advances along with their actual and potential application and resultant consequences. Only in this manner can informed discussion take place that includes both scientists and nonscientists. Chemists, as good citizens, have an obligation to use their knowledge for the benefit of humankind, and part of this obligation is educating the public on their discipline. In turn, all citizens have an obligation to understand basic science in order to make informed decisions. My hope is that these pages have assisted you in your understanding of chemistry and provided a foundation for you to use this knowledge to enhance your own life, as well as the lives of others.

Glossary

Absolute Zero 0 Kelvins or $-273.15°C$, the lowest possible temperature, the point on the temperature scale where all molecular motion ceases

Acceptable Daily Intake the maximum amount of a substance that can be consumed without damaging health

Acid a substance that yields hydrogen ions in solution or donates protons

Acid Precipitation a collective term for rain, snow, or dry deposition with low pH values

Activated Complex momentary intermediate arrangement of atoms when reactants are converted into products in a chemical reaction, also called transition state

Activation Energy minimum energy needed to initiate a chemical reaction

Active how easily a metal is oxidized

Activity Series a ranking of elements in order of their ability to reduce or oxidize another element

Acyclic an open-chained compound

Addition Polymerization bonding of monomers without the elimination of atoms, formation of polymer by the bonding of unsaturated monomers

Adhesion attraction between the surface of two different bodies

Alcohol organic molecules characterized by containing the $-OH$ group

Aldehyde organic molecules characterized by the $-CHO$ group

Aliphatic Hydrocarbon organic hydrocarbons characterized by a straight chain of carbon atoms

Alkali Metal group I elements in the periodic table: Li, Na, K, Rb, Cs, Fr

Alkaline Earth Metal group II elements in the periodic table: Be, Mg, Ca, Sr, Ba

Alkaloid a nitrogen-containing compound obtained from plants, for example, caffeine, nicotine

Alkane an acyclic saturated hydrocarbon

Alkene an acyclic hydrocarbon that contains at least one carbon-carbon double bond

Alkyl Group the group left when a hydrogen atom is removed from an alkane, for example, methane gives the methyl group

Alkylation chemical process in which an alkyl group is introduced into an organic compound

Alkyne an acyclic hydrocarbon that contains at least one carbon-carbon triple bond

Allotrope different forms of an element characterized by different structures

Alloy a mixture of two or more metals, for example, zinc + copper = brass

Alpha Decay nuclear process in which an alpha particle is emitted by the nucleus

Alpha Particle a radioactive particle consisting of two protons and two neutrons, identical to the nucleus of a helium atom

Amine organic compounds that result when one or more hydrogen atoms in ammonia are replaced by organic radicals

Amino Acid organic acids that contain both an amino group $-NH_2$ and a carboxyl group $-COOH$, the building blocks of proteins

Amorphous Solid solids with a random particle arrangement

Amphoteric a substance that exhibits both acidic and basic properties

Amylopectin branched glucose polymer component of starch

Amylose straight-chain glucose polymer component of starch

Anabolism metabolic reactions in which smaller molecules are built up into larger molecules

Anaerobic without oxygen, condition or process that exists when oxygen is absent

Analgesic a drug that relieves pain

Androgen a male sex hormone

Anion a negatively charged ion

Anneal to heat a metal or glass to a high temperature and slowly cool reducing brittleness of material

Anode the positively charged electrode in an electrolytic cell where oxidation takes place

Anoxic condition that exists when oxygen is absent

Antimatter all particles having complementary properties of matter, for example, positrons

Antipyretic substance that reduces fever

Aromatic Hydrocarbon an unsaturated cyclic hydrocarbon that does not readily undergo additional reaction, benzene and related compounds

Arteriosclerosis condition known as hardening of the arteries due to plaque deposits in arteries

Aryl Group an aromatic ring from which a single hydrogen has been removed

Assimilate to incorporate into the body

Atmosphere envelope of gas surrounding Earth's surface composed primarily of nitrogen and oxygen

Atomic Number the number of protons contained in the nucleus of an atom of an element

Avogadro's Law law stating that equal volumes of gases at the same temperature and pressure contain an equal number of particles

Avogadro's Number 6.02×10^{23}, the number of particles defining one mole of a substance

Background Radiation amount of natural radiation detected in the absence of nonnatural radioactive sources

Base a substance that yields hydroxide ions in solution or accepts protons

Becquerel SI unit for activity equal to one disintegration per second, abbreviated Bq

Beta Decay nuclear process in which a beta particle is emitted

Beta Particle in nuclear processes, particle equivalent to an electron

Binding Energy the amount of energy holding the nucleus of element together, equivalent to the mass defect $\times c^2$

Bioaccumulate the accumulation of a chemical in an organism

Biological Magnification an increase in the concentration of a chemical moving up the food chain

Biomolecule molecules associated with life, for example, carbohydrates, fats, proteins

Biosphere total of living and decaying organisms

Bipyridyliums class of herbicides that includes paraquat

Boiling Point Elevation increase in normal boiling point of a pure liquid due to the presence of solute added to the liquid

Bond Energy the heat of formation of a molecule and its constituent atoms or the energy necessary to form or break apart a molecule

Bose-Einstein Condensate phase of matter that is created just above absolute zero when atoms lose their individual identity

Boyle's Law law that states volume of a gas is inversely related to its pressure

Breeder Reactor type of nuclear reactor that creates or "breeds" fissionable plutonium from nonfissionable U-238

Buckministerfullerene C_{60}, allotrope of carbon consisting of spherical arrangement of carbon, named after architect Buckminister Fuller, Invertor of geodesic dome

Buffer a solution that resists a change in pH

Calcination slow heating of a substance without causing it to melt

Calorie unit of energy equivalent to the energy required to raise the temperature of one gram of water from 14.5°C to 15.5°C, equivalent to 4.18 J

Calorimetry experimental technique used to determine the specific heat capacity of a substance or various heats of chemical reactions

Calx oxide of metal formed during calcination process

Carbamates class of insecticides derived from carbamic acid

Carbohydrate large class of biomolecules consisting of carbon, hydrogen, and oxygen and having the general formula $C_x(H_2O)_y$, more specifically a polyhydroxyl ketone or polyhydroxyl aldhyde

Carbonate an ion with the formula CO_3^{2-} or a compound that contains this ion

Carbonate Hardness condition of water when magnesium and calcium are combined with carbonates and bicarbonates, also called temporary hardness because it can be removed by boiling, responsible for scale in pipes

Carbonyl a carbon atom and oxygen atom joined by a double bond C=O, present in aldehydes and ketones

Carboxyl Functional Group a carbonyl with a hydroxyl (–OH) attached to the carbon atom

Carboxylic Acid group of organic compounds characterized by the presence of carboxyl group

Carcinogen a substance that causes cancer

Catabolism metabolic processes where molecules are broken down

Catalysis chemical process in which a reaction rate is accelerated

Catalyst a substance used to accelerate a chemical reaction without taking part in the reaction

Cathode the negatively charged electrode in an electrolytic cell where reduction takes place

Cathodic Protection method where a more active metal is connected to a metal structure such as a tank or a ship protecting the structure because the active metal is oxidized rather than the structure

Cation a positively charged ion

Cellulase a group of enzymes that hydrolyze cellulose

Cellulose common carbohydrate that is the primary constituent of cell walls in plants

Chapman Reactions series of reactions responsible for the destruction of stratospheric ozone

Charles's Law law that states volume of a gas is directly related to its absolute temperature

Chemical Affinity the tendency of particular atoms to bond to each other

Chemical Change a transformation in which one substance changes into another, as opposed to a physical change

Chemical Element a pure substance that cannot be broken down into a simpler substance by chemical means

Chemical Family a group of elements that share similar chemical properties and share the same column in the periodic table, for example, halogens, alkali earth

Chirality condition that describes the "handedness" of a molecule or whether a molecule exists in forms that can be superimposed on each other

Chlorofluorocarbons also called CFCs, compounds consisting of chorine, fluorine, and carbon that are responsible for stratospheric ozone destruction

Coagulation precipitation or separation from a dispersed state

Coefficient of Thermal Expansion measure of the rate at which a substance will expand when heated

Cohesion intermolecular attractive force between particles within a substance

Colligative Property a property dependent on the number of particles in solution and not on the type of particles, for example, boiling point elevation and freezing point depression

Colloid particles having dimensions of 1–1000 nm or a suspension containing particles of this size

Combination Reaction also called synthesis, reaction in which two or more substances combine to form a compound

Combustion burning of a fuel to produce heat, oxidation of a fuel source

Complete Ionic Equation equation used to express reactions in aqueous solutions where reactants and products are written as ions rather than molecules or compounds

Complete Reaction a chemical reaction in which one of the reactants is completely consumed

Complex Carbohydrate polysaccharides such as cellulose and starch

Compound a substance consisting of atoms of two or more different elements chemically bonded

Condensation transformation from the gas to liquid state

Conjugate Acid in the Brønsted theory, the substance formed when a base accepts a proton

Conjugate Acid-Base Pair in the Brønsted theory, an acid and its conjugate base or a base and its conjugate acid

Conjugate Base in the Brønsted theory, the substance that remains when an acid donates a proton

Constructive Interference when waves combine to reinforce each other

Control to remove or account for the effect of a variable in an experiment

Control Rod rods used in nuclear reactor to absorb neutrons and control fission of radioactive fuel

Controlled Experiment method in which all variables except one independent variable are held constant or accounted for in some fashion

Coriolis Force pseudo-force introduced to account for motion that takes place on a rotating Earth, to the right in the Northern Hemisphere and left in the Southern Hemisphere

Cosmic Radiation charged atomic particles originating from space

Covalent Bond chemical bond in which electrons are shared between atoms

Covalent Crystal crystal in which atoms are held together by covalent bonds in a rigid three-dimensional network, for example, diamond

Cracking process by which a compound is broken down into simpler substances, typically employed in petroleum industry to break carbon-carbon bonds

Crenation condition that results when cells lose water and shrivel up

Critical Mass quantity of fissionable material necessary to give a self-sustaining nuclear reaction

Crystalline Solid solid in which atoms are arranged in definite three-dimensional pattern

Cultural Eutrophication accelerated aging of water body due to human influence and water pollution

Curie measure of radiation activity equal to 3.7×10^{10} disintegrations per second

Cyclic arrangement of molecule in which carbons are bonded together in a ring-pattern

Dalton's Law of Partial Pressure law that states the total pressure in a mixture of gases is equal to the sum of pressures each gas would exert in absence of other gases

Daughter the by-product in a radioactive decay

D-Block Element a transition element, valence electrons are in the d orbitals

Decay Activity the rate of decay of a radioactive substance

Decomposition Reaction reaction in which a compound is broken down into its components

Delocalized not associated with a particular region, in metallic bonding it refers to the fact that electrons are shared by the metal as a whole

Denatured to make unsuitable for human consumption

Dephlogisticated Air term Priestley used for oxygen, air devoid of phlogiston

Dephlogisticated Nitrous Air term Priestley used for nitrous oxide, N_2O, laughing gas

Destructive Interference when waves combine to cancel each other

Deuterium isotope of hydrogen containing 1 proton and 1 neutron

Dialysis the separation of particles from a colloid suspension by the passage of suspension solution through a semipermeable membrane

Diffraction the spreading of light or other waves after passing through an opening or passing by an obstacle

Dihydric an alcohol with two hydroxyl groups

Dioxins chlorinated cyclic compounds

Dipeptide two amino acids joined a peptide bond

Dipole a difference in the centers of positive and negative charge in a molecule

Dipole Moment condition when the centers of positive and negative charge in a molecule differ

Dipole-Dipole Force type of intermolecular force existing between molecules possessing dipole moments

Disaccharide carbohydrates in which two monosaccharides are bonded to each other

Dispersing Medium in a colloid suspension, the substance in which the particles are suspended

Dispersing Phase in a colloid suspension, the particles that are suspended in the dispersing medium

Dispersion Force intermolecular force that results from continuous temporary dipoles formed in molecules not possessing permanent dipole moments

Dose the amount of substance administered to an organism during a treatment or experiment

Dose Equivalent a means of comparing different forms of radiation on different body tissue

Double Replacement Reaction a reaction between two compounds in which elements or ions replace each other

Effluent discharge from a pipe or another source, associated with water quality

Electrical Potential the electrical potential energy of charged body above ground measured in volts

Electrochemical Cell a cell in which electrical energy is used to cause a nonspontaneous chemical reaction to occur

Electrode a terminal that conducts current in an electrochemical cell

Electrolysis process involving the passage of current through an electrochemical cell forcing a nonspontaneous reaction to occur

Electrolyte a substance that conducts current when it is dissolved in water

Electrolyze to bring about a chemical change by passing an electrical current through an electrolyte

Electromotive Force force due to a difference in electrical potential causing current to flow in a circuit

Electron Configuration distribution of electrons into different shells and orbitals from the lower to higher energy levels

Electronegativity measure of the attraction of an element for a bonding pair of electrons

Electroplating process where a metal is reduced on to the surface of an object, which serves as the cathode in an electrochemical cell

Electrorefining purification of a metal using electrolysis

Electrostatic charge at rest

Element *see* Chemical Element

Elementary Particle collectively the smallest units of matter, electrons, protons, quarks, and so on.

Elementary Steps single reaction in a series of reactions that produce a net chemical equation

Empirical experimental, to determine experimentally

Emulsifier a substance that allows two phases to mix and form a solution or suspension, used in food processing

Endothermic a chemical reaction in which energy is absorbed from the surroundings

Energy the ability to do work

Enzymes a biological catalyst

Equilibrium state of chemical reaction where the rates of forward and reverse reactions are equal, causing concentrations of reactants and products to remain constant

Equilibrium Constant a number equal to the ratio of the concentration of products at equilibrium over the concentration of reactants at equilibrium all raised to a power equal to the stoichiometric coefficient in the chemical equation

Essential Amino Acid an amino acid that cannot be produced by our bodies and must be ingested

Ester class of organic compounds that results from the reaction of carboxylic acid and an alcohol

Estrogen female sex hormones

Ether class of organic compound containing an oxygen atom singly bonded to two carbon atoms C-O-C

Eutrophication the natural aging of a water body

Exothermic a chemical reaction that liberates energy to the surroundings

Fats biomolecules consisting of glycerols and fatty acids (triglycerides) that are solid at room temperature

Fatty Acid long unbranched carboxylic acid chain

F-Block Element the lanthanides and actinides, valence electrons in the f orbitals

Feedstock a process chemical used to produce other chemicals or products

Fine Chemicals chemicals produced in relatively low volumes and at higher prices as compared to bulk chemicals such as sulfuric acid, includes flavorings, perfumes, pharmaceuticals, and dyes

First Law of Thermodynamics law that states energy in universe is constant, energy cannot be created or destroyed

First Order Reaction reaction in which the rate is dependent on the concentration of reactant to the first power

Fission *see* Nuclear Fission

Fixed Air carbon dioxide, term used by Joseph Black for CO_2

Fossil Fuels a fuel derived from the remains of ancient organisms: oil, natural gas, and coal

Fractional Distillation process in which a mixture of chemicals is separated by their difference in boiling points, used extensively to refine petroleum into its different components or fractions

Free Radical a species possessing at least one unpaired electron

Freezing Point Depression decrease in freezing point of a pure liquid due to the presence of solute added to the liquid

Fuel Cell type of electrochemical cell in which reactants are continually supplied to create electrical energy

Fullerene pure carbon molecule consisting of at least 60 carbon atoms arranged spherically like a soccer ball

Functional Group atom or group of atoms that characterize a class of compounds

Fusion *see* Nuclear Fusion

Galvanize a method used to protect metals by plating them with another metal, for example, coating iron with zinc

Gamma Ray form of high energy electromagnetic radiation with no mass and no charge

Gamma Decay radioactive process in which gamma ray is emitted from the nucleus

Gay-Lussac's Law law that states the pressure of a gas is directly proportional to the gas's absolute temperature

Gay-Lussac's Principle law that states when a chemical reaction takes place involving gases, the volumes of reactants and products are in small whole number ratios

Gel colloid in which a liquid is dispersed in a solid

Gray SI unit to measure the amount of radiation dose equal to deposition of one joule of energy into one kilogram of matter

Greenhouse Effect warming of the Earth due to absorption of infrared radiation by particular atmospheric gases such as H_2O, CO_2, and CH_4

Groups columns or families in the periodic table

Haber Process method used to produce ammonia from nitrogen and hydrogen at high temperature and pressure under the presence of a catalyst

Half Reaction the oxidation or reduction reaction in an oxidation-reduction reaction

Half-Life time it takes for substance to be reduced to half its original amount or activity in a chemical or nuclear reaction

Halogen family consisting of group 17 elements in the periodic table: F, Cl, Be, I, At

Halogenation chemical reaction involving the addition of a halogen

Hard Water water containing a high concentration of divalent metal ions particularly calcium and magnesium

Heat flow of energy from a region of higher temperature to a region of lower temperature

Heat of Fusion energy necessary to change one kilogram of a substance from a liquid to solid at the substance's freezing point

Heat of Vaporization energy necessary to change one kilogram of a substance from a liquid to gas at the substance's boiling point

Heating Curve graph that shows how the temperature of a substance changes as heat is added to the substance

Henry's Law principle that states the amount of gas dissolved in a solution is directly proportional to the partial pressure of the gas above the solution

Heterocyclic Compound a cyclic compound in which at least one carbon atom in the ring has been replaced by another atom or group of atoms

Homogeneous a solution or mixture that is well mixed with a constant composition throughout

Hormone chemical messengers released by glands

Humidity the amount of moisture contained in the air expressed in mass of water vapor per unit volume of air

Hydration process by which solute particles become surrounded by water molecules in an aqueous solution

Hydrocarbon organic compound consisting of only carbon and hydrogen

Hydrogen Bond intermolecular force formed between hydrogen of one molecule and highly electronegative atom such as nitrogen, oxygen, or fluorine on another molecule

Hydrogenation reaction in which hydrogen is added to unsaturated organic compounds in presence of a catalyst, used to turn liquid vegetable oils into solids

Hydrolysis decomposition or change of a substance by its reaction with water

Hydronium the H_3O^+ ion

Hydrosphere the sum total of Earth's water including the oceans, polar ice caps, groundwater, atmospheric water, rivers, and lakes

Hyperbaric a condition of elevated pressure

Hypothesis a tentative statement used to explain some aspect of nature and tested through an experiment

Iatrochemical medical school of thought that attributed disease to chemical imbalances in the body

Ideal Gas gas in which it is assumed gas molecules occupy no appreciable portion of the total volume, do not interact with one another, and undergo elastic collision with one another

Ideal Gas Law relationship describing the behavior of an ideal gas, $PV = nRT$ where P is the pressure, V is volume, n is number of moles, R is ideal gas law constant, and T is absolute temperature

Immiscible when two liquids do not mix, for example, oil and water

Independent Variable the factor which a researcher varies in an experiment

Indicator a substance that exhibits a color change dependent on pH

Industrial Hygiene dealing with the health and safety of workers, a discipline that deals with the maintenance of a healthy work environment

Inhibitor a substance that slows down or stops a reaction

Inorganic Chemistry chemistry of compounds excluding hydrocarbons and other compounds based on carbon

Intermolecular between molecules

Internal Energy the total potential and kinetic energy of all particles comprising a system

Intramolecular within a molecule

Invert Sugar a sugar mixture consisting of glucose and fructose

Iodine Number measure of the amount of iodine absorbed by an unsaturated fat and used to measure the degree of saturation

Ion a charged species created when an atom or group of atoms gains or loses electrons

Ion-Dipole Force intermolecular force between an ion and a dipole

Ion Pair in a solution when a positive and negative ion exist as a single particle

Ion Product Constant in an ionic reaction, the product of each ion's concentration in solution raised to a power equal to the coefficient in the net ionic equation, for water it equals $[H^+][OH^-]$

Ionic Bond intramolecular force created when electrons are transferred from one atom to another creating ions that possess electrostatic attraction for one another

Ionic Solid a solid composed of anions and cations such as NaCl

Ionization process by which an atom or molecule gains or loses electrons and acquires a charge becoming an ion

Ionizing Radiation alpha, beta, or gamma radiation possessing sufficient energy to dislodge electrons from atoms in tissue forming ions that can cause cell damage

Isomer different compounds that have the same molecular formula

Isotope forms of the same element that have different mass numbers

Joule SI unit for energy and work equivalent to 0.24 calorie

Ketone class of organic compounds containing a carbonyl group bonded to two other carbon atoms

Kilojoules unit of energy equal to 1,000 joules

Kinetic Energy energy associated with motion equal to $\frac{1}{2}mv^2$

Kinetic Molecular Theory model that defines an ideal gas and assumes the average kinetic energy of gas molecules is directly proportional to the absolute temperature

Kinetics that area of chemistry that deals with how fast reactions take place

Latent Heat energy transferred to material without a change in temperature, associated with phase changes such as melting and boiling

Lattice Points positions in a unit cell occupied by atom, molecules, or ions

Law of Definite Proportion law that states that different samples of the same compound always contain elemental mass percentages that are constant

Law of Mass Action mathematical expression based on the ratio between products and reactants at equilibrium, an equation to determine the equilibrium rate constant

Law of Multiple Proportions law that states when two elements combine to form more than one compound that the mass of one element compared to the fixed mass of the other element is in the ratio of small whole numbers

Le Chatelier's Principle principle that says when a system is at equilibrium and a change is imposed on the system that the system will shift to reduce the change

Lewis Electron Dot Formula a diagram showing how the valence electrons are distributed around an atom or distributed in a molecule

Lipid a class of biomolecules including fats and oils characterized by their insolubility in water

Liquid Crystals materials that have properties of both solids and liquids used extensively in digital displays

Lithosphere outer surface of Earth including the crust and upper mantle

Lock-and-Key Model model to explain how enzymes catalyze reactions with specific enzymes acting as locks that only certain substrates which act as keys can fit

London Force intermolecular force that arises when momentary temporary dipoles form in nonpolar molecules

Long-Range Order general term to describe regular repeating structure found in crystals

Longwave Radiation infrared radiation emitted by Earth

Main Group Element elements in groups 1, 2, and 13–18, members in each group have the same general electron configuration

Mass Defect the change in mass when a nucleus is formed from its constituent nucleons

Mass Number sum of protons and neutrons in the nucleus of an atom

Matter the material that comprises the universe, anything that has mass and occupies space

Maximum Contaminant Level the maximum concentration of a substance dissolved in water deemed safe for consumption

Meniscus the curved upper surface of a narrow column of water

Metabolism the total of chemical reactions and processes used by organisms to transform energy, reproduce, and maintain themselves

Metal class of elements characterized by their ability to form positive ions, conduct heat and electricity, malleability, and luster

Metallic Bond bond present in metals due to delocalized electrons moving throughout the metal lattice

Metallic Solid type of solid characterized by delocalized electrons and metal atoms occupying lattice points

Metalloid elements have properties intermediate between metals and nonmetals

Mixture combination of two or more substances where the individual substances maintain their identity

Moderator a material such as graphite or deuterium used to slow down neutrons in nuclear reactors

Molality a way to express concentration of a solution, moles of solute per kilogram of solvent

Molarity a way to express concentration of a solution, moles of solute per liter of solution

Mole number of C-12 atoms in exactly 12 grams of C-12 equal to 6.022×10^{23}, a quantity of substance equal to the molar mass expressed in grams

Molecular Equation a reaction in which the reactants and products are expressed as molecules or whole units rather than as ions

Molecular Orbital Theory a model that uses wave functions to describe the position of electrons in a molecule, assuming electrons are delocalized within the molecule

Molecular Solid a solid that contains molecules at the lattice points

Molecule a group of atoms that exist as a unit and are held together by covalent bonds

Monomer discrete molecules that join together to form a polymer

Monosaccharide a simple sugar molecule containing only a single aldehyde or ketone group, for example, glucose

Nanometer one billionth of a meter, 10^{-9} m

Nebula cloud of dust and gases, principally hydrogen, distributed throughout the universe

Net Ionic Equation a chemical equation that shows only the ionic species that actually take part in the reaction

Neutralization process that occurs when an acid reacts with a base, a type of reaction involving an acid and base

Newton SI unit for force equal to 1 kg-m/s^2

Nonelectrolyte a substance that does not conduct current when it is dissolved in water

Nonionizing Radiation electromagnetic radiation with insufficient energy to dislodge electrons and cause ionization in human tissue, for example, radio waves, microwave, visible light

Nonmetal elements found on the right side of the periodic table that conduct heat and electricity poorly

Nonpoint Source Pollution pollution that does not originate from one specific location, such as a sewer pipe, but comes from multiple locations spread over a wide area, for example, runoff from city streets

Nonrenewable Resource a resource in fixed supply or one that it is replenished at a rate so slow that it is exhausted before it is replenished, for example, oil

Normal Alkane an alkane in which all the carbon atoms in the molecule are attached in a continuous chain

Normality a way to express the concentration of a solution, the number of equivalents per liter of solution

Nuclear Fission the splitting of the nucleus of an atom of a heavier element into smaller nuclei to produce energy in a nuclear reaction

Nuclear Fusion the combination of atomic nuclei of lighter elements into heavier nuclei to produce energy

Nucleic Acid complex biomolecules responsible for passing on genetic information and for protein synthesis, include DNA and RNA

Nucleon a proton or a neutron, the number of nucleons in an atom equals the sum of protons and neutrons in the nucleus

Octet Rule general rule that states that the most stable electron configuration occurs when an atom surrounds itself with eight valence electrons

Oil biomolecules consisting of glycerols and fatty acids (triglycerides) that are liquid at room temperature

Oil of Vitriol concentrated sulfuric acid

Optical Isomer isomers that differ in their ability to rotate light in opposite directions

Organic Chemistry chemistry of carbon-based compounds

Organochloride chlorine substituted hydrocarbon molecule widely used in industry, pesticides, CFCs

Organophosphate organic compounds containing phosphorus, widely used in certain pesticides

Osmosis process in which a solvent flows from a diluted solution to a more concentrated solution across a semipermeable membrane

Osmotic Pressure pressure required to stop the movement of the solvent across a semipermeable membrane during osmosis

Oxidation process that describes the loss of electrons

Oxidation Number the amount of charge transferred by or to an atom when it is assumed an ionic bond is formed as electrons are donated to the more electronegative atom

Oxidation Reaction a chemical reaction in which an atom, molecule, or ion loses electrons

Oxidizing Agent a substance that is reduced and hence causes oxidation to occur in a redox reaction

Parent in a nuclear reaction the original substance that emits some form of radiation

Partial Pressure pressure exerted by an individual gas in a mixture of gases

Parts Per Million way to express concentration by giving the number of units of mass or volume out of one million units of mass or volume

Pascal SI unit for pressure equal to 1 newton per square meter

Peptide a sequence of 50 or fewer amino acids joined by peptide bonds

Peptide Bond bond formed between the carboxyl group of one amino acid and the amino group of another amino acid in a peptide

Percent by Mass proportion of the mass of each element in a compound expressed as a percent or proportion of solute mass as total mass of solution

Perfect Gas *see* Ideal Gas

Period a horizontal row in the periodic table

Permanent Hardness condition of water when magnesium and calcium are combined with chlorides and sulfates rather than carbonates, cannot be removed by heating

Petrochemical chemicals derived from oil or natural gas

pH the negative of the base 10 logarithm of the molar concentration of the hydrogen ion concentration of a solution, $-\log[H^+]$

Phenol C_6H_5OH, the molecule formed when a hydroxyl group is substituted for a hydrogen in benzene or a compound that contains the phenol group

Phenyl C_6H_5, group formed when a hydrogen is removed from benzene

Pheromones compounds produced in animals, especially insects, to communicate messages to members of the same species

Philosopher's Stone imaginary substance that alchemists believed could turn base metal into gold or was an elixir to cure diseases

Phlogiston a material once thought to be an element responsible for combustion

Phospholipid lipid containing phosphorus derived from phosphoric acid

Photochemical Oxidants air pollutants produced when hydrocarbons, nitrogen oxides, and other chemicals react under the influence of sunlight, for example, ozone, peroxyacylnitrates (PAN)

Physical Changes transformations in the natural world, such as freezing and boiling, that do not involve a change in substances or in which a chemical reaction does not take place

Phytoplankton marine and freshwater plants that float in the surface waters, many are microscopic and include different forms of algae

Plasma state of matter in which atoms or molecules have been ionized to form positive ions and electrons

Plasmolysis condition that results when cells absorb water and rupture

Pneumatic Chemistry study of gases and air, important in the development of modern chemistry

Pneumatic Trough device used to collect and measure the volume of gases invented in 1700s by Stephen Hales, consisted of glass container submerged in water in which gas displaced water

pOH the negative of the base 10 logarithm of the molar concentration of the hydroxide ion concentration of a solution, $-\log[OH^-]$

Point Source Pollution pollution originating from a specific location, for example, a smokestack

Polar Covalent Bond a covalent bond in which the bonding pair of electrons are not shared equally between atoms, the atom with the greater electronegativity has a greater affinity for the bonding electrons

Polar Molecule a molecule where the centers of positive and negative charge differ, creating a permanent dipole moment

Polarizability ability of an electron cloud in a neutral atom to be distorted

Polarized Light light in which the electromagnetic wave vibrates in only one plane

Polyatomic Ion an ion consisting of more than one atom

Polymer giant molecule formed by the linking of simple molecules called monomers

Polypeptide a chain of more than 50 amino acids

Polysaccharide a carbohydrate polymer consisting of monosaccharides bonded together

Polyunsaturated Fatty Acid fatty acid containing multiple carbon-carbon double bonds

Positron a subatomic particle with the mass of an electron and a charge of $+1$

Potential Energy the ability to do work by virtue of the position or arrangement of a system

Pressure force per unit area

Primary Amine an amine in which the nitrogen atom is bonded to a single carbon atom, when only one hydrogen in ammonia has been replaced by an organic R group

Primary Cells batteries that cannot be recharged

Primary Production conversion of matter into biomass using sunlight by plants in an ecosystem

Primary Structure the sequence of amino acids bonded to each other in a peptide chain

Principal Quantum Number a number that designates an electron's energy level and its average distance from the nucleus of an atom

Products the resulting substances in a chemical reaction

Progestin pregnancy hormones

Protein biomolecules consisting of polypeptide chain with large molecular mass

Protostar early stage in the formation of a star when gases and dust start to contract due to gravitational forces

P-V Work work associated with the expansion or compression of a gas

Pyrethroids synthetic forms of pyrethrins, insecticides based on extracts from chrysanthemums

Qualitative Analysis general method used to determine the composition of a compound

Quality Control a system for maintaining proper standards in a product or collection of data

Quantum Mechanics theory that explains the behavior of matter using wave functions to characterize the energy of electrons in atoms

Quark a fundamental particle hypothesized to form the basic building blocks of all matter; there are six different quark types: up, down, strange, charm, top, and bottom

Quaternary Structure describes the association of the different chains in multichain proteins

Rad a unit for an absorbed dose of radiation equal to 100 ergs of energy absorbed by a gram of matter

Radioactive unstable atomic nuclei spontaneously emitting particles and energy

Radioactive Tracer a radioactive substance used to monitor the movement and behavior of a chemical in biological processes and chemical reactions

Rate Determining Step the slowest reaction in the reaction mechanism

Rate Law general mathematical expression that gives the rate of reaction, depends on rate constant and concentrations of reactants

Reactants substances initially present in a chemical reaction

Reaction Mechanism series of reactions that shows how reactants are converted into products in a chemical reaction

Redox Reaction reaction involving the transfer of electrons, an oxidation-reduction reaction

Reducing Agent a substance that is oxidized and causes reduction to occur in a redox reaction

Reduction process that describes the gain of electrons

Reduction Reaction a chemical reaction in which an atom, molecule, or ion gains electrons

Relative Humidity ratio of the vapor pressure of water in air compared to the saturated vapor pressure of pure water at the same temperature, measure of the amount of water vapor in an air mass expressed as percent of how much water vapor that air mass can hold

Relative Mass mass measured with respect to a standard, atomic masses based on C-12 standard

Rem unit used to measure radiation dose equivalents based on the quantity of radiation absorbed and type of radiation

Residual Chlorine chlorine that has not combined with organic matter

Resins sticky, liquid organic substances exuded from plants that harden upon exposure to air

Resonance Structures any of two or more structures used to represent the overall structure and behavior of a compound even though the compound does not exist in any of the resonance forms

Reverse Osmosis process in which pressure is used to force a solvent of a concentrated solution through a semipermeable membrane toward a diluted solution, often used to purify water

Reversible Reaction a reaction that occurs in both directions: reactants to products and products to reactants

Rod component of nuclear reactor holding fuel (fuel rod) or substances used to control nuclear reaction (control rod)

Saccharides carbohydrates

Sacrificial Anode in cathodic protection, the metal connected to the structure to be protected that is more readily oxidized than the structure

Salt Bridge concentrated solution of electrolyte used to complete the circuit in an electrochemical cell that helps to equalize charge distribution in each half cell

Saltpeter potassium nitrate, KNO_3

Saponification conversion of a fat to soap by reacting with an alkali

Saturated solution that contains the maximum amount of solute under a given set of conditions

Saturated Fatty Acid fatty acids in which the carbon chain contains only single carbon-carbon bonds

Saturated Hydrocarbon a hydrocarbon containing only single carbon-carbon bonds

Scintillation process producing discrete flashes or sparks of light

Scrubbing process used to remove air pollutants from industrial emissions

Secondary Amine an amine in which the nitrogen atom is bonded to two carbon atoms, when two hydrogen atoms in ammonia have been replaced by two organic R groups

Secondary Cell batteries capable of being recharged

Secondary Production conversion of plant matter into biomass by organisms feeding at the second trophic level and above

Secondary Structure designates how atoms in a protein's backbone structure are arranged in space

Semipermeable Membrane media that allows the passage of solvent molecules but blocks solute molecules

Shortwave Radiation radiation emitted by the Sun that passes through and interacts with the atmosphere

Simple Carbohydrate monosaccharides such as glucose, fructose, and galactose that cannot be broken down by water

Single Replacement Reaction type of chemical reaction in which one element replaces another in a compound

Smelting to melt an ore in order to refine a metal from the ore

Solute in a solution the component present in the smaller amount

Solution a homogeneous mixture of two or more substances

Solvation process by which a solute dissolves in a solvent

Solvay Process method used to produce sodium carbonate using sodium chloride, ammonia, and carbon dioxide

Solvent in a solution the component present in the larger amount

Specific Heat Capacity the amount of heat required to raise the temperature of an object by 1 degree Celsius

Spectator Ions ions that do not take part in a chemical reaction

Standard Hydrogen Electrode standard electrode consisting of hydrogen gas at 1 atmosphere pressure, a platinum electrode, and 1 molar hydrochloric acid, the potential of this electrode is defined as 0 and is used to determine the potential of other half-reactions

Standard Reduction Potential the voltage measured for a half-cell under standard conditions when in reference to the standard hydrogen electrode

Steady-State a system that does not change with time

Stereoisomer isomers in which the atoms differ in their three-dimensional spatial arrangement

Steroids lipids that contain four fused carbon rings, three containing six carbons and one containing five carbons, include cholesterol and the sex hormones

Sterols *see* Steroids

Stoichiometry mass relationship in a chemical reaction

Straight Chain hydrocarbon structure in which carbon is bonded in a linear fashion as opposed to branching

Strong Electrolyte a substance that is a good conductor of electricity when dissolved in water

Strong Nuclear Force force responsible for binding the nucleons of an atom's nucleus together

Structural Isomer compounds that have the same formula but have different atomic arrangement

Subcritical Mass when the quantity of radioactive fuel is insufficient to produce a self-sustaining chain reaction

Sublimation process where a substance passes directly from the solid to gaseous phase without going through the liquid phase

Substrate the compound an enzyme acts upon

Supercritical Mass when the quantity of a radioactive fuel is sufficient to produce a self-sustaining chain reaction

Supernova massive explosion of giant star at the end of its life

Supersaturated a solution containing more solute than is present in its saturated state

Surface Tension property of liquid causing it to contract to the smallest possible area due to the unbalance of forces at the surface of the liquid

Surroundings the remainder of the universe outside a defined system

Suspension a mixture of nonsettling solid particles in a liquid, that is, gels and aerosols

Synthesis Reaction see combination reaction

System that part of the universe of concern to us

Temper heat treatment of metals

Temporary Hardness *see* Carbonate hardness

Tertiary Amine an amine in which the nitrogen atom is bonded to three carbon atoms, when three hydrogen atoms in ammonia have been replaced by three organic R groups

Tertiary Structure three-dimensional structure in protein resulting from interaction of amino acids in coiled chain

Thermochemistry area of chemistry that deals with energy changes that take place in chemical processes

Thermoplastics plastics that can be repeatedly heated and reformed

Thermosetting Plastics type of plastics that harden into final form after being heated once

Trace Gas gaseous components of the atmosphere that occur in low concentrations, for example, methane, ozone

Transmutation nuclear transformation in which the atomic nucleus of one element is converted into another element

Triazine group of herbicides that work by disrupting photosynthesis

Triglyceride principal component of fats and oils formed from three molecules of fatty acids and one glycerol molecule

Trihalomethane class of suspected carcinogens that form when chlorine and other halogens react with humic acids, chloroform, and bromoform

Trihydric an alcohol with three hydroxyl groups

Tyndall Effect scattering of light by fine particles in a suspension

Unit Cell most basic repeating unit of a lattice structure

Unsaturated solution that has not reached the point of saturation and can hold more solute

Unsaturated Hydrocarbon a hydrocarbon that contains at least one double or triple bond

Valence a positive number indicating the ability of an atom or radical to combine with other atoms or radicals, the number of electrons an atom can donate or accept in a chemical reaction

Valence Bond Theory theory of bonding based on overlapping valence orbitals

Valence Electron Configuration quantum numbers of electrons that reside in the outermost shell of an atom

Van Der Waal Force intermolecular force that can include dipole-dipole, ion-dipole, or London force

Van't Hoff Factor ratio of the number of actual particles dissolved in solution to the theoretical number of particles predicted when a substance dissolves

Vapor the gaseous state of a substance that normally exists in the solid or liquid phase

Vapor Pressure pressure exerted by a liquid or solid's vapor when the liquid or solid is in equilibrium with its vapor

Vaporization evaporation, process that occurs when molecules pass from liquid to gaseous state

Variable one of any number of factors that change and may influence an observation or experimental outcome

Viscosity the resistance to flow

Vitalism theory that a vital force associated with life was associated with all organic substances

VSEPR Model valence shell electron pair repulsion model, model used to predict the geometry of molecule based on distribution of shared and unshared electron pairs distributed around central atom of a molecule

Vulcanization heating rubber in the presence of sulfur to remove its tackiness and improve its quality

Weak Electrolyte a substance that is a poor conductor of electricity when dissolved in water

White Dwarf old star that no longer undergoes thermonuclear reaction but continues to radiate light

Wien's Law law that states the wavelength of maximum energy radiation from an object is inversely proportional to the surface temperature of the object

Work product of force times displacement

X-Ray Crystallography method used to determine structure of crystal by examining diffraction pattern produced by x-rays directed at crystal

Zooplankton marine and freshwater animals that float in the surface waters, many are microscopic

Zwitterion a molecule with a negative charge on one atom and a positive charge on another atom

Brief Timeline of Chemistry

30000 B.C.   cave paintings

8000 B.C.   agriculture develops

6000 B.C.   copper smelting

4000 B.C.   copper alloys

2000 B.C.   bronze

600 B.C.   Anaximander posits air as primary element; Thasles posits water as primary element

450 B.C.   Empedocies posits air, earth, fire and water as four major elements

400 B.C.   Democritus leads atomists school, atoms basic form of matter

350 B.C.   Aristotle's *Meterologica*

A.D. 300   Emperor Diocletian outlaws chemistry in Roman Empire

700–1100   height of Arab period; work of Geber, Al Razi, Avicenna

1000   gunpowder invented in China

1250   Albertus Magnus's alchemical studies

1250   Roger Bacon studies on gunpowder

1525   Paracelsus uses chemistry in medicinal treatment

1540   *Pirotechnia,* Biringuccio's work on metallurgy

1556   Agricola's *Of Metallaurgy,* treatise on mining

1620   Van Helmont's concept of gas

1640   Torricelli invents barometer

1660   Boyle's law

1661	Boyle publishes *The Skeptical Chymist*
1670	Becher and Stahl lay groundwork for phlogiston theory
1750	Hales invents pneumatic trough
1766	Cavendish isolates hydrogen
1770	Priestley and Scheele isolate oxygen
1789	Lavoisier's *Elements of Chemistry*
1794	Lavoisier beheaded
1770–1800	Volta studies electricity
1802	DuPont chemical company founded
1808	Dalton proproses atoms as basic building blocks of matter
1808	Humphrey Davy uses electrolysis to discover metallic elements
1810	Avogadro's hypothesis
1828	Wöhler synthesizes urea
1830	Faraday's studies on electricity
1845	Kolbe synthesizes acetic acid
1846	ether is used as anesthetic
1852	idea of valence is introduced by Frankland
1856	Perkin perfects synthetic dyes
1859	first commercial oil well is drilled by Drake in Titusville, Pennsylvania
1863	Solvay process is discovered for making sodium bicarbonate
1865	Kekulé determines structure of benzene
1866	Nobel invents dynamite
1870	Mendeleev's periodic table
1876	American Chemical Society is founded
1884	Arrhenius's theory that acids produce hydrogen ions
1893	synthesis of aspirin by Hoffman
1895	Roentgen discoveres x-rays
1896	Becquerel discovers process of radioactivity
1897	Thomson discovers electron
1908	Rutherford's model of the atom
1910	rayon production begins in U.S.A.

1913	Bohr's model of hydrogen atom
1919	Rutherford's alpha particles support existence of protons
1920	covalent bond by Lewis and Langmuir
1923	Brønsted definition of acid and base
1927–30	quantum mechanical model of atom developed
1932	neutron discovered by Chadwick
1931	DuPont produces synthetic rubber neoprene
1936	Carothers' DuPont team produces nylon
1939	Müller synthesizes DDT
1940	synthetic rubber is used in tires
1953	Crick and Watson present structure for DNA
1963	Rachael Carson publishes *Silent Spring*
1969	EPA is established
1976	evidence that chlorine compounds destroy ozone
1985	Smalley and Kroto discover buckminsterfullerenes
1989	Human Genome Project commences
1990	PCR method established
1996	cloning of Dolly the sheep
2000	conclusion of human genome sequencing

Nobel Laureates in Chemistry

Nobel laureates are listed consecutively by year starting with 1901. The country indicates where the chemist did major work. If the chemist's country of birth is different than where major work is done, this is indicated after the "b." In many years, several chemists are recognized. When several chemists are recognized in the same year, it may be either for work in similar areas of chemistry, working collaboratively or independently, or for work in entirely different areas. When several laureates have been recognized for work in similar areas, the work is cited for the first laureate in the list.

Year	Recipient(s)	Country	Work
1901	Jacobus Henricus van't Hoff	Germany b. Netherlands	laws of chemical dynamics and osmotic pressure in solutions
1902	Hermann Emil Fischer	Germany	sugar and purine syntheses
1903	Svante August Arrhenius	Sweden	electrolytic dissociation theory
1904	William Ramsay	Great Britain	discovery of noble gases
1905	Adolf von Baeyer	Germany	organic dyes and hydroaromatic compounds
1906	Henri Moissan	France	isolation of fluorine and electric furnace
1907	Eduard Buchner	Germany	fermentation in absence of cells and biochemistry
1908	Ernest Rutherford	Great Britain b. New Zealand	radioactive decay
1909	Wilhelm Ostwald	Germany b. Russia	chemical equilibrium, kinetics, and catalysis
1910	Otto Wallach	Germany	pioneering work with alicyclic compounds
1911	Marie Curie	France b. Poland	discovery of radium and polonium
1912	Victor Grignard	France	discovery of Grignard's reagent
	Paul Sabatier	France	hydrogenation with metal catalysts
1913	Alfred Werner	Switzerland b. Germany	bonding of inorganic compounds
1914	Theodore Richards	United States	determination of atomic weights
1915	Richard Willstätter	Germany	studies of plant pigments, especially chlorophyll

Year	Recipient(s)	Country	Work
1918	Fritz Haber	Germany	synthesis of ammonia
1920	Walter H. Nernst	Germany	thermochemistry
1921	Frederick Soddy	Great Britain	radioactive substances and isotopes
1922	Francis W. Aston	Great Britain	mass spectrography and discovery of isotopes
1923	Fritz Perl	Austria	organic microanalysis
1925	Richard A. Zsigmondy	Germany b. Austria	colloid chemistry
1926	Theodor Svedberg	Sweden	disperse systems
1927	Heinrich Otto Wieland	Germany	bile acids
1928	Adolf Otto Windaus	Germany	sterols relationship with vitamins
1929	Arthur Harden	Great Britain	fermentation of sugar and sugar enzymes
	Hans von Euler-Chelpin	Sweden b. Germany	
1930	Hans Fischer	Germany	synthesis of hemin
1931	Carl Bosch	Germany	high pressure chemical processing
	Friedrich Bergins	Germany	
1932	Irving Langmuir	United States	surface chemistry
1934	Harold Urey	United States	discovery of heavy hydrogen
1935	Frédéric Joliot	France	synthesis of new radioactive elements
	Irène Joliot-Curie	France	
1936	Peter Debye	Germany b.Netherlands	dipole moments and x-ray diffraction
1937	Walter Norman Hayworth	Great Britain	carbohydrates and vitamin C
	Paul Karrer	Switzerland	vitamins A and B12
1938	Richard Kuhn	Germany b. Austria	carotenoids and vitamins
1939	Adolf F. J. Butenandt	Germany	sex hormones
	Leopold Ruzicka	Switzerland b. Hungary	polymethylenes and terpenes
1943	George DeHevesy	Sweden b. Hungary	isotope tracers
1944	Otto Hahn	Germany	fission of heavy nuclei
1945	Atturi I. Virtanen	Finland	agricultural and food chemistry and preservation of fodder
1946	James B. Sumner	United States	crystallation of enzymes
	John H. Northrop	United States	preparation of proteins and enzymes in pure form
	Wendell M. Stanley	United States	
1947	Robert Robinson	Great Britain	alkaloids
1948	Arne W. K. Tiselius	Sweden	electrophoresis and serum proteins
1949	William F. Giauque	United States	low temperature thermodynamics
1950	Otto Diels	Germany	diene synthesis
	Kurt Alder	Germany	
1951	Edwin McMillan	United States	chemistry of transuranium elements
	Glenn Seaborg	United States	

Year	Recipient(s)	Country	Work
1952	Archer J. P. Martin	Great Britain	invention of partition chromatography
	Richard L. M. Synge	Great Britain	
1953	Hermann Staudinger	Germany	macromolecular chemistry
1954	Linus Pauling	United States	chemical bonding and molecular structure of proteins
1955	Vincent Du Vigneaud	United States	sulfur compounds of biological importance, synthesis of polypeptide hormone
1956	Cyril Hinshelwood	Great Britain	mechanisms of chemical reaction
	Nikolay Semenov	USSR	
1957	Alexander Todd	Great Britain	nucleotides and their co-enzymes
1958	Frederick Sanger	Great Britain	protein structure, insulin
1959	Jaroslav Heyrovsky	Czechoslovakia	polarographic methods of analysis
1960	Willard Libby	United States	carbon-14 dating
1961	Melvin Calvin	United States	CO_2 assimilation in plants
1962	Max Preutz	Great Britain b. Austria	structure of globular proteins
	John Kendrew	Great Britain	
1963	Giulio Natta	Italy	high polymers
	Karl Ziegler	Germany	
1964	Dorothy Crowfoot Hodgkin	Great Britain	x-ray techniques of the structures of biochemical substances
1965	Robert Woodward	United States	organic synthesis
1966	Robert Mulliken	United States	chemical bonds and electronic structure of molecules
1967	Manfred Eigen	Germany	study of very fast chemical reactions
	Ronald Norrish	Great Britain	
	George Porter	Great Britain	
1968	Lars Onsager	United States b. Norway	thermodynamic of irreversible processes
1969	Derek Barton	Great Britain	conformation
	Odd Hensel	Norway	
1970	Luis Leloir	Argentina	sugar nucleotides and carbohydrate biosynthesis
1971	Gerhard Herzberg	Canada b. Germany	structure and geometry of free radicals
1972	Christian Anfinsen	United States	ribonuclease, amino acid sequencing and biological activity
	Stanford Moore	United States	chemical structure and catalytic activity of ribonuclease
	William Stein	United States	
1973	Ernst Fischer	Germany	organometallic sandwich compounds
	Geoffrey Wilkinson	Great Britain	
1974	Paul Flory	United States	macromolecules
1975	John Cornforth	Great Britain	stereochemisitry of enzyme-catalyzed reactions
	Vladmir Prelog	Switzerland b. Bosnia	stereochemistry of organic molecules
1976	William Lipscomb	United States	structure of borane and bonding

Year	Recipient(s)	Country	Work
1977	Ilya Prigogine	Belgium b. Russia	theory of dissipative structures
1978	Peter Mitchell	Great Britain	chemiosmotic theory
1979	Herbert Brown	United States b. Great Britain	organic synthesis of boron and phosphorus compound
	George Wittig	Germany	
1980	Paul Berg	United States	recombinant DNA
	Walter Gilbert	United States	nucleic acid base sequences
	Frederick Sanger	Great Britain	
1981	Kenichi Fukui	Japan	chemical reactions and orbital theory
	Roald Hoffman	United States b. Poland	
1982	Aaron Klug	Great Britain	crystallographic electron microscopy applied to nucleic acids and proteins
1983	Henry Taube	United States b. Canada	electron transfer in metal complexes
1984	Robert Merrifield	United States	chemical synthesis
1985	Herbert Hauptman	United States	crystal structures
	Jerome Karle	United States	
1986	Dudley Herschbach	United States	chemical elementary processes
	Yuan Lee	United States b. Taiwan	
	John Polanyi	Canada	
1987	Donald Cram	United States	development of molecules with highly selective structure specific interactions
	Jean-Marie Lehn	France	
	Charles Pedersen	United States b. Korea, of Norwegian descent	
1988	Johann Deisenhofer	Germany and United States b. Germany	photosynthesis
	Robert Huber	Germany	
	Michel Hartmut	Germany	
1989	Sidney Altman	United States b. Canada	catalytic properties of RNA
	Thomas Cech	United States	
1990	Elias James Corey	United States	organic synthesis
1991	Richard Ernst	Switzerland	nuclear resonance spectroscopy
1992	Rudolph Marcus	United States b. Canada	electron transfer in chemical systems
1993	Kary Mullis	United States	invention of PCR method
	Michael Smith	Canada b. Great Britain	mutagenesis and protein studies
1994	George Olah	United States b. Hungary	carbocation chemistry

Year	Recipient(s)	Country	Work
1995	Paul Crutzen	Germany b. Netherlands	atmospheric chemistry
	Mario Molina	United States b. Mexico	stratospheric ozone depletion
	Rowland F. Sherwood	United States	
1996	Robert Kurl	United States	discovery of fullerenes
	Harold Kroto	Great Britain	
	Richard Smalley	United States	
1997	Paul Boyer	United States	enzyme mechanism of ATP
	John Walker	Great Britain	
	Jens Skou	Denmark	discovery of ion transport enzyme Na^+, K^+-ATPase
1998	Walter Kohn	United States	density function theory
	John Pople	Great Britain	computational methods in quantum chemistry
1999	Ahmed Zewail	United States b. Egypt	transition states using femto spectroscopy
2000	Alan J. Heeger	United States	discovery of conductive polymers
	Alan MacDiarmid	United States	
	Hideki Shirakawa	Japan	
2001	William Knowles	United States	chirally catalyzed hydrogenation reactions
	Ryoji Noyori	Japan	
	K. Barry Sharpless	United States	chirally catalyzed oxidation reactions
2002	John B. Fenn	United States	identical of biological macromolecules
	Koichi Tanaka	Japan	
	Kurt Wüthrich	Switzerland	NMR analysis of biological macromolecules

Table of the Elements

Element	Symbol	Atomic Number	Atomic Mass	Element	Symbol	Atomic Number	Atomic Mass
Actinium	Ac	89	227	Mercury	Hg	80	200.6
Aluminum	Al	13	26.98	Molybdenum	Mo	42	95.94
Americium	Am	95	243	Neodymium	Nd	60	144.2
Antimony	Sb	51	121.8	Neon	Ne	10	20.18
Argon	Ar	18	39.95	Neptunium	Np	93	237
Arsenic	As	33	74.92	Nickel	Ni	28	58.69
Astatine	At	85	210	Niobium	Nb	41	92.91
Barium	Ba	56	137.3	Nitrogen	N	7	14.01
Berkelium	Bk	97	247	Nobelium	No	102	259
Beryllium	Be	4	9.012	Osmium	Os	76	190.2
Bismuth	Bi	83	209.0	Oxygen	O	8	16.00
Bohrium	Bh	107	264	Palladium	Pd	46	106.4
Boron	B	5	10.81	Phosphorus	P	15	30.97
Bromine	Br	35	79.90	Platinum	Pt	78	195.1
Cadmium	Cd	48	112.4	Plutonium	Pu	94	242
Calcium	Ca	20	40.08	Polonium	Po	84	210
Californium	Cf	98	251	Potassium	K	19	39.10
Carbon	C	6	12.01	Praseodymium	Pr	59	140.9
Cerium	Ce	58	140.0	Promethium	Pm	61	147
Cesium	Cs	55	132.9	Protactinium	Pa	91	231
Chlorine	Cl	17	35.45	Radium	Ra	88	226
Chromium	Cr	24	52.00	Radon	Rn	86	222
Cobalt	Co	27	58.93	Rhenium	Re	75	186.2
Copper	Cu	29	63.55	Rhodium	Rh	45	102.9
Curium	Cm	96	247	Rubidium	Rb	37	85.47
Dubnium	Db	105	262	Ruthenium	Ru	44	101.1
Dysprosium	Dy	66	162.5	Rutherfordium	Rf	104	261

Element	Symbol	Atomic Number	Atomic Mass	Element	Symbol	Atomic Number	Atomic Mass
Einsteinium	Es	99	254	Samarium	Sm	62	150.4
Erbium	Er	68	167.3	Scandium	Sc	21	44.96
Europium	Eu	63	152.0	Seaborgium	Sg	106	263
Fermium	Fm	100	253	Selenium	Se	34	78.96
Flourine	F	9	19.00	Silicon	Si	14	28.09
Francium	Fr	87	223	Silver	Ag	47	107.9
Gadolinium	Gd	64	157.3	Sodium	Na	11	22.99
Gallium	Ga	31	69.72	Strontium	Sr	38	87.62
Germanium	Ge	32	72.59	Sulfur	S	16	32.07
Gold	Au	79	197.0	Tantalum	Ta	73	180.9
Hafnium	Hf	72	178.5	Technetium	Tc	43	99
Hassium	Hs	108	265	Tellurium	Te	52	127.6
Helium	He	2	4.003	Terbium	Tb	65	158.9
Holmium	Ho	67	164.9	Thallium	Tl	81	204.4
Hydrogen	H	1	1.008	Thorium	Th	90	232.0
Indium	In	49	114.8	Thulium	Tm	69	168.9
Iodine	I	53	126.9	Tin	Sn	50	118.7
Iridium	Ir	77	192.2	Titanium	Ti	22	47.88
Iron	Fe	26	55.85	Tungsten	W	74	183.9
Krypton	Kr	36	83.80	Ununnilium	Uun	110	269
Lanthanum	La	57	138.9	Unununium	Uun	111	269
Lawrencium	Lr	103	257	Unumbium	Uub	112	277
Lead	Pb	82	207.2	Uranium	U	92	238.0
Lithium	Li	3	6.941	Vanadium	V	23	50.94
Lutetium	Lu	71	175.0	Xenon	Xe	54	131.3
Magnesium	Mg	12	24.31	Ytterbium	Yb	70	1773.0
Manganese	Mn	25	54.94	Yttrium	Y	39	88.91
Meitnerium	Mt	109	268	Zinc	Zn	30	65.39
Mendelevium	Md	101	256	Zirconium	Zr	40	91.22

Selected Bibliography

Aftalion, Fred. *A History of the International Chemistry Industry*, 2nd ed. Philadelphia: Chemical Heritage Press, 2001.

Amato, Evan. *Stuff: The Materials the World Is Made Of*. New York: Basic Books, 1997.

Asimov, Isaac. *A Short History of Chemistry*. Garden City, NY: Anchor Books, 1965.

Baird, Colin. *Environmental Chemistry*. New York: W.H. Freeman and Company, 1995.

Brock, William H. *The Norton History of Chemistry*. New York: W.W. Norton and Company, 1993.

Bureau of Labor Statistics. *Occupational Outlook Handbook*. *http://stats.bls.gov/oco/ocos029.htm*. 25 February 2002.

Campbell, Mary K. *Biochemistry*. Philadelphia: Saunders Publishing, 1991.

Chang, Raymond. *Chemistry*, 7th ed. Boston: McGraw-Hill, Inc., 2002.

Chemical Heritage Foundation. *http://www.chemheritage.org/*. 24 February 2002.

Cobb, Cathy, and Goldwhite, Harold. *Creations of Fire: Chemistry's Lively History from Alchemy to the Atomic Age*. New York: Plenum Press, 1995.

Cox, P. A. *The Elements on Earth*. Oxford: Oxford University Press, 1995.

Emsley, John. *The Elements*, 3rd ed. Oxford: Clarendon Press, 1998.

Encarta Encyclopedia 99 (1999). [CD-ROM]. Redmond, WA: Microsoft Inc.

Farber, Eduard. *Great Chemists*. New York: Interscience, 1961.

Finucane, Edward W. *Definitions, Conversions and Calculations for Occupational Safety and Health Professionals*, 2nd ed. Boca Raton: Lewis Publishers, 1998.

Gibbs, F. W. *Joseph Priestley*. Garden City, NY: Doubleday and Company, Inc., 1967.

Giunta, Carmen. "Selected Classic Papers from the History of Chemistry." *http://webserver.lemoyne.edu/faculty/giunta/papers/html*. 23 February 2002.

Goran, Morris. *The Story of Fritz Haber*. Norman: University of Oklahoma Press, 1967.

Greenaway, Frank. *John Dalton and the Atom*. Ithaca: Cornell University Press, 1966.

Greenberg, Arthur. *A Chemical History Tour*. New York: Wiley Interscience, 2000.

Heiserman, David L. *Exploring Chemical Elements and Their Compounds*. Blue Ridge Summit, PA: TAB Books, 1992.

Herron, J. Dudley, Kukla, David A, Schrader, Clifford, DiSpezio, Michael A., and Erickson, Julia Lee. *Heath Chemistry*. Lexington, MA: D.C. Heath and Company, 1987.

Hill, John W. *Chemistry for Changing Times*, 6th ed. New York: Macmillan Publishing, 1992.

Hill, John W., and Feigl, Dorothy M. *Chemistry and Life: An Introduction to General, Organic, and Biological Chemistry*, 3rd ed. New York: Macmillan Publishing Company, 1978.

Hoffmann, Roald. *The Same and Not The Same*. New York: Columbia University Press, 1995.

Hoover's Handbook of American Business 2002. Austin, TX: Hoover's Business Press, 2002.

HowStuffWorks, Inc. *http://www.howstuffworks.com/*. 20 December 2001.

Hudson, John. *The History of Chemistry*. New York: Chapman and Hall, 1992.

Idhe, Aaron J. *The Development of Modern Chemistry*. New York: Harper and Row, 1964.

International Directory of Company Histories. Detroit: St. James Press, 1988.

Jacobs, Bob. "Wilton High School Chemistry." *http://www.chemistrycoach.com/home.htm.* 23 February 2002.

Jaffe, Bernard. *Crucibles: The Story of Chemistry From Ancient Alchemy to Nuclear Fission,* 4th ed. New York: Dover Publications, Inc., 1976.

Kent, James A., ed. *Riegel's Handbook of Industrial Chemistry,* 7th ed. New York: Van Nostrand Reinhold Company, 1969.

Knight, David. *Humphry Davy: Science and Power.* Oxford: Blackwell Publishers, 1992.

Krebs, Robert E. *The History and Uses of Our Earth's Chemical Elements: A Reference Guide.* Westport, CT: Greenwood Press, 1998.

Lagowski, J.J. *The Chemical Bond.* Boston: Houghton Mifflin Company, 1966.

Landis, Walter S. *Your Servant the Molecule.* New York: Macmillan, 1945.

Lavoisier, Antoine. *Elements of Chemistry, with an introduction by Douglas McKie.* Translated by Robert Kerr. New York: Dover Publications, 1984.

Leicester, Henry M, and Klickstein, Herbert S., *A Source Book in Chemistry, 1400–1900.* Cambridge: Harvard University Press, 1952.

Levere, Trevor H. *Affinity and Matter.* Oxford: Oxford University Press, 1971.

Mathews, Harry, Freeland, Richard, and Miesfeld, Roger L. *Biochemistry: A Short Course.* New York: Wiley Sons, 1997.

McMurry, John, and Fay, Robert C. *Chemistry,* 2nd ed. Upper Saddle River, NJ: Prentice Hall, 1998.

Miller, G. Tyler. *Environmental Science,* 4th ed. Belmont, CA: Wadsworth Publishing Company, 1993.

Morrison, Robert Thornton, and Boyd, Robert. *Organic Chemistry.* Boston: Allyn and Bacon Inc., 1959.

Multhauf, Robert P. *Neptune's Gift.* Baltimore: The Johns Hopkins University Press, 1978.

Nobel Foundation. "Nobel e-Museum." *http://www.nobel.se/index.html.* 1 February 2002.

O'Neil, Maryadale J., Budavari, Susan, Smith, Ann, Heckeleman, Patricia E., and Kinneary, Joanne F., eds. *The Merck Index,* 12th ed. Rahway, NJ: The Merck Publishing Group, 1996.

Park, John L. "The Chem Team, A Tutorial for High School Chemistry." *http://www.dbhs.wvusd.k12ca.us/.* 15 January 2002

Parker, Sybil P., ed. *McGraw-Hill Dictionary of Chemical Terms.* New York: McGraw-Hill Book Company, 1985.

Patterson, Elizabeth C. *John Dalton and the Atomic Theory.* Garden City, NY: Anchor Books, 1970.

Pauling, Linus. *The Nature of the Chemical Bond,* 3rd ed. Ithaca: Cornell University Press, 1960.

Pera, Marcelo. *The Ambiguous Frog: The Galvani-Volta Controversy.* Translated by Jonathan Mandelbaum. Princeton, NJ: Princeton University Press, 1992.

Redgrove, H. Stanley. *Alchemy: Ancient and Modern.* New York: Harper and Row, 1911; reprint New York: Harper and Row, 1973.

Rowh, Mark. *Great Jobs for Chemistry Majors.* Chicago: VGM Career Horizons, 1999.

Sae, Andy S. W. *Chemical Magic from the Grocery Store.* Dubuque, IA: Kendall/Hunt Publishing Company, 1996.

Salzberg, Hugh W. *From Cavemen to Chemist: Circumstances and Achievements.* Washington D.C.: American Chemical Society, 1991.

Sass, Stephen L. *The Substance of Civilization.* New York: Arcade Publishing, 1998.

Shreve, R. Norris, and Brink, Joseph A. *Chemical Process Industries.* New York: McGraw-Hill Book Company, 1977.

Snyder, Carl. *The Extraordinary Chemistry of Ordinary Things.* New York: John Wiley and Sons, 1992.

Stoker, H. Stephen. *General, Organic, and Biological Chemistry.* Boston: Houghton Mifflin Company, 2001.

Summerlin, Lee R., Borgford, Christie L., and Ealy, Julie. Chemical Demonstrations, A Sourcebook for Teachers Volume 2, 2nd ed. Washington, D.C.: American Chemical Society, 1988.

Summerlin, Lee R., and Ealy, James L. Jr. *Chemical Demonstrations: A Sourcebook for Teachers Volume 1,* 2nd ed. Washington, D.C.: American Chemical Society, 1988.

Taylor, F. Sherwood. *The Alchemists.* New York: Henry Schuman, 1949.

Thomas, John Meurig. *Michael Faraday and The Royal Institution: The Genius of Man and Place.* Bristol: Adam Hilger, 1991.

Ware, George W. *The Pesticide Book,* 5th ed. Fresno, CA: Thomson Publications, 1999.

Wendland, Ray T. *Petrochemicals, The New World of Synthetics.* Garden City, NY: Doubleday and Company, 1969.

Woodburn, John H. *Chemistry Careers.* Chicago: VGM Career Horizons, 1997.

Zumdahl, Steven S., and Zumdahl, Susan A., 5th ed. Boston: Houghton Mifflin Company, 2000.

Author Index

Abegg, Richard, 74
Agricola, Georgius, 16
Al-Razi, 13
Anaximander of Miletus, 9
Andrews, Samuel, 305
Aquinas, Thomas, 14
Aristotle, 10, 13, 14, 18
Arrhenius, Svante August, 141, 156, 157, 158, 159
Avicenna, 13
Avogadro, Amedeo, 35, 106

Bacon, Roger, 14
Baekeland, Leo Hendrik, 298
Bayen, Pierre, 26
Bayer, Friedrich, 168
Bayer, Karl Josef, 192
Becher, John, 19, 20
Becquerel, Antoine Henri, 38
Bergman, Tobern, 27, 72
Bernigaut, Louis Marie Hilaire, 297, 298
Berthelot, Marcellin, 196
Berthollet, Claude Louis, 27, 32, 33, 72
Berzelius, Jöns Jacob, 34, 51, 73, 196, 197, 198
Biringuccio, Vannoccio, 16
Black, Joseph, 20, 21, 22, 25, 26
Bohr, Niels, 40
Boltzmann, Ludwig, 103
Bosch, Karl, 153
Bose, Satyendra Nath, 85
Boullay, Pierre Francois Guillaume, 196
Boyle, Robert, 18, 19, 21, 25, 102
Bragg, William Henry, 91
Bragg, William Lawrence, 91

Brønsted, Johannes, 158
Bunsen, Robert, 296

Cannizzaro, Stanislao, 62, 63
Carlisle, Anthony, 176
Carothers, Wallace Hume, 298, 299, 300
Carson, Rachael, 283
Cavendish, Henry, 20, 25, 26, 27, 36
Celsius, Anders, 104
Chadwick, James, 37, 38
Chapman, Sydney, 262
Charles, Jacques Alexandre, 105, 107
Chesebrough, Robert, 218
Colgate, William, 306
Couper, Archibald Scott, 198
Crick, Francis, 234, 236
Crookes, William, 35, 36
Crutzen, Paul, 266
Curie, Marie, 38
Curie, Pierre, 38

Dalton, John, 10, 33, 34, 35, 38, 45, 46, 51, 56, 107, 131, 176
Daniell, John Fredrick, 180
Davy, Humphrey, 72, 73, 156, 176
de Broglie, Louis, 40
de Buffon, Comte, 72
de Fourcroy, Antoine Francois, 27
de Mestral, George, 299
de Morveau, Guyton, 27
Democritus, 9, 10, 47, 71
Diocletian, 12
Döbereiner, Johann, 61, 62

Dow, Herbert Henry, 302, 303
Drake, Edwin L., 216
Du Fay, Charles, 172
Dumas, Jean-Baptiste André, 196, 197, 198
DuPont, Henry, 292, 293
DuPont, LaMont, 293
DuPont, Pierre Samuel, 292
DuPont de Nemours, Eleuthère Irènèe, 292

Eastman, George, 298
Ehrlich, Paul, 296
Einstein, Albert, 85, 246
Empedocles of Agrigentum, 9
Epicurus, 10

Fahrenheit, Daniel, 104
Faraday, Michael, 73, 176, 205, 294
Fermi, Enrico, 46
Firestone, Harvey Samuel, 305
Fischer, Ernest Gottfried, 32
Ford, Henry, 305
Frankland, Edward, 73, 198
Franklin, Benjamin, 23, 172
Fresenius, Karl, 296
Fuller, F. Buckminster, 96

Galen, 14
Galilei, Galileo, 99
Galvani, Luigi, 172, 173, 174, 180, 193
Gamble, James, 305, 306
Gamble, William, 306
Gay-Lussac, Joseph Louis, 34, 35, 105, 107, 196,
    290
Geiger, Johannes Wilhelm, 39
Gell-Mann, Murray, 44
Geoffroy, Etienne Francois, 72
Gerhardt, Charles Fredric, 168, 197, 198
Gilbert, William, 171
Glover, John, 290
Goodrich, Benjamin Franklin, 304, 305
Goodyear, Charles, 300, 304
Gore, Wilbert, 299
Graham, Thomas, 136
Grove, Sir William, 188
Guldberg, Cato Maximilian, 151

Haber, Fritz, 153, 154
Hales, Stephen, 21, 22
Hall, Charles Martin, 191, 192, 193
Heisenberg, Werner, 40
Henry, William, 33, 131
Heraclitus of Ephesus, 9
Héroult, Paul L. T., 191, 192, 193

Hippocrates, 167
Hoffman, Felix, 168
Hooke, Robert, 6, 18, 19, 297
Hyatt, John Wesley, 298

Jabir ibn Hayyan (Geber), 13, 14
Joule, James Prescott, 115

Kekulé, Friedrich August, 73, 198, 206, 296
Kirwan, Richard, 27
Koch, Robert, 296
Kolbe, Adolph Wilhelm Hermann, 168, 196
Kossel, Walther, 74
Kroto, Harold, 96

Langmuir, Irving, 74
Laurent, Auguste, 73, 197, 198
Lavoisier, Antoine, 18, 20, 21, 23, 25, 26, 27, 28, 29,
    31, 32, 35, 70, 156, 193, 196, 292
Le Châtelier, Henri Louis, 149, 153
LeBlanc, Nicholas, 291
Leclanché, George, 186
Leucippus, 9, 71
Lewis, Gilbert Newton, 74, 75, 76, 159
Libavius, Andreas, 16
Libby, Williard, 246
Liebig, Justus, 156, 197, 198, 295, 296
Lister, Joseph, 296
London, Fritz, 89
Lowry, Thomas, 158, 159
Lucretius, 10

Magnus, Albertus, 14
Marsden, Ernest, 39
Maxwell, James Clerk, 103
Mayow, John, 19
Mendeleev, Dmitri, 63, 64, 65
Meyer, Julius Lothar, 63
Millikan, Robert Andrew, 37
Mitscherlich, Eilhardt, 205
Molina, Mario, 265, 266
Morehead, James T., 304
Morton, William T. J., 209
Moseley, Henry, 64
Müller, Paul, 282, 283
Mullis, Kary B., 236

Newlands, John, 62
Newton, Isaac, 6, 25, 71, 72
Nicholson, William, 176
Nieuwland, Julius, 300
Nobel, Alfred, 293
Nobel, Immanuel, 293

Paracelsus, 14, 15, 18
Parmenides, 9
Pasteur, Louis, 296
Pauli, Wolfgang, 42
Pauling, Linus, 77
Perkins, William Henry, 295
Perrin, Jean Baptiste, 56
Planté, Gaston, 186
Priestley, Joseph, 20, 23, 24, 25, 26, 27, 31, 172
Procter, Harley, 306
Procter, William, 305, 306
Proust, Joseph Louis, 32, 33

Queeny, John Francis, 303

Richter, Jermias Benjamin, 31, 32
Rockefeller, John D., 305
Roebuck, John, 290
Roentgen, Wilhelm Conrad, 38, 46
Rouelle, Giullame Francois, 25
Rowland, F. Sherwood, 265, 266
Rutherford, Daniel, 22
Rutherford, Ernest, 37, 39

Scheele, Carl, 24, 25, 26
Schrödinger, Edwin, 40
Schweppe, Jacob, 23
Seaborg, Glenn, 69
Seiberling, Charles, 304
Seiberling, Frank A., 304
Sennert, Daniel, 18

Smalley, Richard, 96
Smith, Robert Angus, 267
Sobrero, Ascanio, 293
Solvay, Ernest, 292
Sorenson, Soren Peer Lauritz, 162
Stahl, George, 20, 21
Stoney, George, 37

Thompson, William (Lord Kelvin), 104, 105
Thomson, Joseph John, 36, 37, 38, 73, 176
Topham, Charles, 297
Torricelli, Evangelista, 99, 100
Tyndall, John, 137

van der Waals, Johannes Diederick, 90, 91
van Helmont, Jan Baptista, 17, 18, 21
van Musschenbroek, Pieter, 172
van't Hoff, Jacobus Hendricus, 142
Volta, Alessandro, 72, 172, 174, 175, 176, 180, 193
von Guericke, Otto, 171
von Hofmann, Wilhelm, 295, 296
von Laue, Max Theodor Felix, 91

Waage, Peter, 151
Watson, James, 234, 236
Watt, James, 22, 26
Williams, Charles Greville, 300
Willson, Thomas L., 304
Wöhler, Friedrich, 195, 196, 197, 294

Zweig, George, 44

Subject Index

Absolute zero, 105
Acetone, 210
Acid ionization constant (K_a), 160
Acid precipitation, 164, 266
Acid rain, 266–70, 314; effect on aquatic organisms, 269
Acids: common, 157; definitions of, 156–59; general properties of, 156; history of 155–56, 290–91; stock solutions, 299
Actinides, 67
Activated complex, 141
Activation energy, 141
Activity series, 181
Addition polymerization, 203
Adhesion, 95
Alchemy, 11
Alcohols, 208, 217; common, 386; denatured, 386
Aldehyde, 209–10
Alizarin, 294
Alkali metal, 64
Alkaline earth metal, 64
Alkaloid, 215
Alkanes, 199–201, 205
Alkenes, 202–3, 205
Alkyl group, 207
Alkynes, 202–3, 205
Allotrope, 495
Alloy, 8, 317
Alpha: decay, 39, 243–44; particle, 39, 243–44
Alpha helix, 231–32
Aluminum, 191–93
American Chemical Society, 332–33
Amine, 215–17

Amino acid, 229; essential, 230
Ammonia, 152–54
Ammonium ion, 52
Amorphous solids, 92
Amylopectin, 224
Amylose, 224
Anabolism, 221
Androgen, 228
Anion, 46
Annealing, 8
Antacids, 167
Antimatter, 254
Aqua regia, 155
Aromatic hydrocarbons, 205–7, 217
Arsenic, 276; use as pesticide, 282
Arteriosclerosis, 228
Aryl group, 207
Aspirin, 168–69
Atmospheric pressure, 99–100, 318–20; composition of, 100
Atom: Bohr model, 40; history of, 9, 33–34; planetary model, 39
Atomic mass, 45
Atomic number, 42, 45
Atomic orbitals, 40–43
Atomic theory, 31–34
Avogadro's law, 106
Avogadro's number, 56

Bakelite, 298
Base ionization constant (K_b), 160
BASF, 306–7
Bayer Company, 168, 306–7

Becquerel: radiation unit, 244
Benzene, 205–7
Beta: decay, 39, 243–44; particle, 39, 243–44
B. F. Goodrich Company, 301–4
Binding energy, 246–47
Biological magnification, 283
Bipyridyliums, 286
Boiling point elevation, 132–33
Bond energies, 79
Bonds: covalent, 76–78; early ides, 71–74; ionic, 74–76; metallic, 80; polar covalent, 76–77
Bose-Einstein condensate, 85
Boyle's Law, 102
Breeder reactor, 249
Buckminsterfullerenes, 96–97
Buffer, 166–67

Caffeine: structure, 215
Calcination, 19–20
Calorimetry, 118–20
Calx, 20
Carbamates, 284
Carbohydrates, 221–26
Carbon dioxide: emissions by country, 273
Carbon monoxide, 279–80
Carbon-14 dating, 246
Carbonyl, 209
Carboxyl, 210
Carboxylic acids, 210–12, 217
Catabolism, 221
Catalysts, 144–47
Catalytic converter, 146
Cathode-ray tube, 35–36
Cathode rays, 35–36
Cathodic protection, 190
Cation, 46
Cellulase, 223–24
Cellulose, 223–24
Celsius temperature scale, 104
Chain reaction, 248
Chapman reactions, 262
Charles' law, 105
Chemical change, 4
Chemical composition: of Earth, 70; of human body, 70
Chemical compound: naming, 51–53
Chemical equilibrium, 147–49
Chemical formula, 54
Chemist's code of conduct, 333
Chirality, 222–23
Chlorine: as water disinfectant, 275
Cohesion, 95
Colligative properties, 131–34
Collision theory, 140–42

Colloids, 136–37
Compounds, 49; common names, 53; cyclic, 205; naming, 51–53
Concentration units: of solutions, 129
Conjugate acid-base pairs, 158
Corrosion, 189–90
Cracking process, 218
Crenation, 133
Crookes tube, 35
Crystalline solid, 90–92
Curie: radiation unit, 255
Cyclic compounds, 205

Dalton's Law of Partial Pressure, 107
d-Block elements, 67
DDT, 282–84
Dephlogisticated air, 23
Diabetes, 224–25
Dialysis, 133
Diamond: structure, 4; synthetic, 97
Dioxins, 285
Dipole-dipole force, 89
Dipole moment, 271
Dispersion force, 89
DNA, 2; replication, 235; structure, 233–35
Dry cell battery, 185–86
Dry ice, 324
Dupont Chemical Company, 292–93, 302
Dyes, 293–97
Dynamite, 293

Electrochemical cell, 179–84
Electrolysis: of water, 185, 315
Electrolyte, 128
Electromagnetic radiation, 262; spectrum of, 263
Electromotive force, 182
Electron: charge to mass ratio, 36–37; discovery of, 36–37; size, 45
Electron configuration, 42–45
Electronegativity, 76–78
Electroplating, 190
Element, 3, 28
Elementary steps, 145
Elements: discovery of, 61–62; periodic trends, 63; table of, 361–62
Emulsifier, 227
Endothermic reaction, 121
Energy content: of selected fuels, 122
Enzymes, 146–47
Error: random, 312; systematic, 312
Esters, 211–12, 217
Estrogens, 228
Ether, 208–9, 217; as anesthetic, 209
Eutrophication, 274

Exothermic reaction, 120
Exxon, 301, 305

Fahrenheit temperature scale, 104
Fatty acids, 211, 226–27; common, 212; in common fats and oils, 227; in soap, 213
Fermentation, 8
Firestone Tire and Rubber Company, 305
First law of thermodynamics, 121
Fission, 246–47
Fixed air, 22
Formaldehyde, 209
Fossil fuels, 216
Fractional distillation, 217
Free radical, 265
Freezing point depression, 132
Fructose, 222
Fuel cells, 188
Fullerene, 96–97
Functional group, 198, 207, 217
Fundamental particle, 44–45
Fusion, 247

Gamma: decay, 39, 243; energy, 39, 243
Gasoline, 218; combustion of, 279
Gay Lussac's law, 105–6
Geiger counter, 256
Genetic engineering, 238
Glass, 92–93
Glucose, 222–23
Glycogen, 224
Goodyear Tire and Rubber Company, 300, 304
Gore-Tex, 299
Gunpowder, 292–93

Haber process, 152–53
Half-Life: chemical, 144; of common isotopes, 245; radioactive, 244
Half reactions, 177
Halogen, 64
Halogenation, 202
Hard water, 274
Heat capacity, 116
Heat of: fusion, 117, 119; reaction, 120; vaporization, 117, 119
Henry's law, 110, 131
Histamine, 215
Human genome project, 236
Hydration, 127
Hydrocarbons, 199; saturated, 199; unsaturated, 199
Hydrogen, 25
Hydrogen bond, 87–88
Hydrogenation, 202
Hydrolysis, 213

Ideal gas, 102
Ideal gas law, 106–7
Indigo, 294
Internal energy, 114
Invert sugar, 223
Iodine number, 226; of common fats and oils, 227
Ion product constant (K_w), 162
Ion-dipole force, 89
Ionization, 46
Ionizing radiation, 256
Ions, 45–46; present in seawater, 125
Iron: corrosion of, 189
Isotope, 45; half-lives of, 245
IUPAC, 69
Ivory soap, 305–6

Joule: unit of energy, 115

K_a, 160
K_b, 160
Kelvin temperature scale, 104–6
Ketone, 209–10, 217

Latex, 299
Lathanides, 67
Law of Definite Proportions, 33–34
Law of mass action, 151
Law of Multiple Proportions, 33–34
Lead: toxic effects, 266–67
Lead-chamber process, 290
Lead storage battery, 186–87
LeBlanc method, 291
LeChâtelier's Principle, 149–50
Lewis acid, 159
Lewis base, 159
Lewis electron dot formula, 74–76
Limewater, 321
Lipids, 226
Liquid crystals, 95–96
Lock-and-key model, 147
London force, 89
Long-range order, 93

Main group elements, 64–67
Mass defect, 246
Mass number, 45
Matter, 3, 10
Mauve, 295
Mercury: toxic effects, 276–77
Metallic bond, 80
Metalloids, 67
Metals, 67–69
Mixture, 49
Mole, 56

Molecular orbital theory, 83
Molecule, 35, 49–50
Monsanto, 303–4
MTBE, 209

Neutron, 37, 45
Nicotine: as pesticide, 282; structure, 215
Nitroglycerin, 293
Nitrous gas, 34
Nobel prize, 293
Nobel prizes: in chemistry, 355–59
Noble gas, 64, 68
Nomenclature: ancient symbols, 50; common
    names, 53; history of, 49–52; for naming com-
    pounds, 51–54
Non-ionizing radiation, 256
Nonmetals, 67
Nuclear fission, 246–47
Nuclear fusion, 247; and stellar evolution, 251–52
Nuclear power, 248–51
Nuclear stability, 241–42
Nucleon, 241
Nylon, 298–99

Octane number, 201–2
Octet rule, 74
Oil, 94
Optical isomer, 296
Organic Foods Production Act, 219
Organic substitution, 197
Organochloride, 284
Organophosphates, 284
Osmosis, 133
Osmotic pressure, 133
Oxidation numbers, 177–79
Oxidation reaction, 176–77
Oxidizing agent, 177
Oxygen: discovery of, 23–25
Ozone, 262; depletion, 262–66

Partial pressure, 107; and diving, 109–10
Pauli Exclusion Principle, 42
PCBs, 277–78
Peptide, 229
Perfect gas, 102
Periodic table, inside cover; history of, 63–64; mod-
    ern, 64; recently discovered elements and, 69
Pesticides, 281–87; history of, 281–82
Petrochemicals, 218
pH, 161–63; of common substances, 163; indicator,
    163, 310
Phase changes, 113–16
Phenol, 208
Phenyl, 207

Pheromones, 212
Philosopher's stone, 11
Phlogiston theory, 19–20
Physical change, 4
Plasmolysis, 133
Plastics, 297–99; recycling symbols, 204
Pneumatic trough, 21
pOH, 161–63
Polarizability: and intermolecular forces, 90
Polarized light, 96
Polyatomic ions, 52
Polyethylene terephthalate (PET), 214
Polymerase chain reaction, 236–37
Polymerization, 203
Polysaccharide, 223
Potash, 290–91
Potential energy, 114; chemical, 120
Pressure, 99; units, 101
Procter and Gamble, 305–6; and Ivory soap, 306
Progestin, 228
Protein, 229; structure, 229–32; synthesis, 235
Proton, 44–45; discovery of, 37
Pyrethroids, 284–85

Quantum numbers, 40–42
Qualitative analysis, 136
Quarks, 44
Quaternary structure, 231–32

Radiation: background, 259; biological effects,
    257–59; detection, 256–57; units, 255–56
Radical theory, 196
Radiometric dating, 244–46
Random error, 312
Rate law, 143
Reaction order, 143
Reaction rates: factors affecting, 142
Reactions, 54; balancing, 55; complete, 147; oxida-
    tion/reduction, 176–77; reversible; 147; types
    of, 55
Recombinant DNA, 238
Recycling symbols, 204
Reducing agent, 177
Relative humidity, 108
Relative mass, 45; Dalton's, 33
Resins, 209
Resonance structures, 206
Reverse osmosis, 134
RNA, 235–36
Rubber: synthetic, 299–301

Saccharides, 221, 223
Saccharin, 2
SAE number, 94

Saponification, 213–14
Scientific method, 310–13
Sickle-cell anemia, 231
Smog, 280–81
Soap, 213
Soda ash, 291. *See also* LeBlanc method; Solvay process
Solids: amorphous, 92; crystalline, 90; ionic, 91; metallic, 91; molecular, 91
Solubility: of gases, 130–131; rules for ionic compounds, 128; and temperature, 130
Solutions: concentration units, 129; saturated, 129
Solvation, 127
Solvay process, 292
Specific heat, 116; of common substances, 116
Specific heat capacity, 116
Standard hydrogen electrode, 182
Standard reduction potential, 182; of common half-reactions, 184
Starch, 224
States of matter, 85–87
Stereoisomers, 223
Steroids, 228
Stoichiometry, 32
Structural isomers, 201
Sublimation, 113–14
Sulfuric acid, 155, 157; lead chamber process, 290
Surface tension, 94

Temperature scales, 104
Tokamak, 251

Transmutation, 252
Triazines, 285–86
Trihalomethanes, 275
Tyndall effect, 137

Ultraviolet radiation, 263; in ozone depletion, 262–63
Union Carbide, 304
Urea: synthesis of, 195

Valence, 73
Valence bond theory, 83
Valence electron configuration, 43–44
Valence shell electron pair repulsion (VSEPR), 80–83
Vapor pressure, 107
Vaporization, 113–14; heat of, 117
Vaseline, 218
Velcro, 299
Viscosity, 93
Vitalism, 195
Vulcanization, 300

Water: hydrogen bonding in, 87; quality, 273–78
Wien's Law, 270
Work: P-V, 115

X-Ray crystallography, 91–92

Zwitterion, 231

About the Author

RICHARD MYERS is Professor of Environmental Science at Alaska Pacific University. He teaches undergraduate and graduate courses in chemistry, statistics, and environmental science. Dr. Myers' published research includes work on air quality, water quality, science education, and science and the humanities.